W9-CSP-646

Fluidization Technology

SERIES IN THERMAL AND FLUIDS ENGINEERING

EDITORS

JAMES P. HARTNETT and THOMAS F. IRVINE, JR.

Chang
- **Control of Flow Separation: Energy Conservation, Operational Efficiency, and Safety**

Keairns
- **Fluidization Technology**

IN PREPARATION

Begell
- **Glossary of Terms in Heat Transfer and Related Topics: English, Russian, German, French, and Japanese**

Brzustowski and Wojcicki
- **Aerosols in Combustion and Explosions**

Chi
- **Heat Pipe Theory and Practice: A Sourcebook**

Denton
- **Future Energy Production Systems: Heat and Mass Transfer Processes***

Eckert and Goldstein
- **Measurements in Heat Transfer, 2nd edition**

Ginoux
- **Two-Phase Flows with Application to Nuclear Reactor Design Problems†**

Goulard
- **Combustion Measurements: Modern Techniques and Instrumentation**

Hahne
- **Heat Transfer in Boiling**

Hartnett
- **Alternative Energy Sources***

Hsu and Graham
- **Transport Processes in Boiling and Two-Phase Systems, Including Near-Critical Fluids**

Moore and Sieverding
- **Two-Phase Steam Flow in Turbines and Separators: Theory, Instrumentation, Engineering†**

Pfender
- **High-Temperature Phenomena in Electric Arcs**

Richards
- **Measurement of Unsteady Fluid Dynamic Phenomena†**

van Stralen and Cole
- **Boiling Phenomena**

Yovanovich
- **Advanced Heat Conduction**

*A publication of the International Centre for Heat and Mass Transfer, Belgrade.
†A von Karman Institute Book, Brussels.

Fluidization Technology

VOLUME I

Bubble Phenomena
Gas Exchange and Fluid Bed Modeling
Liquid Phase Fluidization
Three Phase Fluidization

EDITOR

Dale L. Keairns

IN COOPERATION WITH

M. A. Bergougnou
J. F. Davidson
J. M. Matsen
C. Y. Wen

HEMISPHERE PUBLISHING CORPORATION

Washington

IN ASSOCIATION WITH

McGRAW-HILL INTERNATIONAL BOOK COMPANY

New York   St. Louis   San Francisco   Auckland   Düsseldorf   Johannesburg
London   Mexico   Montreal   New Delhi   Panama
Paris   São Paulo   Singapore   Sydney   Tokyo   Toronto

6166-7602v

CHEMISTRY

"These proceedings are based upon the International Fluidization Conference which was sponsored by the Engineering Foundation at Asilomar Conference Grounds, Pacific Grove, California on June 15-20, 1975. The views presented here are not necessarily those of the Engineering Foundation, 345 East 47th Street, New York, New York 10017."

FLUIDIZATION TECHNOLOGY, Volume I

Copyright © 1976 by Hemisphere Publishing Corporation. All rights reserved. Printed in the United States of America. No part of this publication may be reproduced, stored in a retrieval system, or transmitted, in any form or by any means, electronic, mechanical, photocopying, recording, or otherwise, without prior written permission of the publisher.

1 2 3 4 5 6 7 8 9 0   D O D O   7 8 4 3 2 1 0 9 8 7 6

Library of Congress Cataloging in Publication Data

International Fluidization Conference, Pacific Grove,
   Calif., 1975.
   Fluidization technology.

   An Engineering Foundation conference, co-sponsored by
American Institute of Chemical Engineers and other
organizations.
   (Series in thermal and fluids engineering)
   1. Fluidization—Congresses.   I. Keairns, D. L.,
1940-   II. Bergougnou, Maurice Amédée.   III. Engineering Foundation, New York.   IV. American Institute
of Chemical Engineers.   V. Title.
TP156.F65I48 1975   660.2'8429   75-40106
ISBN 0-89116-005-1 (v. I)

T P
156
F65
I48
1975
v.1
CHEM

ENGINEERING FOUNDATION CONFERENCE

Proceedings of the International Fluidization Conference
held at
Asilomar Conference Grounds
Pacific Grove, California
June 15–20, 1975

Co-Sponsored by

American Institute of Chemical Engineers
Canadian Society of Chemical Engineers
Institution of Chemical Engineers, United Kingdom
Société de Chimie Industrielle, France
Society of Chemical Engineers, Japan

Financial Assistance was Provided by

Engineering Foundation
National Science Foundation

19562

CONTENTS

PREFACE

The proceedings of the International Fluidization Conference held at Asilomar Conference Grounds, Pacific Grove, California, June 15–20, 1975 are presented in these two volumes. Volume I contains reports of investigations into fluidization fundamentals—bubble phenomena, gas exchange and fluid bed modeling, liquid phase fluidization, and three phase fluidization. The primary emphasis in Volume II is on fluidization processing—fluidized bed performance with internals, solids mixing and transport, fossil fuel processing, and applications.

The International Fluidization Conference was organized to disseminate present knowledge and to further the development of fluidized bed technology and applications of the technology. Investigators and practitioners from industry, universities, governments and research institutions throughout the world participated in the Conference.

The need for the international conference on fluidization arose out of a number of concerns. Fluidization technology is being applied in increasing numbers of new applications. However, the understanding of fluidization phenomena is far from adequate and the design of fluidization processes is primarily based on past experience. The extensive research work being carried out is illustrated by the thousands of papers and patents which have been issued. Previous international conferences, in particular the Toulouse meeting in 1973 and the Eindhoven meeting in 1967, provided effective means to exchange information. While there continue to be numerous symposia on fluidization fundamentals and applications, a major conference dealing exclusively with fluidization was considered timely. Discussions held with research investigators, personnel involved with large scale development programs, government contract monitors, and individuals responsible for fluidized bed production units resulted in this conference. The conference consisted of eight technical sessions, corresponding to the eight divisions in these two volumes. Parallel workshops were held on the day following the seven sessions on specific fluidization areas. The applications session was held on the final day followed by reports from the workshop discussion leaders. The conference site was selected to provide an atmosphere where informal discussion and exchange of ideas could be most readily fostered.

In order to achieve the conference objectives, the organizing committee adopted a rapporteur format. This was adopted to permit more information to be available for discussion and provided an opportunity to assimilate information through the use of reporters. Summary papers containing the scope, objectives, results and significance for each piece of work were provided to all conference participants. Extended manuscripts were available to the session organizers and reporters. The extended manuscripts form the basis for these proceedings.

It is my pleasure to record the important role of my colleagues on the organizing committee: M. A. Bergougnou, J. F. Davidson, J. M. Matsen and C. Y. Wen. The success of the conference was made possible by their work on the program organization, cooperative efforts, ideas, suggestions, and knowledge of investigators in the field. I personally express my appreciation to the following session chairpersons, rapporteurs, and workshop discussion leaders who provided the environment which resulted in a stimulating meeting with valuable technical insights.

Bubble Phenomena
    Chairperson   I. O. Molerus
    Rapporteur   P. N. Rowe
    Workshop discussion leader   P. H. Calderbank

Fluidized Bed Performance With Internals
    Chairperson   H. Angelino
    Rapporteur   J. S. Halow
    Workshop discussion leader   D. F. Wells

Gas Exchange and Fluid Bed Modeling
    Chairperson   D. Kunii
    Rapporteur   D. L. Pyle
    Workshop discussion leader   O. Levenspiel

Solids Mixing and Transport
    Chairperson   L. Massimilla
    Rapporteur   L. S. Leung
    Workshop discussion leader   J. S. M. Botterill

Liquid Phase Fluidization
    Chairperson   P. Le Goff
    Rapporteur   N. Epstein
    Workshop discussion leader   H. Littman

Three Phase Fluidization
    Chairperson   D. Harrison
    Rapporteur   C. G. J. Baker
    Workshop discussion leader   K. Ostergaard

Fossil Fuel Processing
    Chairperson   S. Freedman
    Rapporteur   D. H. Archer
    Workshop discussion leader   B. S. Lee

Applications
    Chairperson   F. A. Zenz
    Rapporteur   L. Reh

The session chairpersons were effective at establishing a positive format and were sensitive to potentially difficult situations. The technically thorough and comprehensive analyses presented by the rapporteurs provided an excellent basis for discussion. Their comments were effective at assimilating a diversity of information. The discussion leaders for the workshops enabled the prior discussion of specific investigations in each area to be extended and integrated with other experience and information.

On behalf of the organizing committee, I express our appreciation to the Engineering Foundation for enabling this conference to be held and to S. S. Cole, M. Keenberg, and W. E. Reaser for their efforts to implement the planning and arrangements. Financial support from the Engineering Foundation and the National Science Foundation which enabled many of the participants to attend is gratefully acknowledged. Also, I acknowledge the cooperation of N. Del Gobbo of the National Science Foundation. The technical societies and institutions who cosponsored this conference are acknowledged for their support and their assistance with publicity and providing contact with members.

The experience at Asilomar demonstrated to many participants the need for future exchange of information among those working on fluidization technology. As a result of this interest, a second Engineering Foundation conference will be held the first week of April in 1978 at the University of Cambridge, Cambridge, England.

It is our hope that the International Fluidization Conference and these proceedings effectively contribute to the concerns which led to the conference and provide for the exchange of information and ideas on fluidization technology.

Dale L. Keairns
Conference Chairperson

PART I
Bubble Phenomena

A LOW-REYNOLDS-NUMBER ANALYSIS
OF GAS BUBBLES IN FLUIDIZED BEDS

R. H. WEILAND

The gas and solids flow through and around bubbles in fluidized beds is treated under the assumption that such bubbles rise at low Reynolds numbers. In the case of three-dimensional bubbles this enables the calculation of bubble shape. The bubbles are found to be slightly elongated and akin in shape to a gas slug in a tube and the large bubbles described by Goldsmith and Rowe (1975). The analysis suggests that a detailed examination of the transition from bubble to suspension phase will be needed for any precise description of bubbles, which rise naturally at intermediate Reynolds numbers.

1. INTRODUCTION

When gas or liquid is forced to flow upwards through a porous bed of solid particles a pressure drop is created across the bed. If this pressure drop is sufficient to support the weight of the particles, they become separated and free to move about; such a bed is said to be incipiently fluidized. Further increases in flow rate are generally accommodated by bed expansion in the case of liquid fluidization but if the ratio of particle density to fluid density is large as in fluidization by gases, then a large proportion of the excess flow over that needed for incipient fluidization passes through the bed in the form of bubbles of fluid virtually devoid of particles. This is termed aggregative fluidization. It is characterized by the presence of bubbles or voids and is the usual form in most practical applications.

The properties and behavior of single steadily-rising low-Reynolds-number bubbles in gas fluidized beds is the subject of the present work. The development given here was motivated by the need that some consideration be given to the influence of bed viscosity on the phenomena associated with bubble rise. All previous analyses have ignored viscous stresses in the equations of motion but shear viscosities of such suspensions have been measured using Couette viscometers and found to be some tens of poise, So little is known about the effective viscosity of fluidized beds however, (and even less about their rheological properties) that the Reynolds number of a bubble can hardly be estimated although it is usually larger than unity. All previous analyses have been based on inviscid theory in that they have totally neglected viscous stress terms in the equations of motion. The opposite and no-less-extreme point-of-view is taken here that these stresses are dominant features of the motion. The real situation is almost certainly one in which viscous stresses are quite important but a low-Reynolds-number approximation is an oversimplification and probably of very limited importance to practical bubbles. Its relevance lies in the much needed consideration of the effect of bed viscosity on bubble behavior. Along with the recent work of Batchelor (1973), it should be taken as only a first attempt at the inclusion of viscous stresses in a description of the motion around bubbles in incipiently fluidized beds.

Our present theoretical understanding of bubble behavior has generally come from the analogy with bubbles rising in inviscid liquids. Although the bubble rise velocities in these two cases are similar, it is well known that the basic physics is different. The main dissimilarity is that bubbles in liquids involve two immiscible phases wherein the gaseous contents of the bubble remain firmly identified with it—it has certainly been recognized in all previous theoretical developments that this is not the case for fluidized beds.

As so clearly described by Rowe (1971) a bubble in a fluidized bed is not a discontinuity between two immiscible phases or a structured shell of particles. It is a void that sets up a pattern of flow in an otherwise uniform field. This in turn brings about a pattern of forces on the particles and causes them to move in just that way which perpetuates the void and causes it to move upwards. The bubbles are regions of extremely high permeability in surroundings of comparatively low permeability. The gas, which is moving through the dense region in streamline flow converges towards the bottom of the bubble, flows through it, then diverges again as it flows through the bubble's roof and returns to the dense phase. The flow of gas through the bubble is the prime cause of the whole phenomenon.

Davidson et al. (1959) suggested that the well-known relation of Davies and Taylor (1950) for the rise velocity of large gas bubbles in inviscid liquids might also be applicable to bubbles in fluidized beds. Within the range of observational errors involved, experimental evidence (Harrison and Leung, 1962; Rowe and Partridge, 1965; Pyle and Harrison, 1967; Littman and Homolka, 1970, 1973) indicates that such bubbles rise at more or less constant Froude number and thereby lends support to the proposal of Davidson et al. (1959). As far as bubble rise velocity is concerned, fluidized beds give the appearance of being inviscid.

The extension of this observation to the use of inviscid theory as the basis for a description of the fluid and particle flow around bubbles in gas fluidized beds has given rise to several models each with its own set of further assumptions and approximations and each with its own points of agreement and disagreement with experiment. However, these theories are in surprisingly good accord with each other and with experiment on certain major points and have contributed greatly to the understanding of aggregative fluidization. For example they all accurately reproduce experimental data on bubble rise velocity and at worst differ only slightly from the observed size of the recirculating gas cloud which accompanies bubbles rising at higher velocity than the fluidizing gas in a uniform bed. Some other features however, are not well described by these theories and so although they are useful they must be regarded as incomplete.

An analysis of the flow around rising bubbles was first presented in the elegantly simple theory of Davidson (1961). It accurately describes observed bubble rise velocities (Collins, 1965), cloud size and throughflow velocities and also meets the constant pressure condition on the bubble surface which it maintains as a streamline for the flow of particles. Unfortunately particle momentum is totally neglected and as a result this theory fails to account for the particle flow pattern in a physically satisfying way.

In attempting a more complete description of the flow, Jackson (1963) naturally included particle momentum considerations in his analysis and in addition permitted the volume fraction of particles to vary near the bubble. The resulting estimates of cloud size are in slightly better agreement with experiment than Davidson's and in addition it was found that the particle number density should decrease as the bubble is approached. The latter prediction has been confirmed experimentally by Lockett and Harrison (1967) and Nguyen, Leung and Weiland (1973) although the agreement with theory is by no means perfect. A less happy outcome of this model however, is that it gives a highly variable pressure distribution on the surface of the bubble whereas, within the range of measurement errors involved, it is an experimental fact that the pressure is constant (the gas density is about three decades smaller than that of the solids so the convergence of fluid streamlines inside the bubble gives rise to quite negligible variations in pressure). Murray's (1965b) analysis introduces some features of Jackson's treatment into the Davidson theory but it too gives a varying pressure at the bubble surface. Recently Nguyen (1974) repeated the work of Jackson and adjusted the shape of the bubble to achieve constancy of pressure. The results are generally similar to Jackson's but the surface pressure could be made constant only at the expense of admitting a large component of solids velocity normal to the bubble surface. At worst there is only a small amount of rainthrough of solids observed experimentally so in this regard the model sharply contradicts reality.

Despite the vast differences in the assumptions which have been made in the various theories, striking agreement with certain important experimental results has been found. It is equally true however, that each model either is in sharp disagreement with certain other and no-less-important experimental findings or, by not meeting conservation laws, provides no fluid dynamical explanation for the phenomena involved.

It seems likely that resolution of the problem lies in the viscous stresses which result from the mutual proximity of particles and the concomitant denseness of the suspension. There is no *a priori* justification for neglecting the stress tensor of the solids phase—its omission appears to have been merely a matter of expediency. Indeed the well-documented evidence of Schügerl et al. (1961) and Hagyard and Sacerdote (1966) that the viscosities of fluidized beds are enormous suggests that viscous stresses play an important role in determining the flow around bubbles in gas fluidized beds.

The present work is an enquiry into the influence of viscous stresses on the motion around bubbles. It has already been mentioned that inertial and viscous forces are probably both important in determining the flow, i.e., the Reynolds numbers of rising bubbles are neither large nor small, but at the present time the case of such intermediate Reynolds number flows is not amenable to analysis by other than numerical means. All previous theories have assumed $R \rightarrow \infty$. In the present work we take the point-of-view that bubble motion in an incipiently fluidized bed is dominated by viscous stresses so that the problem is a low-Reynolds-number one. Although such an assumption is not representative of real bubbles it has the advantage of being open to analytical investigation. The flow around two- and three-dimensional bubbles is treated as a singular perturbations problem using the now-classical matched asymptotic expansions technique of Proudman and Pearson (1957) and Kaplun and Lagerstrom (1957). In the three-dimensional case we obtain the first approximation to the distortion of the bubble from a sphere.

2. EQUATIONS OF MOTION AND BOUNDARY CONDITIONS

The fluid mechanical theory of fluidized beds has recently been reviewed by Jackson (1971). The basic assumption common to all proposed equations describing the fluidized state is that the fluid and particulate phases behave as though they were interpenetrating continua: point variables have been smoothed by averaging over regions large compared with the particle spacing but small compared with the physical scale of the gross phenomena of interest. Thus we are concerned with the relative motion of fluid and particles as two distinct phases rather than as a single rheologically complex continuum.

An infinite bed of equally-sized particles of the same density ρ_s is uniformly fluidized by an incompressible fluid of density ρ_f so that the volume fraction of particles is Z. If v and V are the average velocities of the particulate and fluid phases respectively then the corresponding continuity equations are (Murray, 1965a)

$$\frac{\partial Z}{\partial t} + \text{div } Zv = 0 \qquad (2.1)$$

$$\frac{\partial Z}{\partial t} - \text{div } (1 - Z)V = 0 \qquad (2.2)$$

In the incipiently fluidized state with which we are concerned, particles lie very near each other; hence, it is reasonable to assume that the gas stress tensor is negligible compared with the solids stress tensor. For systems in which $\rho_f/\rho_s \ll 1$ the momentum equations are

$$\frac{\rho_s Z \, Dv}{Dt} = - g\rho_s Z_j + Z \, \text{div} \, \sigma - \text{grad} \, p \tag{2.3}$$

$$\text{grad} \, p = - \beta(Z)(V - v) \tag{2.4}$$

where $\beta(Z)$ is the fluid-particle drag coefficient per unit bed volume, σ is the stress tensor associated with the presence of the solid particles and p is the fluid pressure. It is worth noting that in deriving these equations Murray *assumed* that the pressure of the solid phase which would arise from particle collisions is negligible by comparison with the fluid pressure. By way of justification for this assumption he offered the evidence of low noise levels and attrition rates in fluidized beds, thereby concluding that frequent particle collisions do not occur. It is difficult to believe that particles in the incipiently fluidized state (in which they would be in very close proximity to each other) would not be in an almost constant state of collision. If however one does not make some assumption about the distribution of pressure between the two phases, then the governing equations are indeterminate in that there is one more variable than equations. To neglect the solids pressure altogether is the simplest approximation one can make; whether or not it is correct is another matter. This also raises the question of just what is being measured when pressure transducers and manometers are used as pressure sensors. For example the pressure distribution above and below bubbles found by Nguyen (1974) using water-filled manometers in a two-dimensional fluidized bed are very much lower than those found by Littman and Homolka (1973) using pressure transducers. In fact Nguyen's results correspond more closely to Reuter's (1963) measurements of three-dimensional bubbles, taken using pressure transducers. Great care must be exercised in comparing theoretical and experimental pressures because the experimental results depend very much on the means used to obtain them.

Schügerl et al. (1961) and Hagyard and Sacerdote (1966) have found that gas fluidized beds in Couette viscometers exhibit roughly Newtonian behavior except at very low shear rates. Theirs are the only data available so we approximate the particulate phase by a Navier-Stokes fluid with the constitutive relation

$$\sigma_{ij} = \mu_s \left(\frac{\partial v_i}{\partial x_j} + \frac{\partial v_j}{\partial x_i} \right) + \left(\zeta_s - \frac{2}{3} \mu_s \right) \delta_{ij} \, \text{div} \, v \tag{2.5}$$

in which μ_s and ζ_s are respectively the shear and bulk viscosities associated with the solids phase. One expects both viscosities to be dependent on the volume fraction of solids in the suspension and relationships of the form

$$\mu_s = \mu_f A D_s \tag{2.6}$$

and $$\zeta_s = \mu_f B D_s^3 \tag{2.7}$$

have been postulated by Murray (1965a). Here A and B are 0(1) constants, $D_s = Z/(Z_s - Z)$ and Z_s is the volume fraction of solids in the unfluidized state. Clearly near incipient fluidization μ_s and ζ_s are large and show very strong dependence on Z.

As it stands, the system (2.1) to (2.4) provides six entirely coupled equations for p, Z and the four velocity components u, v, U, and V. For the analysis to proceed we must make a further approximation. If we take Z to be uniform throughout the flow field and equal to its value in the undisturbed stream at infinity then the equations decouple and the solids motion can be determined without reference to the motion of the fluid. The steady flow equations then become

$$\text{div } v = 0 \tag{2.8}$$

$$\text{div } V = 0 \tag{2.9}$$

$$v \cdot \text{grad } v = - gj + \rho_s^{-1} \text{ div } \sigma - (\rho_s Z_\infty)^{-1} \text{ grad } P \tag{2.10}$$

$$\text{grad } P = - \beta(Z_\infty)(V - v) \tag{2.11}$$

The effect of bulk viscosity vanishes and the constitutive relation becomes

$$\sigma_{ij} = \mu_s \left(\frac{\partial v_i}{\partial x_j} + \frac{\partial v_j}{\partial x_i} \right) \tag{2.12}$$

where μ_s is now a constant independent of spatial coordinates.

In obtaining these governing equations we have neglected the gas density compared with that of the solids hence the gas flow possesses no momentum. As a result the gas pressure inside the bubble is constant despite the fact that the motion is not rectilinear. This constant internal pressure must be balanced by the total normal stress acting on the exterior of the bubble, or equivalently, on the surface we have $-p + 2\mu e_{nn}$ = constant where e_{nn} is the component of the rate-of-strain tensor normal to the boundary. The normal stress boundary condition (at least as it is expressed in terms of Murray's model of a suspension) is troublesome however because we have a pressure jump at the interface but no mechanism with which to generate it; the gas has been assumed devoid of momentum, the solids pressure-free and the simplification of constant solids fraction has excluded the action of bulk viscous effects. In reality of course the gas velocity roughly doubles in moving from the bubble to the dense phase thereby creating a sudden decrease in pressure. Nevertheless, the change from this source would be small so our neglect of gas density cannot be held to account for the rather large effect with which we are concerned. It is plausible to associate with the solids phase a pressure quite distinct from that in the fluid, due to the very frequent particle-particle collisions which *must* occur in an agitated suspension whose average particle spacing is considerably less than a diameter. On the bubble side of the interface there is a particulate vacuum while on the suspension side this pressure is non-zero and must be balanced by the viscous normal stress there. Finally, because the velocity field has been taken as solenoidal there is no contribution of bulk viscous effects to the normal stress. This too could contribute in a significant way to resolving the problem, especially in view of the rapid variation in suspension density which takes place near the bubble (see below). It must be pointed out that the moment one wishes to take viscous effects into account, this problem will arise no matter how small one takes the effective viscosity to be. The result of inviscid theory is a non-uniform surface pressure or in lieu of that, a large component of solids velocity normal to the bubble surface; on the other hand, the inclusion of viscosity in the formulation gives rise to an imbalance of normal stress at the interface—the same type of problem reappears in different form. It would seem paramount to further theoretical progress that the physics of this situation be quantitatively examined in some detail.

Of course if we insist that on both sides of the surface, fluid pressure alone is constant then we are left with a large unbalanced and ignored viscous normal stress and this would be in direct contradiction to our postulate that the stress tensor is a significant factor in determining the motion. The available theoretical (Jackson, 1963) and experimental (Lockett and Harrison, 1967; Nguyen, Leung and Weiland, 1973) evidence is that the suspension becomes less concentrated near the bubble surface so that in view of (2.6) and (2.7) its viscosity is significantly reduced there. The difficulty encountered in dealing with the normal stress condition and the absence of any mechanism for generating the pressure jump have arisen solely from a consideration of viscosity and it has been suggested that further progress of a fundamental nature may well hinge on answering these questions. So little is known about solids pressure, bulk viscosity and the way viscosity varies with the concentration of the suspension that any quantitative discussion of conditions to be met on the bubble surface is necessarily very limited at present.

3. CIRCULAR CYLINDRICAL BUBBLE

We begin by considering a circular cylindrical bubble of radius a, rising in a fluidized bed by unlimited depth and breadth. The coordinate system is taken stationary with respect to the bubble, having origin at its center. Non-dimensionalizing the solids and fluid velocities with the free stream velocities $U_{s\infty}$ and $U_{f\infty}$, respectively, taking the bubble radius as the relevant length and scaling the pressure with $Z_\infty \rho_s U_{s\infty}{}^2$ we write the equations of motion in circular cylindrical coordinates:

$$(ru)_r + v_\theta = 0 \tag{3.1}$$

$$(rU)_r + V_\theta = 0 \tag{3.2}$$

$$uu_r + r^{-1} vu_\theta - r^{-1} v^2 = -p_r - F^{-1} \cos \theta +$$
$$R^{-1} (u_{rr} + r^{-1} u_r + r^{-2} u_{\theta\theta} - r^{-2} u - 2r^{-2} v_\theta) \tag{3.3}$$

$$uv_r + r^{-1} vv_\theta + r^{-1} uv = - r^{-1} P_\theta + F^{-1} \sin \theta +$$
$$R^{-1} (v_{rr} + r^{-1} v_r + r^{-2} v_{\theta\theta} + 2r^{-2} u_\theta - r^{-2} v) \tag{3.4}$$

$$r^{-1} P_\theta = \frac{-a\beta(Z_\infty)}{Z_\infty \rho_s U_{s\infty}} \left(\frac{U_{g\infty}}{U_{s\infty}} V - v \right) \tag{3.5}$$

where $F = U_{s\infty}{}^2 /ga$ is the Froude number and $R = \rho_s a U_{s\infty}/\mu_s$ is the Reynolds number for the flow of solids around the void. We shall assume $R \ll 1$.

If pressure is eliminated between (3.3) and (3.4) the gravitational terms also disappear: introducing the stream function through $u = -r^{-1} \psi_\theta$, $v = \psi_r$ we have

$$\nabla_r^4 \psi = - \frac{R}{r} \frac{\partial (\psi, \nabla_r^2 \psi)}{\partial (r,\theta)} \tag{3.6}$$

where
$$\nabla_r^2 = \frac{\partial^2}{\partial r^2} + \frac{1}{r} \frac{\partial}{\partial r} + \frac{1}{r^2} \frac{\partial^2}{\partial \theta^2} \tag{3.7}$$

The kinematic condition and the requirement that the bubble surface be shear free give

$$\psi(1,\theta) = \psi_{rr}(1,\theta) - r^{-1} \psi_r (1,\theta) = 0 \tag{3.8}$$

while the condition relative to the bubble center of free streaming at infinity is

$$\psi \to r \sin \theta \quad \text{as } r \to \infty \tag{3.9}$$

A further condition is that there be no unbalanced normal stresses on the bubble surface. This condition must be met regardless of the shape of the bubble surface.

3.1 Solution for the Stream Function

Proudman and Pearson (1957) and Kaplun and Lagerstrom (1957) have provided asymptotic solutions to the problem of flow past a solid sphere and circular cylinder at small but non-zero Reynolds numbers. Apart from the no-slip condition rather than the zero-shear-stress condition, this problem is nearly identical to the one at present under consideration. The same approach will be adopted here; details of the analysis can be found in Proudman and Pearson.

The first term of the Stokes solution meeting the conditions (3.8) is

$$\psi_0 = Br \log r \sin \theta \tag{3.10}$$

where we have assumed an inner expansion in the usual form

$$\psi(r, \theta) = \sum_0^\infty f_n(R) \psi_n(r, \theta). \tag{3.11}$$

As shown by Proudman and Pearson, this solution fails when $r = 0(1/R)$. We therefore seek an outer solution in terms of Oseen variables

$$\rho = Rr, \ \Psi = R\psi. \tag{3.12}$$

The required solution is

$$\Psi_0 = \rho \sin \theta \tag{3.13}$$

where an expansion of the form

$$\Psi(\rho, \theta) = \sum_0^\infty F_n(R) \Psi_n(\rho, \theta)$$

has been assumed with $F_0(R) = 1$. Because of matching requirements we get

$$B = 1 \tag{3.14}$$

in (3.10). If $f_0(R)\psi_0$ is not to contain any terms of greater order than unity we must have

$$f_0(R) = -1/\log R. \tag{3.15}$$

The second term in the Oseen expansion satisfies

$$\left(\nabla_\rho{}^2 - \frac{\partial}{\partial \xi} \right) \nabla_\rho{}^2 \ \Psi_1 = 0 \tag{3.16}$$

where $\xi = \rho \cos \theta$. A first integral bounded at infinity is

$$\nabla_\rho{}^2 \Psi_1 = e^{1/2\xi} \sum_1^\infty X_n K_n (1/2\rho) \sin n\theta \tag{3.17}$$

where $K_n(\tfrac{1}{2}\rho)$ is a modified Bessel function. We apply the matching requirement directly to $\Delta_\rho^2 \psi_1$. It can be readily shown from the Stokes solution that

$$f_0(R)\nabla_r^2 \Psi_0 = \frac{-2}{r \log R} \sin \theta \qquad (3.18)$$

which is the same as the result obtained using the no-slip condition. The subsequent analysis is identical to that of Proudman and Pearson and leads to the solution

$$\Psi_1 = \sum_1^\infty \phi_n(1/2\rho) \, \frac{\rho \sin n\theta}{n} \qquad (3.19)$$

where $\qquad \phi_n = 2K_1 I_n + K_0 (I_{n-1} + I_{n+1}) \qquad (3.20)$

the K_m and I_m being modified Bessel functions. Following their analysis further, it can be shown that inertial terms never become important enough to be considered in the governing equations so that subsequent terms in the Stokes expansion only differ from each other by a numerical factor. Calculation of the next term in the Oseen expansion allows us to write $F_2(R)\psi_2$ in terms of the Stokes coordinate r and matching requirements give

$$f_2(R) = \frac{-(1/2 - \gamma + \log 4)}{(\log R)^2} \qquad (3.21)$$

where γ is Euler's constant.

This process can be continued to give all of the ψ_n and $f_n(R)$. The solution for the stream function valid near the cylindrical void is then

$$\psi(r,\theta) = -\frac{r \log r}{\log R} \left[1 + \frac{1/2 - \gamma + \log 4}{\log R} + 0(\log R)^{-2} \right] \sin \theta \qquad (3.22)$$

3.2 The Normal Stress Condition

Since inertial terms never alter the form of the Stokes solution they cause no distortion in the shape of the bubble and the condition of zero unbalanced normal stress on the surface can be written in the form

$$P_\theta(1,\theta) - 2R^{-1} u_{r\theta}(1,\theta) = 0. \qquad (3.23)$$

The tangential pressure gradient can be easily calculated from (3.4), and (3.23) becomes

$$\left[\frac{1}{F} + \frac{4}{R \log R} \left(1 + \frac{1/2 - \gamma + \log 4}{\log R} \right) \right] \sin \theta = 0. \qquad (3.24)$$

If this is to be identically zero the coefficient of $\sin \theta$ must be zero and we are led to the relationship

$$\frac{1}{F} = \frac{4}{R} \left(\log \frac{4}{R} - \gamma + \frac{1}{2} \right)^{-1} \qquad (3.25)$$

which has been rewritten in the form of Kaplun and Lagerstrom (1957).

The resulting pressure field valid near the bubble is easily calculated as

$$p(r,\theta) = -\frac{2}{R}\left(\log\frac{4}{R} - \gamma + \frac{1}{2}\right)^{-1}\left(2r - \frac{1}{r}\right)\cos\theta + \text{const.} \tag{3.26}$$

in which the constant is arbitrary and merely sets the reference pressure.

Finally, the stream function ψ for the gas flow is found from (3.5) to be

$$\Psi = \frac{U_{S\infty}}{U_{g\infty}}\left[\psi + \frac{Z_\infty \rho_S U_{S\infty}}{a\beta(Z_\infty)}\int_r^{-1} p_\theta \, dr\right]$$

$$= \frac{U_{S\infty}}{U_{g\infty}}\left[\psi + \frac{2Z_\infty \rho_S U_{S\infty}}{a\beta(Z_\infty)R}\left(\log\frac{4}{R} - \gamma + \frac{1}{2}\right)^{-1}\left(2r + \frac{1}{r}\right)\sin\theta\right] \tag{3.27}$$

for the inner region; ψ meets the kinematic condition $\psi(r, 0) = \psi(r, \pi) = 0$.

We are not particularly interested in events far from the bubble so we have not performed the tedious calculations needed to describe the pressure and gas flow fields there.

4. THREE-DIMENSIONAL BUBBLE

The three-dimensional case exhibits a significant difference from the circular cylindrical void in that inertial terms enter into the Stokes solution and therefore into the expressions for the fluid pressure and the normal stress exerted by the solids flow on the bubble surface. If the pressure there is to be constant the bubble must be allowed to distort from the spherical; the condition of constant surface pressure determines the shape of the void.

We begin with a bubble whose volume is equal to that of a sphere of radius a. Following the non-dimensionalisation of §3 and using the equivalent radius as the characteristic length, the equations of motion in spherical coordinates stationary with respect to the bubble center are

$$\frac{1}{r}(r^2 u)_r + \frac{1}{\sin\theta}(v\sin\theta)_\theta = 0 \tag{4.1}$$

$$\frac{1}{r}(r^2 U)_r + \frac{1}{\sin\theta}(V\sin\theta)_\theta = 0 \tag{4.2}$$

$$uu_r + \frac{vu_\theta}{r} - \frac{v^2}{r} = -p_r - \frac{\cos\theta}{F} + \frac{1}{R}\left[\frac{1}{r^2}(r^2 u_r)_r\right.$$

$$\left. + \frac{1}{r^2\sin\theta}(u_\theta\sin\theta)_\theta - \frac{2u}{r^2} - \frac{2v_\theta}{r^2} - \frac{2v\cot\theta}{r^2}\right] \tag{4.3}$$

$$uv_r - \frac{vv_\theta}{r} + \frac{uv}{r} = -\frac{1}{r}p_\theta + \frac{\sin\theta}{F}$$

$$+ \frac{1}{R}\left[\frac{1}{r^2}(r^2 v_r)_r + \frac{1}{r^2\sin\theta}(v_\theta\sin\theta)_\theta + \frac{2u_\theta}{r^2} - \frac{v}{r^2\sin^2\theta}\right] \tag{4.4}$$

$$r^{-1}P_\theta = \frac{-a\beta(Z_\infty)}{Z_\infty \rho_s U_{s\infty}} \left(\frac{U_{g\infty}}{U_{s\infty}} V - v\right) \tag{4.5}$$

where the Froude and Reynolds numbers have the same definitions as before. We again assume $R \ll 1$ and eliminate pressure from (4.3) and (4.4). The resulting equation in terms of the stream function, $u = -\psi_\theta/r^2 \sin\theta$, $v = \psi_r/r \sin\theta$ is

$$\frac{1}{r^2}\frac{\partial(\psi, D_r^2 \psi)}{\partial(r,\mu)} + \frac{2}{r^2} D_r^2 \psi L_r \psi = \frac{1}{R} D_r^4 \psi \tag{4.6}$$

where $\mu = \cos\theta$

$$D_r^2 = \frac{\partial^2}{\partial r^2} + \frac{1-\mu^2}{r^2}\frac{\partial^2}{\partial\mu^2}$$

and $\qquad L_r = \frac{\mu}{1-\mu^2}\frac{\partial}{\partial r} + \frac{1}{r}\frac{\partial}{\partial\mu}$

Take $r = 1 + \zeta(\mu)$ as the equation of the surface of the bubble with the requirement that $|\zeta(\mu)| \ll 1$ since we shall deal with small departures from spherical shape. Boundary conditions on the stream function are then

$$\psi[1 + \zeta(\mu), \mu] = 0$$

$$\left.\psi_{rr}[1 + \zeta(\mu), \mu] - 2r^{-1}\psi_r[1 + \zeta(\mu), \mu] = 0\right\} \tag{4.7}$$

$$\psi \to \frac{1}{2}r^2(1 - \mu^2) \text{ as } \quad r \to \infty \tag{4.8}$$

while the normal stress condition is

$$-p[1 + \zeta(\mu), \mu] + 2R^{-1}u_r[1 + \zeta(\mu), \mu] = \text{constant.} \tag{4.9}$$

4.1 Solution for the Stream Function

The analysis closely follows that of Proudman and Pearson (1957) and Taylor and Acrivos (1964), so much of the detail will be omitted from what follows.

The expansion in the Stokes region is assumed to take the form

$$\psi(r, \mu) = \sum_0^\infty f_n(R)\psi_n(r, \mu) \tag{4.10}$$

and the boundedness requirement on the magnitude of the velocity allows us to write $f_0(R) = 1$. The fir first term of the Stokes solution which meets the conditions (4.7) and (4.8) is

$$\psi_0 = \frac{1}{2}r(r - 1)(1 - \mu^2) + 0(\zeta). \tag{4.11}$$

In terms of Oseen variables, $\rho = Rr$ and $\psi = R^2 \psi$, the leading term in the outer solution is the uniform stream

$$\Psi_0 = \frac{1}{2} \rho^2 (1 - \mu^2).$$ (4.12)

Here we have assumed an expansion of the form

$$\Psi(\rho, \mu) = \sum_0^\infty F_n(R) \Psi_n(\rho, \mu)$$ (4.13)

with $F_0(R) = 1$.

The second term in the Oseen expansion satisfies

$$\frac{1 - \mu^2}{\rho} \frac{\partial D_\rho{}^2 \Psi_1}{\partial \mu} + \frac{\partial D_\rho{}^2 \Psi_1}{\partial \rho} = D_\rho{}^4 \Psi_1.$$ (4.14)

As is well known (Proudman and Pearson, 1957) the most general solution vanishing at infinity and along $\mu = \pm 1$ is

$$D_\rho{}^2 \Psi_1 = e^{1/2 \rho \mu} \sum_1^\infty A_n \left(\frac{1}{2} \rho\right)^{1/2} K_{n+1/2} \left(\frac{1}{2} \rho\right) Q_n(\mu)$$ (4.15)

where $K_{n+1/2}(\frac{1}{2}\rho)$ is a modified Bessel function and $Q_n(\mu)$ is the integrated Legendre polynomial of degree n,

$$Q_n(\mu) = \int_{-1}^\mu P_n(x)\, dx.$$ (4.16)

By immediately applying boundary conditions it can be shown that

$$A_1 = -\left(\frac{\pi}{2}\right)^{-1/2}$$

$$A_n = 0, n \geqslant 2$$ (4.17)

and this gives

$$D_\rho{}^2 \Psi_1 = \frac{1}{2} \left(1 + \frac{2}{\rho}\right) e^{-1/2\rho(1-\mu)} (1 - \mu^2).$$ (4.18)

The particular integral which is not of greater order than unity in the Stokes region and whose terms of this order match the Stokes solution is

$$\Psi_1 = -(1 + \mu)(1 - e^{1/2\rho(1-\mu)}).$$ (4.19)

The inertial terms in the Stokes form (4.6) can be estimated from the leading Stokes term. Thus the equation for ψ_1 is

$$D_r^4 \psi_1 = 6\left(\frac{1}{r^2} - \frac{1}{r^3}\right) Q_2(\mu) \tag{4.20}$$

where we have put $f_1(R) = R$ but retained the possibility of constants of integration being functions of R. A particular integral of (4.20) is

$$\frac{1}{4}(r-1)Q_2(\mu) + 0(\zeta).$$

It can be shown that the only term in the general solution of the biharmonic equation which meets matching requirements and the boundary conditions is just a multiple of ψ_0; the full solution to (4.20) is then

$$\psi_1 = Br(r-1)(1-\mu^2) - \frac{1}{8}r(r-1)\mu(1-\mu^2). \tag{4.21}$$

Writing this in terms of Oseen variables and taking $\rho = 0(1)$ for $R \ll 1$, (4.21) becomes

$$f_1(R)\psi_1 = R\left[\frac{1}{4}\rho^2 Q_2(\mu) - 2B\rho^2 Q_1(\mu)\right] + 0(R). \tag{4.22}$$

If $B = 1/8$ this expansion agrees with the Oseen term Ψ_1 for small ρ and the solution for the second term in the Stokes expansion is therefore

$$\psi_1 = \frac{1}{8}r(r-1)(1-\mu^2)(1-\mu). \tag{4.23}$$

We now return to the first term in the Stokes expansion. Because of the distortion of the bubble from spherical shape the constants of integration are in error by an amount $0(\zeta)$ or equivalently $0(R)$. After adjusting the constants to meet conditions on $r = 1 + \zeta(\mu)$ we have

$$\psi_0 = \frac{1}{2}r\{r - [1 + \zeta(\mu)]\}(1-\mu^2). \tag{4.24}$$

No adjustment is required at this stage in ψ_1 because it contributes terms $0(R)$ already.

It is obvious how the calculations can be continued to reveal higher order terms but we shall stop here; the effect of such terms will be discussed in section 5.

4.2 The Normal Stress Condition

It is convenient to apply the condition (4.9) on the surface of a sphere rather than on $r = 1 + \zeta(\mu)$. Expanding (4.9) about $r = 1$ and differentiating with respect to θ we obtain

$$p_\theta + \zeta p_{r\theta} - \zeta' p_r \sin\theta - 2R^{-1}(U_{r\theta} + \zeta U_{rr\theta} - \zeta' U_{rr} \sin\theta) = 0(\zeta^2) \tag{4.25}$$

on $r = 1$. The pressure terms in (4.25) can be readily calculated from (4.3) and (4.4). Then the requirement of balanced normal stress becomes

$$\left(\frac{1}{F} - \frac{3}{R}\right) - \frac{3}{4} + 2\mu + \left(\frac{1}{F} + \frac{3}{R}\right)\zeta + \frac{\mu}{F}\zeta' + \frac{3(1 - \mu^2)}{2R}\zeta'' = 0. \tag{4.26}$$

The $0(1/R)$ terms give the usual Hadamard (1911) relationship

$$F = \frac{1}{3}R \tag{4.27}$$

while the $0(1)$ terms give an equation for the first approximation to the distortion of the bubble. This is

$$(1 - \mu^2)\zeta'' + 2\mu\zeta' + 4\zeta = \frac{1}{2}R\left(1 - \frac{8}{3}\mu\right). \tag{4.28}$$

A particular integral is

$$\frac{R}{8}\left(1 - \frac{16\mu}{9}\right).$$

We note that the homogeneous part of (4.28) is of the form

$$(1 - x^2)w'' + 2(n - 1)xw' + 2nw = 0 \tag{4.29}$$

with $n = 2$. The complimentary function is therefore (Whittaker and Watson, 1963) the doubly integrated Legendre function of the second order, namely

$$C(1 - \mu^2)^2.$$

The constant of integration is found by requiring the bubble to have a volume equal to that of the unit sphere. This condition may be written

$$\int_{-1}^{1} \zeta(\mu)\, d\mu = 0 \tag{4.30}$$

and gives $C = -\frac{15}{64}R$. The bubble surface is therefore the line

$$r = 1 + \frac{R}{8}\left[\left(1 - \frac{16\mu}{9}\right) - \frac{15}{8}(1 - \mu^2)^2\right] + 0(\zeta^2). \tag{4.31}$$

The pressure distribution around the bubble in the Stokes region is readily calculated from (4.4) and is found to be

$$p(r, \mu) = -\frac{\mu}{R}\left(3r - \frac{1}{r^2}\right) + \frac{\mu}{2r^2}\left[\left(\frac{1}{4} - \frac{2\mu}{3}\right) - \frac{45}{32}(1 - \mu^2)^2\right] + \text{constant}. \tag{4.32}$$

in which the constant sets the reference pressure.

Finally the gas flow field in the linear region is found from (4.5) to be

$$\Psi(r,\mu) = \frac{U_{s\infty}}{U_{g\infty}} \left\{ -\frac{Z_\infty \rho_S U_{s\infty}}{a\beta(Z_\infty)}(1-\mu^2)\left[\frac{1}{2R}\left(3r^2 + \frac{2}{r}\right)\right.\right.$$

$$\left.\left. -\frac{1}{r}\left(\frac{37}{64} + \frac{2\mu}{9} - \frac{45}{64}\mu^2 + \frac{15}{64}\mu^4\right)\right] + \psi(r,\mu) \right\}$$

$$(4.33)$$

and we note that $\Psi(r,\mu)$ meets the kinematic condition on $\mu = \pm1$.

5. DISCUSSION

The expansions we have obtained for the circular cylindrical void are quite different from those found for the three-dimensional bubble. The entire Stokes expansion for the cylinder corresponds to just the first term in the Stokes expansion for the axisymmetric case. Since inertial terms never enter into the expansion for the cylinder we can gain no information whatever about its distortion from the basic circular shape.

In the case of the three-dimensional void (4.31) gives the first approximation to the distortion from spherical shape, and it is interesting to note that unlike the immiscible drops described by Taylor and Acrivos (1964) which were found to be oblate spheroidal in shape, in the present case of bubbles in fluidized beds, only axial symmetry is observed. This shape may well be due to our inability to provide a mechanism for the (non-uniform) pressure jump at the interface but axial symmetry is certainly characteristic of real bubbles which are found to progressively flatten at the rear as the Reynolds number increases. In the current analysis however, the bubble exhibits elongation with a underside of large radius of curvature as shown in Figure 1, more akin to the shape of a gas slug in a tube but also similar to the shape of large two-dimensional bubbles reported by Goldsmith and Rowe (1975). To the order of terms retained in arriving at (4.31) the characteristic indentation at the rear does not emerge. As shown by Taylor and Acrivos (1964), consideration of higher order effects result in an indented rear for liquid drops and a similar result is likely here. It must be kept in mind of course that the low-Reynolds-number approximation is a gross over-simplification made because of the availability of techniques for handling such flows—real bubbles rise at Reynolds numbers greater than unity.

Experimentally the rise velocity of a bubble in a fluidized bed is determined solely by the bubble size. On the other hand, the gas velocity required for fluidization depends for example on the density of the particles and their size. Thus a given bubble may rise at a higher or lower velocity than the fluidizing gas. If the latter occurs the void acts merely as a shortcut for the passage of gas. If the rise velocity is higher however, a recirculating cloud of gas forms around the bubble. This can be most clearly understood by realizing that relative to the bubble the direction of undisturbed gas flow is downwards yet the void offers no resistance to flow so that the pressure gradient causes gas to flow through its center. The upflow is returned down the sides to the base and continues to circulate in this manner.

This feature of real bubbles is partly shown by the stream functions for the gas flow fields given by (3.27) and (4.33) which possess zeros other than along the line of rise of the bubbles. As the ratio of bubble to fludizing gas velocity becomes large the cloud shrinks as one would expect and in qualitative agreement with observation. As the velocity ratio is decreased towards unity the cloud would be expected to become even larger, eventually encompassing the entire fluidized bed but this region is well outside the Stokes limit and equations (3.27) and (4.33) do not apply there.

The assumption of bubble rise at low Reynolds number has allowed us to exploit certain techniques and known solutions for the rise of fluid bubbles and drops in another fluid. It has also served

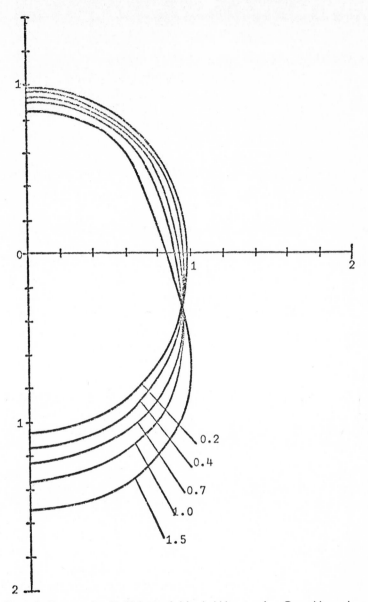

Figure 1. Cross-sectional half-shape of rising bubbles at various Reynolds numbers.

to point out a fundamental problem which arises at the bubble boundary. The physical situation is one of rise at intermediate Reynolds numbers so that a full solution to the real-bubble problem must await consideration of this case, undoubtedly by numerical techniques but the success of such as analysis will almost certainly depend on a consideration of the transition which takes place across the bubble surface.

ACKNOWLEDGEMENTS

I would like to express my thanks to Dr. L. S. Leung and Dr. A. M. Watts for several very useful discussions.

REFERENCES

Batchelor, G. K. 1973. Low-Reynolds-number bubbles in fluidized beds. Proceedings of the XI-th Polish Fluid Dynamics Symposium, in press.

Collins, R. 1965. A simple model of the plane gas bubble in a finite liquid. J. Fluid Mech., *22*, 763.

Davidson, J. F. 1961. Symposium on fluidization—discussion. Trans. Inst. Chem. Engrs., *39*, 230.

Davidson, J. F., Paul, R. C., Smith, M. J. S. & Duxbury, H. A. 1959. The rise of bubbles in a fluidized bed. Trans. Inst. Chem. Engrs., *37*, 323.

Davies, R. M. & Taylor, G. I. 1950. The mechanics of large bubbles rising through extended liquids and through liquids in tubes. Proc. Roy. Soc. A., *200*, 375.

Goldsmith, J. A. & Rowe, P. N. 1975. The shape of the bubbles in a two-dimensional gas fluidized bed. Chem. Eng. Sci. *30*, 439.

Hadamard, M. J. 1911. Mouvement permanent lent d'uni sphere liquide et visqueuse dans un liquide visqueux. Compte Rendu, *152*, 1735.

Hagyard, T. & Sacerdote, A. M. 1966. Viscosity of suspensions of gas-fluidized spheres. 2nd Eng. Chem. Fund., *5*, 500.

Harrison, D. & Leung, L. S. 1962. The rate of rise of bubbles in fluidized beds. Trans. Inst. Chem. Engrs., *40*, 146.

Jackson, R. 1963. The mechanics of fluidized beds. Part II: The motion of fully-developed bubbles. Trans. Inst. Chem. Engrs., *41*, 22.

Jackson, R. 1971. Fluid mechanical theory; in *Fluidization,* J. F. Davidson & D. Harrison (Eds.) Academic Press.

Kaplun, S. & Lagerstrom, P. A. 1957. Asympototic expansions of Navier-Stojes solutions for small Reynolds numbers. J. Math. Mech., *6*, 585.

Littman, H. & Homolka, G. A. J., 1970. Bubble rise velocities in two-dimensional gas-fluidized beds from pressure measurements. Chem. Eng. Prog. Symp. Series No. 105, *66*, 37.

Littman, H. & Homolka, G. A. J. 1973. The pressure field around a two-dimensional gas bubble in a fluidized bed. Chem. Eng. Sci., *28*, 2231.

Lockett, M. J. & Harrison, D. 1967. The distribution of voidage fraction near bubbles in gas fluidized beds. Int. Symposium on Fluidization—Proceedings. Netherlands University Press, Amsterdam.

Murray, J. D. 1965a. On the mathematics of fluidization, Part 1. Fundamental equations and wave propagation. J. Fluid. Mech., *21*, 465.

Murray, J. D. 1965b. On the mathematics of fluidization, Part 2. Steady motion of fully developed bubbles. J. Fluid Mech., *22*, 57.

Nguyen, X. T. 1974. Bubble formation and rise in a two-dimensional fluidized bed. Ph.D. Thesis, University of Queensland.

Nguyen, X. T. Leung, L. S. & Weiland, R. H. 1973. On void fractions around a bubble in a two-dimensional fluidized bed. Proceedings of Int. Congress on Fluidization, Toulouse, in press.

Proudman, I. & Pearson, J. R. A. 1957. Expansions at small Reynolds number for the flow past a sphere and a circular cylinder. J. Fluid Mech., *2*, 237.

Pyle, D. L. & Harrison, D. 1967. The rise velocity of bubbles in two-dimensional fluidized beds. Chem. Eng. Sci., *22*, 531.

Reuter, H. 1963. Druckverteilung um Blasen im Gas-Feststoff-Fleissbett. Chem. Ing. Tech., *35*, 98.

Rowe, P. N. 1971. Experimental properties of bubbles. In *Fluidization*, J. F. Davidson & D. Harrison (Eds.) Academic Press.

Rowe, P. N. & Partridge, B. A. 1965. An x-ray study of bubbles in fluidized beds. Trans. Inst. Chem. Engrs., *43*, 157.

Schügerl, K., Merz, M., & Fetting, F. 1961. Rheologische Eigenschaften von gasdurchströmten Fleissbettsystemen. Chem. Eng. Sci., *15*, 1.

Taylor, T. D. & Acrivos, A. 1964. On the deformation and drag of a falling viscous drop at low Reynolds number. J. Fluid Mech., *18*, 466.

Whittaker, E. T. & Watson, G. N. 1963. A Course of Modern Analysis, 4th ed. Cambridge University Press.

THE FLUID MECHANICS OF SINGLE BUBBLES

A. L. F. MARTIN-GAUTIER AND D. L. PYLE

NOMENCLATURE

a	bubble radius
$\underline{g}$	gravity
$\underline{i}$	unit vector (in the downwards direction)
P	pressure
r	radial coordinate
r_c	cloud radius
t	time
U_b	bubble velocity
U_o	superficial velocity at minimum fluidisation
$\underline{u}$	gas velocity
u_o	interstitial velocity at minimum fluidisation
$\underline{v}$	particle velocity
x	vertical axis
X	drift of particle in x direction
y	horizontal axis
Fr	Froude number, $Fr = \dfrac{U_b^2}{ga}$
Re	bubble Reynolds number, $Re = \dfrac{2aU_b\ \rho_p(1-\varepsilon_o)}{\mu_p}$
α_b	dimensionless ratio $\alpha_b = \dfrac{U_b}{u_o}$
β	fluid particle drag coefficient
γ	dimensionless ratio, $\gamma = \dfrac{\beta_o}{\beta}$
ε	voidage fraction
ζ	component of vorticity
η	coordinate in transformed plane, θ direction
θ	angle
$\mu_f,\ \mu_p$	viscosity
$\nu_f,\ \nu_p$	kinematic viscosity

ξ	coordinate in transformed plane, r-direction
ρ	dimensionless radial component
$\rho_f, \; \rho_p$	density
τ'	dimensionless time on drift profile w.r.t. an undisturbed layer of particles above the bubble
χ	gas stream function
ψ	particle stream function
$\underline{\omega}$	vorticity vector

Subscripts

f	fluid phase
p	particle phase
o	minimum fluidisation conditions
$r, \; \rho, \; \xi$	radial component
$\theta, \; \eta$	tangential component
x	vertical component
y	horizontal component
∞	conditions at the outer boundary

Superscripts

| * | dimensionless quantity |

SYNOPSIS

A detailed theoretical study of the steady motion of a circular bubble in a two-dimensional bed has been carried out; a newtonian model is assumed to describe the behaviour of the particulate phase. Results for the gas and particle motion are obtained for bubble Reynolds numbers in the range 0.01-200.

Experimental data on the gas velocity distribution around the bubble and particle path - and streamlines, drift profiles and velocity distribution are compared with existing theories. It is shown that there is a marked difference between the behaviour of injected and naturally-occurring bubbles. It is concluded from the studies that no single theory can provide an adequate description of the particles and gas. Close to the bubbles, potential flow gives a fair description of the particle movement; further away, the motion appears non-Newtonian. The gas motion is often adequately described by Davidson's theory, but there are significant effects of particle size, and due to the bubble wake.

INTRODUCTION

The first coherent theory of bubble motion in a fluidised bed
developed by Davidson (1,2) assumed that the particles moved around
the bubble in potential flow and that the percolation of gas
through the particulate phase was given by Darcy's Law. Jackson (3)
and Murray (4) introduced a momentum equation for the particle
motion, whilst assuming the absence of any interparticle forces; the
predictions of the gas motion of these theories were quantitatively
different from Davidson's analysis, and measurements on the gas
flow patterns--as in Rowe's studies of the gas clouds (5)--have
tended to favour these latter two theories. On the other hand there
have been various criticisms of the neglect by all these workers of
the possibility of interparticle forces. For example, evidence for
the existence of such forces has been produced by several groups
of workers (e.g. Buysman & Peersman (6), Rietema (7), Verloop
et al (8), Agbim et al (9), Rowe (10)) and Rietema (11) has applied
mechanical stress theory to fluidisation. Anderson & Jackson (12)
have also demonstrated the importance of the stress tensor in the
continuum form of the equations of motion. Before discussing the
problems of incorporating such terms into the analysis of bubble
motion, it is worthwhile to consider briefly some of the experimental evidence on the gas and particle movement in bubbling beds.

Considering first the particle motion, there are essentially two
very different types of information available. First, there are
measurements,both direct and indirect, of the rheological properties
of fluidised systems. There is an excellent review of this field
by Schügerl (13). The most reliable direct measurements appear to
be those of Schügerl et al (14) and of Hagyard & Sacerdote (15).
In all these studies it is clear that the bed behaviour is non-
newtonian around the point of incipient fluidisation; Schügerl
quotes some results showing the existence of a limiting shear stress
below which particle motion is not observed--the particulate phase
thus has the properties of a Bingham plastic. For fully fluidised
beds both sets of workers agree that often the behaviour can be
described in terms of a newtonian shear viscosity. The behaviour
is very sensitive to factors such as particle size, size distribution, the nature of the particle surface etc. Apparent viscosities of the order of 10 poise appear to be typical of such systems;
defining a Reynolds number for the bubble motion by

$$Re = \frac{2a\, U_b\, \rho_p(1-\varepsilon_o)}{\mu_p}$$, these measurements are equivalent to Reynolds
numbers in the range 10-100 for bubbles in
typical experimental situations.

Besides direct viscosity measurements, attempts have been made to
deduce the values from the behaviour of isolated bubbles. For
example, Murray (16) deduces an order of magnitude for μ_p from the
bubble motion, arriving at a value of around 10 poise. Stewart (17)
used the measured pressure distribution along the bubble centre
line, and deduced values which agree fairly well with Schugerl's
experimental values for medium-sized particles (100µm<dp<300µm).

The most satisfactory method was developed by Grace (18), using an
analogy with spherical-cap bubbles in viscous liquids. The values
deduced agree well with other experimental data.

In the second type of study, the influence of rising bubbles on the
movement of particles--either individually or in layers--has been
studied. Various workers (Rowe et al (19 a,b), Raso et al (20),
Woollard & Potter (21), Singh et al (22), Toei (23)) have measured
the movement or 'drift' of layers of particles during the passage
of a bubble. There are certain differences between these measurements and the motion predicted by assuming that the particles
behave like a perfect fluid (Maxwell (24), Darwin (25)). For
example, Toei et al (23) showed that the net drift was much smaller
than the theoretical; the observed downward movement of particles
away from the bubble does not accord with the theory.

Gabor (26,27) studied the movement of individual particles (in a
two-dimensional bed); he failed to observe the predicted particle
'looping' near the bubbles. He also (28) compared his results
with three models for the behaviour of the particulate phase--as
a newtonian fluid (under creeping flow), a power-law fluid, and
Bingham fluid--and concludes that the latter model is the most
appropriate. This accords with Schugerl's measurements on beds
close to incipient fluidisation.

Other measurements of particle velocities (e.g. Baeyens & Geldart
(29), Patureaux et al (30)) show some discrepancies from Davidson's
theory.

In synthesis, the evidence, although partial, suggests that inviscid
theories may be inadequate to describe the particle motion. The
experiments also show the importance of the ratio between the bed
and bubble dimensions, and of particles falling from the bubble
roof.

If anything, there are fewer detailed studies of the gas motion
around rising bubbles. The various studies on the gas clouds
surrounding the bubbles show better agreement with Murray's &
Jackson's theories than with Davidson's (see e.g. Rowe's (5)
summary of the experimental data). However, there are few studies
on the local properties of the gas velocity field. Rowe and
Partridge (31) found reasonable agreement between Davidson's theory
and their measurements using a nitrogen dioxide tracer, but the
method is not very accurate. Anwer & Pyle (32) concluded that the
velocity field ahead of the bubble and cloud is predicted better
by Murray's than by Davidson's theory. Studies of the gas throughflow through the bubble do not discriminate clearly between the
theories.

In this study, then, we set out to investigate two parts of this
problem: first, the predicted effects on particle and gas motion of
incorporating viscous stresses into the equations of motion for
the particle phase, and, second, the experimental evidence on the
movement of both phases around single bubbles rising in fluidised

beds. In this paper we summarise the latter findings: more detailed accounts of the theoretical development and procedures, and discussion of the results will be published elsewhere.

Theory and Numerical Solutions

There are considerable problems in establishing valid constitutive equations for fluidised systems; in this work we have not attempted to solve the problem of their derivation, but rather, given the evidence--however limited--for the newtonian behaviour of fluidised systems, we have chosen to treat the particle phase as a fluid with viscosity μ_p. Given the predictions from the theory and experimental evidence on the gas and particle motion we can then test the adequacy of the theory and of our assumptions. A complete account of the work to be found in Martin-Gautier (33).

The starting point is the set of equations of motion derived by Anderson & Jackson (12) and summarised by Jackson in his review (34). The gas and particles are treated as two interpenetrating continua, and we also make a number of further assumptions. In this work we consider the fluid mechanics of a single, two-dimensional bubble, rising steadily through a gas-fluidised bed. It is assumed that the bubble remains cylindrical in shape, so that solutions for the rear of the bubble have dubious validity. Furthermore, we assume : (i), that viscous stresses from the particulate phase are much more significant than from the fluid (since v_p/v_f is $O(10^2)$), (ii) that the momentum of the fluid phase is neglible and,(iii), that the bouyancy contribution from the fluid is negligible (since ρ_p/ρ_f is $O(10^3)$).

The interaction force per unit volume of bed is taken to be $\beta(\varepsilon)(\underline{u} - \underline{v})$, where $\underline{u}$ and $\underline{v}$ are the gas and particle velocity vectors and ε is the voidage fraction.

Finally the isotropic component of the particle pressure P_p is neglected, since at the present time its magnitude and form is unknown (Jackson (34)).

This leaves four equations of continuity and momentum, with the state variables $\underline{u}$, $\underline{v}$, ε and P_f; the boundary conditions are discussed below. Despite the simplifications, the set is extremely difficult to solve, and the iterative procedure proposed by Jackson (3) is employed. That is, on the first iteration we take $\varepsilon = \varepsilon_o$ everywhere except where it enters the drag coefficient $\beta(\varepsilon)$. The independent variable then becomes β, and the solution for β determines the new estimate of ε, which can be used in further iterations. In this study we limit ourselves to the first iteration, and with these assumptions the equations of continuity and momentum become:

$$\text{div } \underline{u} = 0 \tag{1}$$

$$\text{div } \underline{v} = 0 \tag{2}$$

$$\rho_p(1 - \varepsilon_o)(\underline{v}.\underline{\nabla})\underline{v} = -\underline{\nabla}P_f + \rho_p(1 - \varepsilon_o)\underline{g} + \mu_p\nabla^2\underline{v} \tag{3}$$

$$\underline{\nabla}p_f = -\beta(\varepsilon)(\underline{u} - \underline{v}) \tag{4}$$

It will be recognised immediately that eqns (2) and (3) are identical with the equations of continuity and momentum for an incom-
pressible liquid of viscosity μ_p. Since $\underline{g} = \underline{\nabla} \, \Phi$ it is convenient
to define

$$-\underline{\nabla}P = -\underline{\nabla}P_f + \rho_p(1 - \varepsilon_0)\underline{g} \qquad\qquad (5)$$

where P measures the difference in pressure between two points on
the same horizontal line.

It is also convenient to put the equations into dimensionless form
using the following reference variables:

$$\begin{aligned}
\text{Length} \quad &: \quad a \\
\text{Time} \quad &: \quad a/U_b \\
\text{Mass} \quad &: \quad \rho_p(1 - \varepsilon_0)
\end{aligned}$$

A complete listing of all dimensionless variables is given in
Table I:

Method of Solution

As in the case of Jackson's study of the inviscid problem we first
obtain a solution for the particle motion from eqns (2) and (3).
Unfortunately it is not possible to obtain an analytical solution
as in the case of inviscid flow, and the problem must be solved
numerically. The solution for the pressure distribution and the
Froude number (i.e. bubble rising velocity) is then used to generate
the voidage distribution around the bubble and then the gas
velocity field.

Table 1 Dimensionless variables

Variable or group	Dimensionless Form
Length	$x^* = x/a$
Radial distance	$\rho = r/a$
Velocity	$v^* = v/U_b$
Pressure	$P^* = P/(\tfrac{1}{2}\rho_p(1 - \varepsilon_0)U_b^2)$
Streamfunction	$\psi^* = \Psi/aU_b$
Vorticity	$\zeta^* = a\zeta/U_b$
Drag function	$\gamma = \beta/\beta_0$
Differential operator	$\nabla^* = a\nabla$
Reynolds no.	$Re = \dfrac{2aU_b\rho_p(1 - \varepsilon_0)}{\mu_p}$
Froude no.	$Fr = U_b^2/ga$

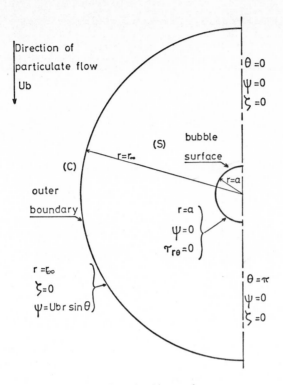

Figure 1. Domain of integration.

In the present problem it is convenient to choose the demain of
integration (s) shown in figure I, with the domain limited by the
curve (c) where:

   (i) along θ = 0 and θ = π, the particle streamfunction (Ψ) and
        vorticity (ζ) are both zero.

   (ii) on the outer boundary of (c) the particle flow is ideal
        and uniform (i.e. is a potential flow).

   (iii) the surface of the bubble is a particle streamline, so
         Ψ = 0 on r = a.

   (iv) finally, the viscous stress on the bubble surface is
        assumed small so that we take $\tau_{r\theta}$ = 0.

The boundary conditions are summarised in Table II:

We solve for the particle motion in terms of the streamfunction and
vorticity; the dimensionless form of the equations derived from
(2) and (3) are:

$$\zeta* = \nabla*^2 \psi*$$ (6)

and $$\frac{1}{\rho}\left[\frac{\partial\psi*}{\partial\rho}\frac{\partial\zeta*}{\partial\theta} - \frac{\partial\psi*}{\partial\theta}\frac{\partial\zeta*}{\partial\rho}\right] = \frac{2}{Re}\nabla*^2\zeta*$$ (7)

Table 2   Boundary conditions for the particulate flow

Boundary Conditions	Dimensionless form of the Conditions
(i) $\theta=0$ $a\leqslant r\leqslant r_\infty$ $\begin{cases}\Psi=0 \quad \dfrac{\partial\Psi}{\partial r}=0 \\[4pt] \zeta=0\end{cases}$ $\theta=\pi$ $a\leqslant r\leqslant r_\infty$	(i) $\theta=0$ $1\leqslant\rho\leqslant\rho_\infty$ $\begin{cases}\Psi^*=0 \quad \dfrac{\partial\Psi^*}{\partial\rho}=0 \\[4pt] \zeta^*=0\end{cases}$ $\theta=0$ $1\leqslant\rho\leqslant\rho_\infty$
(ii) $r=r_\infty$ $0\leqslant\theta\leqslant\pi$ $\Psi=U_b r_\infty\sin\theta$ $\zeta=0$	(ii) $\rho=\rho_\infty$ $0\leqslant\theta\leqslant\pi$ $\Psi^*=\rho_\infty\sin\theta$ $\zeta^*=0$
(iii) $r=a$ $0\leqslant\theta\leqslant\pi$ $\Psi=0$ $\dfrac{\partial\Psi}{\partial\theta}=0$ $\tau_{r\theta}=0:\ \dfrac{\partial^2\Psi}{\partial r^2}=\dfrac{1}{a}\dfrac{\partial\Psi}{\partial r}$	(iii) $\rho=1$ $0\leqslant\theta\leqslant\pi$ $\Psi^*=0$ $\dfrac{\partial\Psi^*}{\partial\theta}=0$ $\tau_{\rho\theta}{}^*=0:\ \dfrac{\partial^2\Psi^*}{\partial\rho^2}=\dfrac{\partial\Psi^*}{\partial\rho}$

These were solved by finite difference methods using a transforma-
tion of the domain S defined by $\rho = e^{\pi\xi}$ and $\theta = \pi\eta$ first discussed
by Apelt (35) and which ensures a rectangular grid and a fine mesh
near the bubble surface where gradients in the vorticity and stream
function are highest. The vorticity equation (7) was put into
finite difference form using the Peaceman-Rachford alternating
direction method, and the equation for the stream function (eqn 6)
was solved by the method of successive overrelaxations. The initial
approximation to the stream function, Ψ^* (o), was taken as the
potential flow solution as given by Davidson & Harrison (2); the
initial vorticity field was assumed zero everywhere except on the
bubble surface, where initial values were approximated by a four-
point formula. The complete calculation procedure is discribed by
Martin-Gautier (33), together with the conditions employed to
ensure convergence. Given the streamfunction Ψ^* and vorticity ζ^*,
the pressure distribution can then be calculated by transformation
and integration of eqns (3) and (5), as can the complete particle
velocity field, particle drift, and the drag coefficient on the
bubble.

Once the solution for the particle streamfunction Ψ^* (and therefore
for $\underline{v}$) has been obtained, it is relatively straightforward to compute the gas velocity $\underline{u}$. Eqn (4) can be put into the form

$$\underline{u}^* = \underline{v}^* - \frac{\gamma}{2}\ \frac{Fr}{\alpha_b}\ \nabla^*P_f^* \tag{8}$$

$$\text{with} \qquad \nabla^*P_f^* = \nabla^*P^* + \frac{2}{Fr}\ \underline{i} \tag{9}$$

where $\gamma = \beta(\epsilon_b)/\beta$, and $\alpha_b = U_b/U_o$.

The Fronde number is calculated from the tangential component of
eqn (3) by satisfying the condition (Davies & Taylor (36), Jackson

(3)) that changes in pressure along the bubble surface are small near $\theta = 0$. From (3) it can be shown that to ensure

$$\frac{\partial^2 p}{\partial \theta^2} = 0 \text{ at } \theta = 0, \rho = 1, \tag{10}$$

$$Fr = \cfrac{1}{\left(\frac{\partial v_\theta^*}{\partial \theta}\right)^2\Bigg|_{\substack{\rho = 1 \\ \theta = 0}} - \frac{2}{Re}\frac{\partial^2 \zeta^*}{\partial \rho \partial \theta}\Bigg|_{\substack{\rho = 1 \\ \theta = 0}}}$$

which reduces correctly to the Davies & Taylor result ($Fr = 0.250$) for irrotational inviscid flows i.e. as $Re \to \infty$.

The variation in voidage fraction is obtained by taking the divergence of eqn (4), which leads to

$$\underline{\nabla}^* \gamma \cdot \underline{\nabla}^* P_f^* + \gamma \nabla^{*2} P_f^* = 0 \tag{11}$$

and which can be solved by the method of characteristics. Using the Carman-Kozeny equation to define the drag coefficient β it is easy to show that

$$\varepsilon = \frac{\sqrt{\gamma}}{\frac{1-\varepsilon_0}{\varepsilon_0} + \sqrt{\gamma}} \tag{12}$$

which finally allows one to calculate the voidage variation around the bubble.

Having solved for γ, Fr, $\underline{\nabla}^* P_f^*$ and $\underline{v}^*$ the solution for the gas velocity $\underline{u}^*$ is thus completely determined. The results can also be represented in terms of a gas stream-function χ.

Thirteen cases were calculated with bubble Reynolds numbers ranging from $Re = 0.01$ to $Re = 200$, requiring in all some 3.2 hours computing time on the Imperial College CDC 6400 computer.

Experimental

The experiments were carried out in a two dimensional bed with faces constructed of Armour plate glass, and with dimensions 128cm high × 48cm wide × 1.27cm deep. The distributor was made of porous plastic (Vyon). Humidified constant temperature air was used to fluidise the particles, whose characteristics are given in Table III. Values for the shear viscosity for those systems are also given in Table III. These values were used to calculate bubble Reynolds numbers in the subsequent tests of the theory.

Artificial bubbles could be injected some 12cm above the distributor using a solenoid valve.

The particle monement was followed using black tracer particles (coloured with Nigrosine alcohol soluble dye). They were filmed with a Milliken D.B.M. 5 High Speed camera at 400 frames/sec. Depending on the particle size, the area of bed filmed varied from 16×12 cm to 10×8 cm.

Table 3   Characteristics of particles

Ballotini grade	Size range μ m	Density $\rho_p (g/cm^3)$	Apparent viscosity μ_p (poise)	Kinematic viscosity $\nu_p (cm^2/sec)$	Minimum of fluid-isation U_o (cm/sec)
7	505-700	2.86	9.5[1]	3.32	34.3*
8	440-530	2.86	12[2]	4.2	25.6*
10	210-325	2.86	9[2]	3.14	8.9*

*   values obtained from a curve : pressure drop v.s. superficial velocity

(1) viscosity value calculated by Grace (18)

(2) viscosity value measured by Schügerl (14)

Two types of experiment were carried out. In the first, single artificial bubbles were injected through the solenoid valve. In all cases it was necessary to maintain the gas velocity through the bed just below the experimentally determined minimum fluidising velocity to prevent elongation or splitting of the bubbles. In such cases only relatively small stable bubbles could be examined. In a second series of experiments in a freely bubbling bed, single bubbles, free from interference, were filmed. In all cases the results were analysed with the aid of a PCD digital motion analyser.

The gas velocity inside the dense phase region of the fluidised bed was measured using a specially constructed shielded hot wire probe. The basic sensor was a DISA 55M10 anemometer with 55 F11 probe (3mm long). A protective cage around the sensor was made of 190μm diam. stainless steel wire, spiral wound, such that the voidage fraction of the entire assembly was close to that of the fluidised bed ($\epsilon \sim 0.4$). The dimensions of the probe were 3mm×3mm. The probe was calibrated in a fixed bed prior to each run, and the bubble injection unit was connected to the U.V. recorder used to record the anemometer output, so that a mark was made on the recorder trace when the solenoid valve opened. The bubbles were also filmed and a flash was illuminated at the same time as the injection so that all events were synchronised.

Results and Comparison with Theory

It is not possible to give here an exhaustive account of both the theoretical predictions and of the experimental results; instead an attempt is made to summarise some of the main features of the experimental results and their relevance to theory.

(a)   Particle motion

Experimental data on particle motion was obtained in terms of the particle pathlines, drift profiles, and velocity distributions around the bubble.

Figures (2) and (3) show typical results for the particle pathlines
in the region of the horizontal axis of the bubble. Since the
theoretical results were only computed at a limited number of Reynolds numbers between 0.01 and 200, here--as in the remaining
examples--predicted values for comparison with the experimental results are taken at the nearest value of the Reynolds number to that
estimated experimentally. The potential flow solution is also shown
since in the case of the particle motion this coincides with Davidson's and Jackson's theory. It will be noted that the theoretical
pathlines are asymmetrical at intermediate Reynolds numbers; it may
be remarked that 'looping' in the form of an ' α ' is predicted to
disappear altogether for Reynolds numbers less than around 10.

The experimental results show two interesting features: first, there
is a quite remarkable difference in behaviour between bubbles
injected into incipiently fluidised beds and those occurring in
freely bubbling beds. In the former case the particles are much
less mobile, and indeed the particle movement only extends to
around two bubble radii, whereas in freely bubbling beds the motion
extends up to around five radii. Secondly, it appears possible to
discuss the results for natural bubbles in two areas: away from the
bubbles the experimental results perhaps appear closer to those
calculated including the viscous terms than the potential flow prediction, although the agreement is not good; close to the bubble the
results are quite close to the potential flow prediction, perhaps

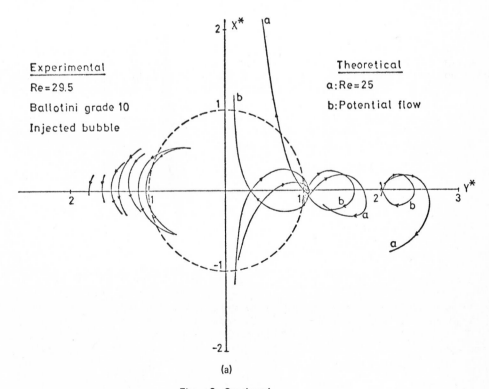

(a)

Figure 2. Continued.

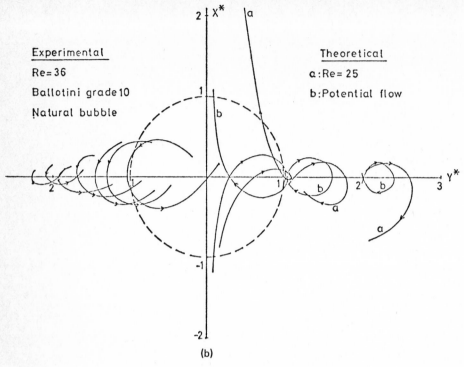

Figure 2. Paths of particles.

because the voidage is rather higher in this region. Nonetheless,
the results do show the predicted looping: it appears likely that
Gabor's failure to observe this looping was due to the small bubble
size, since in this work looping only appeared with larger bubbles.

An alternative way of treating the results is in terms of the particle streamlines and 'drift' with respect to the bubble. The drift
lines represent successive positions of a layer of tracer particles;
these lines are thus lines of time = constant, and in figures (4) &
(5) are plotted at integral values of the dimensionless time
$\tau = tU_b/a$, with respect to the initial undisturbed position of the
layer. The total drift, that is, the net volumetric movement of
particles in the direction of bubble rise, was found to be much less
than predicted by all Newtonian theories (in the case of Stokes
flow, the theoretical drift is infinite).

More conclusive and quantitative results can be obtained by considering the particle velocities. Typical results for natural bubbles
are shown in figures (6) and (7), showing the vertical particle
velocity component along the horizontal axis. Any effect of Reynolds number on the experimental results is masked by the experimental variation, and close to the bubble the best representation of
the particle velocity field is given by potential flow. Away from
the bubble, the theories are close and within the range of experimental error. Measurements of the absolute particle velocity

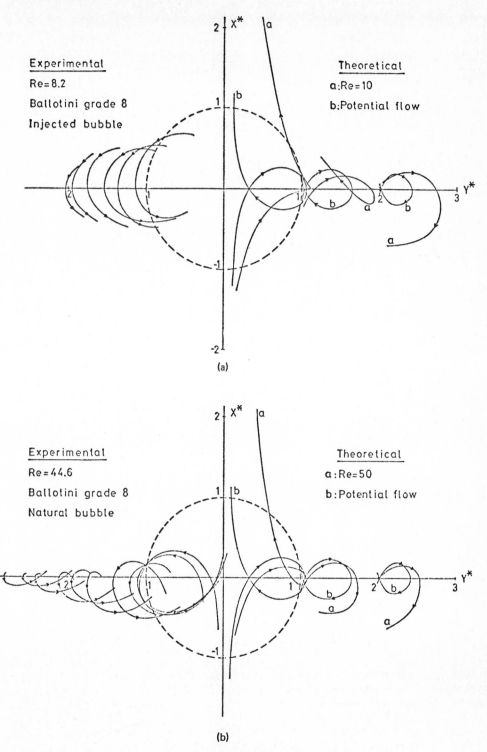

Figure 3. Paths of particles.

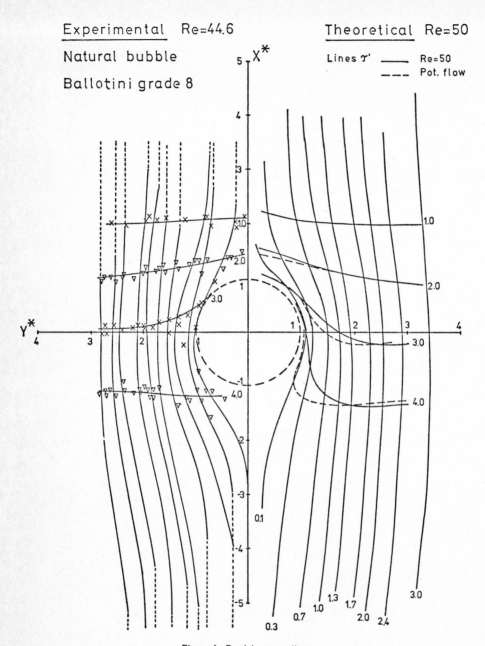

Figure 4. Particle streamlines.

along a path line show that the maximum exists for θ > 90°; theory
predicts maxima at low Reynolds numbers at both θ < 90° and θ > 90°.
The first maximum was not observed.

Corresponding results for injected bubbles show that the velocity
falls to zero very rapidly--within Y* < 2, as opposed to around
Y* < 5 in the case of natural bubbles.

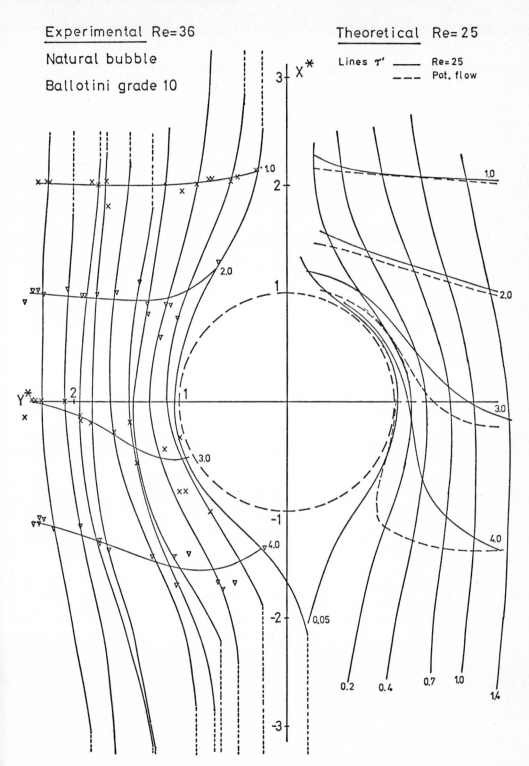

Figure 5. Particle streamlines.

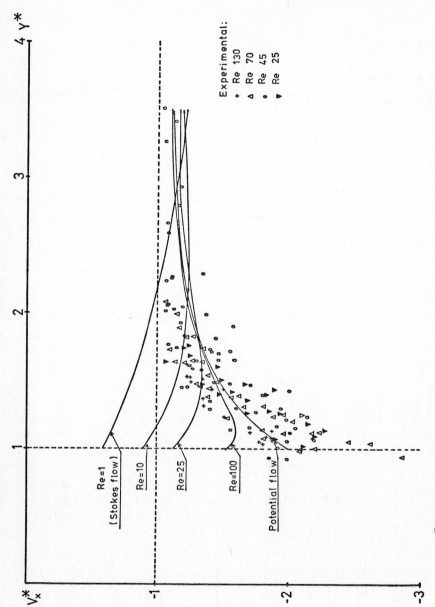

Figure 6. Velocity of particles on the horizontal axis ($\theta = \pi/2$) Ballotini grade 7 natural bubbles.

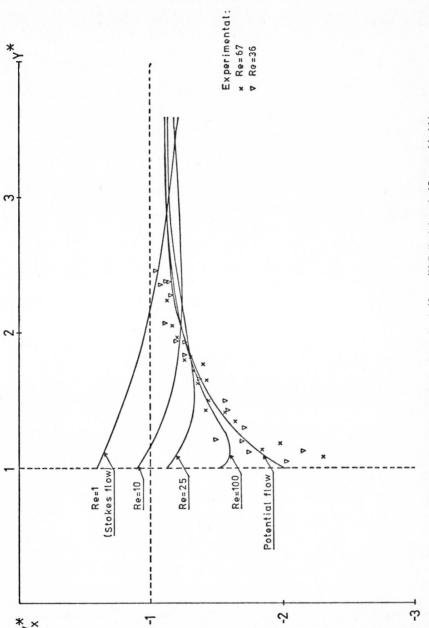

Figure 7. Velocity of particles on the horizontal axis ($\theta = \pi/2$) Ballotini grade 10 natural bubbles.

(b) Gas motion

Qualitatively the predictions of the theory are, as expected,
similar to those of Davidson (1), Murray (4), and Jackson (3), and of
course the present solutions become identical with Jackson's at
high Reynolds numbers (around 100). Figure 8 shows typical gas
streamlines for one Reynolds no. and for comparison, Davidson's solution. The shape of the gas clouds for two values of α_b are shown in
figure 9: it will be seen that the cloud size predicted by the present theory is larger than predicted by Davidson's theory at low
Reynolds numbers (below about five). The cloud becomes more asym-
metric as the Reynolds number increases and approaches Jackson's
solution. Experimental data on the size of gas clouds is plotted in
figure 10 and compared with the range of theoretical values. The
predictions for intermediate Reynolds numbers (10 < Re < 50) agree
quite well with the experimental data. Stokes flow certainly overestimates the cloud dimension.

Measurements of the vertical component of the interstitial gas velocity were obtained from the anemometry studies, and these are
compared with the theoretical predictions in the following series of
figures. Figure (11) shows some typical results for larger particles with $\alpha_b < 1$. The scatter in these results is rather high, but
it appears that the present theory tends to underestimate the velocity close to the bubble. In practice it is not easy to discriminate between the different theories ahead of the bubble, but it
seems that the results lie between Murray's and Davidson's theories.
Behind the bubble, although Davidson's theory overestimates the
velocity it is a much better approximation than Murray's theory.

For $\alpha_b > 1$, the results agree quite well with the present theory
and most satisfactorily with Murray's model ahead of the bubble,
although again the difference between the theories is not very
significant except for very low Reynolds numbers. Behind the bubble
and cloud Murray's theory gives a better prediction of the data, but
it will be seen that the cloud is extremely small in the wake of the
bubble (the cloud limit is defined by $u_x = 0$).

Some results for the velocity distribution around the bubble are
shown in figures (13) – (16). Here the vertical velocity component
is plotted versus the vertical distance from the horizontal axis,
x*; the position of the probe with respect to the bubble axis is
represented by the dimensionless horizontal distance y*.

In practice the influence of a bubble on the gas velocity extends
out to around y* $\sim$ 4. In some cases two theoretical curves for
Reynolds numbers above and below the experimental value are given,
but in any event the results are not very sensitive to Reynolds
number. Most of the experimental results lie between Jackson's
and Davidson's predictions, and very often agree better with the
latter theory. This confirms our own results on the pressure
distribution (33) and those of Littman & Homolka (37).

The viscous theory does not materially improve on existing theories
and in all cases the variations in gas velocity are underestimated.

DISCUSSION AND CONCLUSIONS

The theory developed in this work predicts gas and particle behaviour around rising bubbles which, especially at low Reynolds numbers, is different from the predictions of other theories. In the case of particle drift, it is predicted that at low Reynolds numbers the looping predicted by inviscid theory disappears, and the total volume drift becomes infinite. The cloud sizes predicted for the gas flow when $\alpha_b > 1$ are larger than Davidson's prediction for low Reynolds numbers (Re < 5) and tend to Jackson's theory at higher values (Re $\sim$ 100).

The experiments show the difference between a fully fluidised bed and a bed near incipient fluidisation conditions. For a fully fluidised bed, complete looping of particles was observed for particles starting their motion near the axis of the bubble and within a distance of about $\frac{1}{4}$ of the bubble radius from it. This is similar to predictions obtained from the inviscid theory. For particles starting their motion further away there is a better agreement with the theory including viscous effects. Thus the behaviour of the particulate phase can be divided into two regions:
   - a region very close to the bubble, with a width of about $\frac{1}{4}$ of the bubble radius, where convective terms are more important than viscous effects and which is thus well described by a potential flow solution. We suggest that this may be attributed partly to an increase in voidage fraction of the particulate phase in this region.
   - a region further away where viscous effects are important. In this region, although the present theory describes qualitatively the behaviour of the particles, the quantitative agreement is not very satisfactory. Therefore, presumably, the behaviour is not completely Newtonian.

The complete motion of the particles shows that the net volume drift obtained experimentally is smaller than that predicted by the inviscid theory. This is in agreement with published data.

In the case of natural bubbles, the velocity field of the particulate flow was found to be slightly asymmetrical, the maximum value occurring at $\theta < 90^o$, in qualitative agreement with published data obtained from measurements of the heat transfer coefficient (Clift & Tuot (38)).

Gas velocity results show that there is a difference between the particles which were used in this work. In the case of small particles and with $\alpha_b > 1$, Murray's theory or the present theory give good predictions of the gas velocity. In particular, cloud sizes are well represented by the above theories.

In the case of larger particles and with $\alpha_b < 1$, Davidson's theory is more satisfactory as measurements on the bubble axis and on the bubble sides show.

We can conclude that the theory attempting to describe the rheology of the particulate phase in terms of a Newtonian behaviour developed in this work does not significantly improve on other theories of bubble motion. It is in fact worth asking whether a more compli-cated model should be introduced since this work tends to show that it is probably very difficult to derive a theory which would explain the behaviour of all variables. This question should be asked the more because factors like the true shape of the bubble, the bubble wake, particles falling through the bubble, instability of the roof, all factors which presumably influence to some extent the behaviour of both phases in the vicinity of the bubble, have been completely neglected.

On the other hand, it is hoped that the theory and experiments sum-marised here will be useful in understanding what happens around a bubble rising in a fluidised bed.

ACKNOWLEDGEMENTS

One of use (A.L.F.M-G) gratefully acknowledges financial support from the CNRS and the Royal Society.

REFERENCES

1. DAVIDSON J.F., Trans Inst Chem Eng 39 p.230 (1961).

2. DAVIDSON J.F., HARRISON D., Fluidised Particles.
                              Cambridge University Press (1963).

3. JACKSON R., Trans Inst Chem Eng 41 p.22 (1963).

4. MURRAY J.D., J Fluid Mech 22 p.57 (1965).

5. ROWE P.N., Experimental properties of bubbles. Chap.4 in
          Fluidisation, J.F. Davidson, D. Harrison (Eds)
          Academic Press (1971).

6. BUYSMAN P.J. & PEERSMAN G.A.L., Proc Int Symp on Fluidisation
   Eindhoven, Ed: A.A.H. Drinkenburg,p.38 (1967), Netherlands
   University Press, Amsterdam.

7. RIETEMA K., Chem Eng Sci 28 p.1493 (1973).

8. VERLOOP J., HEERTJES P.M., Powder Technology, 7 p.161 (1973).

9. AGBIM J.A., ROWE P.N., Chem Eng Sci 26 p.1293 (1971).

10.ROWE P.N., Chem Eng Sci 24 p.415 (1969).

11.RIETEMA K., Proc Intl Symp on Fluidisation, Eindhoven.
   Ed. A.A.H. Drinkenburg,p.201 (1967) Netherlands Univ. Press,
   Amsterdam.

12.ANDERSON T.B., JACKSON R., Ind Eng Chem Fund 7 p.12 (1968).

13.SCHÜGERL K., Rheological behaviour of fluidised systems. Chap.6
          in Fluidisation, J.F. Davidson, D. Harrison (Eds)
          Academic Press (1971).

14.SCHÜGERL K., MERZ M., FETTING F., Chem Eng Sci 15 p.1 (1961).

15.HAGYARD T., SACERDOTE A.M., Ind Eng Chem Fund 5 p.500 (1966).

16.MURRAY J.D., Rheologica Acta 6 p.27 (1967).

17.STEWART P.S.B., Trans Inst Chem Eng 46 p.60 (1968).

18.GRACE J.R., Can J Chem Eng 48 p.30 (1970).

19a.ROWE P.N., Chem Eng Prog Symp Ser N°58 p.42 (1962).

19b.ROWE P.N., PARTRIDGE B.A., CHENEY A.G., HENWOOD G.A., LYALL E., Trans Inst Chem Eng 43 p.271 (1965).

20.RASO G., VOLPICELLI G., MAITZ C., Quaderni dell'ingegnere chem. ital, 1 p.157 (1965).

21.WOOLARD I.N., POTTER O.E., Am Inst Chem Eng J, 14 p.388 (1968).

22.SINGH R., FRYER C., POTTER O.E., Powder Technology 6 p.239 (1972)

23. TOEI R., MATSUNO R., NAGAI Y., Mem Fac Eng Kyoto Univ 28 p.428 (1966).

24.MAXWELL J.C., Scientific papers, ed. W.D. Niven, Cambridge University Press, p.208 vol.II (1890).

25.DARWIN C., Proc Camb Phil Soc 49 p.342 (1953).

26.GABOR J.D., Proc Intl Symp on Fluidisation, Eindhoven, Ed. A.A.H. Drinkenburg, p.230 (1967), Netherlands University Press, Amsterdam.

27.GABOR J.D., Chem Eng Sci 26 p.1247 (1971).

28.GABOR J.D., Chem Eng J. 4 p.118 (1972).

29.BAEYENS J., GELDART D., Proc Intl Sym on Fluidisation, Toulouse, p.182, (1973), Cepadues Toulouse.

30.PATUREAUX T., VERGNES F., le GOFF P., Ibid, p. 63 (1973).

31.ROWE P.N., PARTRIDGE B.A., Chem Eng Sci 18 p.511 (1963).

32.ANWER J., PYLE D.L., Proc Intl Symp on Fluidisation, Toulouse, p.240 (1973), Cepadues Toulouse.

33.MARTIN-GAUTIER A.L.F., Ph.D. Dissertation, University of London,(1974).

34.JACKSON R., Fluid mechanical theory – Chap.3 in Fluidisation, Ed. J.F. Davidson, D. Harrison. Academic Press (1971)

35.APELT C.J., Aero Res Counc Rep and Mem N° 3175 (1958).

36.DAVIES R.M., TAYLOR G.I., Proc Roy Soc, A200 p.375 (1950).

37.LITTMAN H., HOMOLKA G.A.J., Chem Eng Prog Symp Series N° 105, 66 p.37 (1970).

38.TUOT J., CLIFT R., Am Inst Chem Eng Symp Ser, N° 128, 69 p.78 (1973).

CHARACTERISTICS OF FLUIDIZED BEDS AT HIGH TEMPERATURES

KUNIO YOSHIDA, SHIGETAKA FUJII AND DAIZO KUNII

SCOPE

A number of investigations have been made on the behavior of rising bubbles in fluidized beds at room temperature. In practical applications, however, fluidized beds are usually operated at elevated temperatures, somewhat higher for physical operations and then much higher for chemical reactions. Despite this fact, little investigations have been reported on the temperature dependence of fluidization. In the present work, therefore, an attempt was made to investigate the effects of temperature on the size and rising velocity of bubbles in a two-dimensional fluidized bed, using several kinds of solid particles.

CONCLUSIONS AND SIGNIFICANCE

Through practical experiences of research and development at high temperatures, some chemical engineers are well aware of the fact that a fluidized bed becomes more fluid at elevated temperatures by comparison with room temperatures, at the same superficial gas velocity. Nevertheless, all procedures to predict conversion and selectivity of fluidized bed reactor, for example, rely on experimental data for the behavior of rising and coalescing bubbles, which were taken at room temperature.

Visual observations of bubbles behavior were made at various temperatures in a two-dimensional fluidized bed with a transparent window of 24cm x 35cm, the cross sectional dimensions of the bed being 24cm x 2.0cm. Experimental results clearly indicated remarkable effects of the operating temperature on bubble size.

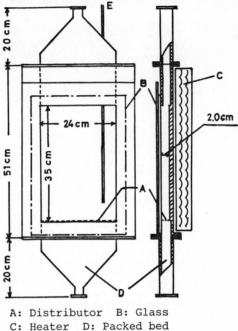

A: Distributor   B: Glass
C: Heater   D: Packed bed
E: Thermocouple

Figure 1. Experimental equipment.

EXPERIMENTAL WORKS

Preceding Results in Three-Dimensional Beds

Preliminary experiments were made in fluidized beds of
10cm I.D. at various temperatures up to 800°C. Graphite
particles of size ranging from 0.2 to 0.4mm were fluidized
by air or nitrogen. Bubble frequency was detected to increase
very much at higher temperatures, for instance, from 2 sec^{-1}
at normal temperatures to 6 sec^{-1} at 800°C at the same super-

Table 1 Particles Used

Micro-spherical catalysts: solid density 1.54g/cm^3,
                             sphericity 1.0
          I. 0.125 - 0.177 mm, mean size 0.151 mm
         II. 0.177 - 0.250 mm, mean size 0.214 mm
        III. 0.062 - 0.250 mm, mean size 0.151 mm
Sands: solid density 2.58g/cm^3, sphericity 0.8
          0.125 - 0.177 mm, mean size 0.151 mm

Table 2   Minimum Fluidization Velocities Measured, U_{mf} (cm/sec)

Micro-spherical catalyst		20°C	200°C	400°C
I.	0.125 - 0.177 mm	6.3	5.6	4.1
II.	0.177 - 0.250 mm	4.5	3.7	2.8
III.	0.062 - 0.250 mm	3.6	2.2	1.8
Sand	0.125 - 0.177 mm	5.5	3.9	3.1

ficial gas velocity. The apparent viscosity of fluidized beds decreased to about one-forth at the same conditions as the above. Details are found in ref.(1).

Bubble frequency was also measured in a bed of micro-spherical catalyst, obtaining similar results to the above, for example, $1.3 sec^{-1}$ at 1000°C by comparison with $0.7 sec^{-1}$ at 600°C. See ref.(2).

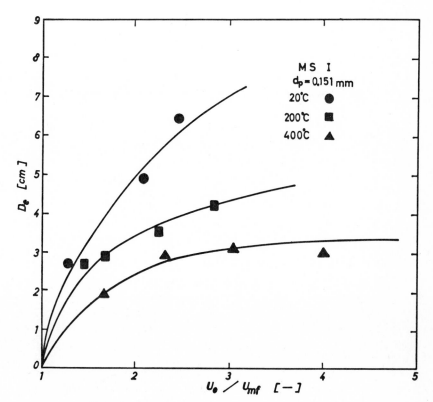

Figure 2. Relationship between D_e and u_0/u_{mf} for MS I.

Experimental Results in a Two-Dimensional Bed

Fig. 1 shows the experimental equipment used to observe bubble behavior at high temperatures. The distributor was made of ceramic fiber p plate supported by perforated metal plate with]mm orifices, the cross sectional dimensions being 24cm x 2.0cm. A front window of 24cm x 35cm was covered with a glass plate. A nichrome wire heater was positioned behind the back wall of the equipment, to keep it at a known temperature.

Minimum fluidization velocity U_{mf} were measured in this two dimensional fluidized bed to give Table 2.

Pictures of rising bubbles were taken by 16 mm cine camera, and both the mean size and rising velocity of bubbles were determined at a level of 20cm from the upper surface of the distributor, using twenty independent pic-

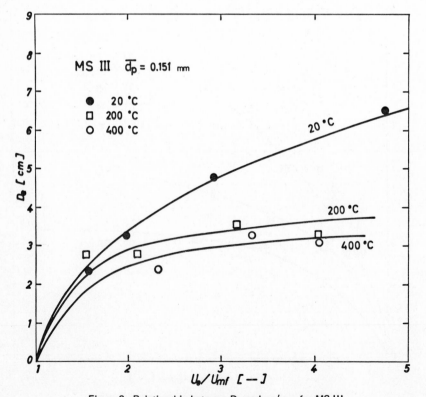

Figure 3. Relationship between D_e and u_0/u_{mf} for MS III.

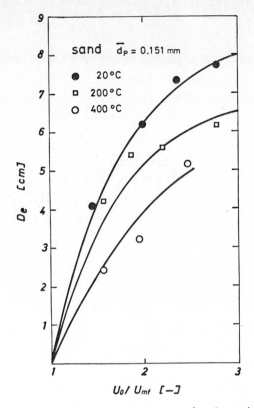

Figure 4. Relationship between D_e and u_0/u_{mf} for sands.

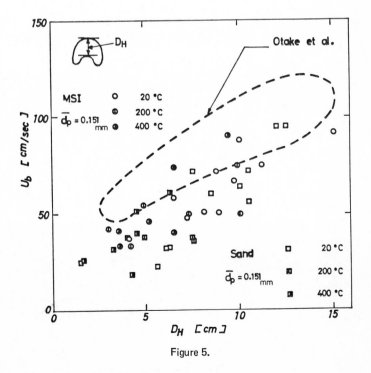

Figure 5.

tures at a given condition.   Since the shape of rising
bubbles seen in pictures was somewhat irregular, the following procedure was adapted to give the equivalent bubble size
D_e.

1) Assume the shape of any bubble seen through the front
   glass to be the central section of a cylindrical bubble.
2) The equivalent bubble size D_e is calculated as the
   diameter of a sphere which possesses the same volume
   of the above cylinder.

Figs. 2, 3 and 4 show the remarkable effect of temperature on equivalent bubble size D_e.   In Fig. 5, rising velocity
of bubbles are shown.

LITERATURE CITED

1) T.Mii, K.Yoshida and D.Kunii, J.Chem.Eng.Japan, Vol.6, 100(1973)
2) K.Yoshida, T.Ueno and D.Kunii, Chem.Eng.Sci. Vol.29, 77(1974)
3) T.Otake et al, Preprint of Autumn Meeting, Society of Chemical
   Engineers, Japan, Tokyo, Oct.(1974)

CONVECTIVE INSTABILITY IN
UNIFORM FLUIDIZED BEDS

R. JACKSON

In several publications (1→4) the stability of an infinite uniformly
fluidized bed has been examined on the basis of a model of the bed
as a pair of interpenetrating and interacting continua, and the
system has been found to be unstable against rising "compression
wave" disturbances. Such disturbances have been observed and
investigated experimentally. However, it is also well known to
workers in fluidization that large scale circulation patterns of
particle movement tend to develop, even in liquid fluidized beds,
when the pressure drop across the distributor is small compared with
that across the bed itself. These look very much like the convective
motion of a fluid heated from below, which is a consequence of the
well known hydrodynamic instability first investigated by Rayleigh (5).
By analogy we might guess that a properly formulated stability
analysis of a fluidized bed of finite extent would yield an instability
of circulatory form. It is also a well established experimental fact
that the observed circulation patterns can be suppressed by a sufficient
increase in the pressure drop across the distributor, and it should be
possible to account for this stabilization through the analysis.

This paper reports such a linearized stability analysis of a fluidized
bed of finite depth. It is based on equations of motion and continuity
for the fluid and particles given by Anderson and Jackson (6), and the
procedure is entirely analogous to Rayleigh's treatment of thermal
convection, though the algebra is very much more complicated because
of the existence of two continuity equations and two interacting
momentum balances. Perturbation modes periodic in the horizontal
direction, with spatial period λ , and exponential in time with growth
factor s are sought.

The distributor pressure drop is expressed as κ , the ratio of the
pressure drop across the distributor to that across the bed in the
uniform unperturbed state, and the distributor thickness is expressed by
by δ , its ratio to the depth of the unperturbed bed.

In addition, the following physical parameters of the bed are important.
(In brackets are the numerical values used in the present computations,
for water and air fluidized beds respectively, in that order.)

(i)	ρ_s	Solid density. (gm/cm^3)	(2.86,2.86)
(ii)	ρ_f	Fluid density. (gm/cm^3)	(1.0,0.0012)
(iii)	μ_s	Particle phase shear viscosity(poise)	(10.0,10.0)
(iv)	λ_s	Particle phase bulk viscosity(poise)	(10.0,10.0)
(v)	ε_o	Void fraction.	(0.42,0.42)
(vi)	d_p	Particle diameter (cm)	(0.086,0.086)
(vii)	u_{mf}	Fluid velocity at minimum fluidization	
(viii)	D	Bed depth (cm)	(86.0,86.0)
(ix)	$\left| dp_s/d\varepsilon \right|$	Derivative of particle phase pressure (dyne/cm^2)	(20.0,50.0)

Fig. 1 shows the computed critical wavelength λ_c , corresponding to
neutral stability, plotted against κ , for various fixed values of δ.
Below and to the right of the curves, the disturbances are damped, while
above and to the left they grow.

It is clear that, for each value of δ , the region of stability
enlarges as κ is increased, so the bed is indeed stabilized by increasing the distributor pressure drop. For each δ , beyond a certain finite
value of κ the bed is stable for all values of λ ; in other words beds
of all widths are stable if the distributor pressure drop is large enough.
For values of κ smaller than this, the bed is stable for small λ and
unstable for large λ , so there exists a critical bed width below
which the bed circulates and above which it does not. Finally, increas-
the distributor thickness δ decreases the effectiveness of the
distributor pressure drop in stabilizing the bed.

Fig. 2 shows corresponding curves for an air fluidized bed. They
are similar in form to those of Fig. 1, but the regions of stability

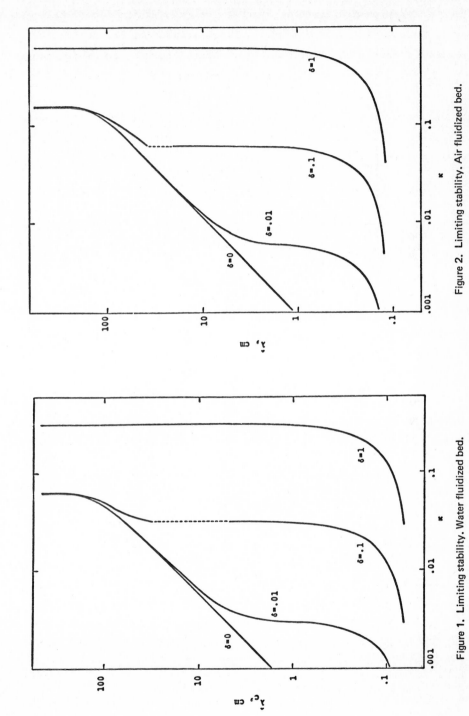

Figure 1. Limiting stability. Water fluidized bed.

Figure 2. Limiting stability. Air fluidized bed.

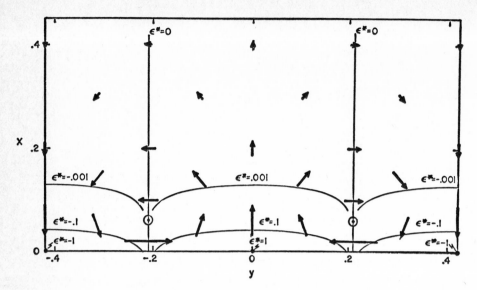

Figure 3. Velocity and voidage perturbations.

are somewhat smaller, showing that air fluidized beds are a little
less stable relative to circulation.

In Fig. 3 the particle velocity field and contours of the perturbation
in voidage are shown for a water fluidized bed with $\delta = 0$ and $\kappa = 0.002$.
The arrows are in the direction of particle motion and their length is
proportional to the magnitude of the particle velocity. There are
seen to be circulation loops centred around points quite low in the
bed. There are also alternating regions of high and low voidage above
the distributor, with the particles ascending in the regions of high
voidage and descending where the voidage is low.

A quantitative experimental investigation of these effects in a two
dimensional bed of variable width is currently being initiated.

REFERENCES

1.   Jackson, R., Trans. Inst. Chem. Eng. 41, 13 (1963)
2.   Pigford, R. L. and Baron, T., Ind.Eng. Chem. Fund. 4, 81 (1965)
3.   Murray, J. D., J. Fluid Mech. 22, 57 (1965)
4.   Anderson, T. B. and Jackson, R., Ind. Eng. Chem. Fund. 7, 12 (1968)
5.   Rayleigh, Lord., Phil. Mag. 32, 529 (1916)
6.   Anderson, T. B. and Jackson, R., Ind. Eng. Chem. Fund. 6 , 528 (1967)

ON THE ELASTICITY OF GAS-FLUIDIZED SYSTEMS

K. RIETEMA AND S. M. P. MUTSERS

At the International Symposium on Fluidization at Eindhoven
in 1967 one of us stated already that in gas-solid fluidization there is permanent contact between the solid particles. At Eindhoven research on this aspect has been continued. At present the experimental proofs of this permanent
contact are so overwhelming that they can no longer be
denied. Furthermore, the stability of homogeneously expanded gas-fluidized beds can only be understood by assuming
a permanent mechanical structure of the dense phase which
must have a certain elasticity. The experimental proofs
which is referred to above are: a) a gas-fluidized bed of
particles which are electrically conductive always maintains
a considerable electrical conductivity which cannot be explained by occasional collisions between particles [1],[2].
b) when a homogeneous gas-solid fluidized bed of fine particles (e.g. cracking catalyst, average particle size =
70 μm) is tilted along a horizontal axis the bed surface
does not shear off but is tilted as well untill a critical
tilting angle is reached which angle depends on the degree
of bed expansion, nature and size of fluidized particles
but also on the viscosity of the fluidization gas [3], [4].
c) exact measurements of the pressure drop over a fluidized bed indicate that it is smaller than the weight of the
particles divided by the cross sectional area of the bed.
This is only possible if the difference is carried by the
wall, which is only possible if there is momentum transport
through the dense phase.
A mechanical structure always has a certain elasticity which
means that any deformation of the structure is opposed by
an elastic force. Although - as we shall see - this elasti-

city appears to be very small a theoretical analysis indicates that it is high enough to stabilize the structure of
a homogeneous expanded bed. In this connection it should
be remembered that the major part of the weight of the particles is carried by the upflowing gas.
According to Wallis [5] the structure is stable if the propagation velocity U_c of continuity waves is smaller than
that of dynamic waves $(=U_d)$. Dynamic waves can occur only
in systems which have elastic properties.
A further analysis (see also [3] shows that $U_c = v_o \frac{3-2\varepsilon}{\varepsilon}$
in which v_o is the superficial gas velocity and ε the bed
porosity, while $U_d = \sqrt{\frac{E}{\rho_d}}$ in which ρ_d is the density of
the particles and E the elasticity modulus of the struc-
ture defined by: $F_e = + E (\frac{\partial \varepsilon}{\partial z})$ where F_e is the force per
unit volume opposing the deformation $\frac{\partial \varepsilon}{\partial z}$, z being the
vertical coordinate. On basis of this analysis it is found
that there is stability (which means that there are no
bubbles) if:

$$N_F = \frac{\rho_d^3 d_p^4 g^2}{\mu^2 E} \leq \left| \frac{150(1-\varepsilon)}{\varepsilon^2 (3-2\varepsilon)} \right|^2 \qquad (1)$$

From experiments in which the maximum bed porosity at which
not yet bubbles appear is determined, the elasticity modulus E can be calculated. In [4] E has been correlated with
the cohesion constant c which exists between the particles
in the packed stat; $E = \gamma c$ in which, therefore, γ is a
dimensionless elasticity coëfficiënt. We also tried to determine this elasticity directly by measuring the interaction between a fluidized bed and a vibrating body dipped
into the bed. As such was chosen a horizontal wire netting
connected by four vertical supports to the coil of an
electro dynamical oscillation system (developed from a small
loudspeaker). The coil was suspended by means of three
small corrugated metal strips which brought the elasticity
coëfficiënt k_v of the system as low as 40 N/m. Electrically the coil was connected via an ohmic resistance to a
frequency generator. The phase difference β between the

electrical current and the electrical voltage was measured
as function of the frequency ω. β depends on ω, the vibrating mass m, the ohmic resistance R of the coil, the self-
induction L, the elasticity coëfficiënt k_v, the damping
coëfficiënt k_d and on the electro mechanical coupling constant K:

$$\tan \beta = \frac{\omega^5 Lm^2 + \omega^3 (Lk_d^2 - mK^2 - 2Lmk_v) + \omega (Lk_v^2 + K^2 k_v)}{\omega^4 Rm^2 + \omega^2 (Rk_d^2 + k_d K^2 - 2Rmk_v) + Rk_v^2} \qquad (2)$$

By means of separate experiments in air the values of m,
R, L, K_v, k_d and K are measured. When the wire netting is
dipped into a fluidized bed it is found that k_v, k_d and
m are increased. By curve fitting (performed by the compu-
tor) the exact values of Δk_v, Δk_d and Δm are determined
(see fig.1).

The increase of k_v points to some elasticity of the bed.
This elasticity might be due to three effects:

a) compressibility of the gas which cannot escape fast
enough

b) hydrodynamical effects as suggested by Verloop [6] and
Oltrogge [7].

c) elasticity of the mechanical dense phase structure. It
was found, however, that decreasing the gas pressure from
1 atm down to 0,05 atm did have no effect on Δk_v which
excludes a contribution of the gas compressibility. Also
no effect of the gas viscosity was found when fluidizing
with hydrogen or neon instead of air which excludes hydrodynamic effects. A strong influence of the nature of the
solid particles (even at the same average particle size)
was found, however. The elasticity Δk_v found therefore
must entirely be due to a mechanical elasticity of the bed
structure.

To calculate the elasticity modulus E from Δk_v we used the
relation:

$$E = \frac{(\Delta k_v)(\Delta m)}{4A_n^2 (1-\varepsilon)\rho_d} \qquad (3)$$

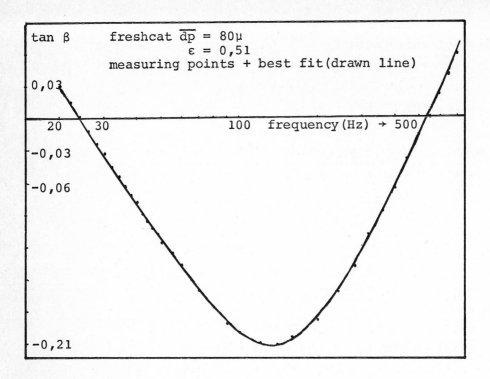

freshcat, HA - HPV, $\overline{dp}$ = 80 μ,
fluidization with air.

ε	ΔK_d N sec/m	ΔK_v N/m	Δm gram	E from (3) N/m^2	E from max.bed expansion N/m^2
0,45	0,239	6,9	0,160	5,45	1,23
0,48	0,141	4,5	0,090	2,11	0,67
0,51	0,104	2,9	0,070	1,12	0,36

Figure 1.

in which A_n is the surface area of the wire netting. Values
of E calculated in this way show the same dependence on the
porosity as determined from the bed expansion experiments
although the numerical values are somewhat too high (see
table, further results will be published in the appendix).
Formula (3), however is only a preliminary relation based
on the assumption that the mass of the dense phase (Δm)
which vibrates with the wire netting is restricted to
two layers (above and below the wire nettingI each with

a thickness = $\frac{1}{4}$ λ where λ is the wave lenght in the solid material

$$(\lambda = \frac{2\pi}{\omega} \sqrt{\frac{E}{\rho_d}}) \tag{4}$$

This relation, however, has further to be investigated. It can be concluded that it has been proven that the dense phase of gas fluidized beds of fine particles like e.g. cracking catalyst has a structure with a certain mechanical strength and a mechanical elasticity. On basis of this mechanical elasticity the critical bed expansion at which bubbling starts can be predicted. The mechanical strength and elasticity must have a strong influence also on the rheology of such systems and therefore on the bubbling behaviour and the understanding of the bubble mechanism.

REFERENCES

1. Goldschmidt, D.; Doctoral Thesis, Université de Nancy,1965

2. Rietema, K; Proceedings Int. Symp. on Fluidization, pag. 38, Eindhoven, Neth.University Press, 1967

3. Rietema, K.; Chem.Eng.Sc., vol 28 (1973) pag.1493-1497

4. Rietema, K., Mutsers, S.M.P.; Fluidization Conference Toulouse, October 1973, Proceedings page 28-40

5. Wallis, G.B.; One dimensional two-phase flow, Mc Graw-Hill Book Comp., 1969

6. Verloop, J., Heertjes, P.M.; Chem.Eng.Sc. 1970, 25, 825

7. Oltrogge, R.D.; Doctoral Thesis, The University of Michigan, 1972.

FLUIDIZATION UNDER PRESSURE

J. R. F. GUEDES DE CARVALHO AND D. HARRISON

INTRODUCTION

Relatively few experimental accounts of fluidized bed behavior under pressure have been published, and quantitative information is scarce. However, there is general agreement concerning the dependence of the minimum fluidizing velocity (U_{mf}) on fluid pressure, which is predictable from equations of the Carman-Kozeny type. In qualitative terms, bubbling fluidization is said to be "smoother" or "better" at higher pressures; which are terms similar to those used to describe the effect of the addition of fines to a fluidized bed.

In theoretical studies, the effect of increased pressure has been taken as that of increased fluid density only, as gas viscosity is effectively independent of pressure. Nevertheless, the importance of experimental work at high pressure has been emphasized because the fluid density reached can be much greater than that of dense gases at atmospheric pressure.

This paper describes experimental work in which the behavior of the fluidized bed under pressure can be observed directly in the range 1 → 18 bar.

EXPERIMENTAL

The fluidized bed was contained in a thick perspex column 2.4m high (I.D. 0.05m) which has been hydraulically tested to 25 bar. Cracking catalyst ($d_{av} = 48\mu$m; $U_{mf} = 0.0021$ m/s at 1 bar) has been fluidized by air.

In each run, using a particle charge of 0.664 kg which gave a settled bed height of about 0.55 m, the bed was first fluidized vigorously and then the fluidizing flowrate was reduced. At each flow rate, the bed-pressure drop was recorded together with the bed heights just before and after bubbles or slugs erupted at the top of the bed. Figure 1 shows a plot of these heights as a function of gas superficial velocity for various pressures.

DISCUSSION

The system is in the slugging regime at and near atmospheric pressure, and quite large oscillations of the bed height have been observed. At higher pressures, however, the bed height oscillates within only 0.5 cm over the full range of flowrate before severe elutriation occurs. The bed behavior is indeed "smoothed". Under pressure, bubbles—rather than slugs—reach the bed surface; in our experiments bubble diameters of about 1-2 cm have been observed. The present observations would support a theory of maximum stable bubble size, and this aspect of our work is being considered. Further experimental work is also in progress.

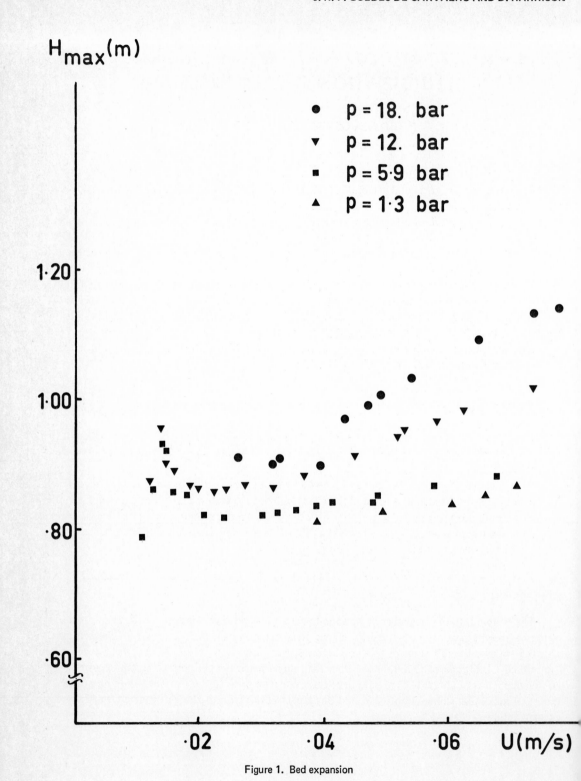

Figure 1. Bed expansion

ACKNOWLEDGEMENT

One of the authors (JRFGC) is grateful for financial support from the Calouste Gulbenkian Foundation during the course of this work.

MEASUREMENT AND SPECTRAL ANALYSIS OF PRESSURE FLUCTUATIONS IN SLUGGING BEDS

T. E. BROADHURST AND H. A. BECKER

NOMENCLATURE

D = column diameter

D_p = particle diameter

f = frequency of pressure oscillation

g = acceleration due to gravity

H = settled bed depth

v = superficial fluid velocity

Y = magnitude of peak in the pressure spectrum

γ = $g(\rho_p - \rho_f)$, specific weight of particle minus their specific buoyancy

ψ = $(6\,V_p/\pi)^{2/3}/(S_p\pi)$, particle sphericity ($V_p$ = volume, S_p = surface area)

ρ = density

μ = viscosity of fluid

Subscripts

f = a fluid property

p = a particle property

ABSTRACT

The visual observation of fluid beds in a laboratory environment is relatively easy. Visual observation of operating, industrial fluid beds is difficult because of high temperatures and opaque vessel walls. This work is concerned with using spectral

analysis of bed pressure fluctuations as a tool for interpreting

fluid bed behaviour. In addition, dimensionless correlations are

provided for predicting the magnitude and frequency of these

pressure fluctuations.

These objectives were realized by performing spectral

analysis of the pressure fluctuations in columns 2.5 to 21 cm

diameter, in beds of uniform particles 70 to 350 μm diameter,

fluidized with gases having a range in density of 0.17 to 5.2 kg/l

at settled bed height to diameter ratios ranging from 2 to 6.

INTRODUCTION

Bubbles rising and bursting at the surface of fluid beds cause

pressure fluctuations which can be measured by a sensitive pressure trans-

ducer, and recorded in a variety of ways. Early investigators (Morse and

Ballou 1951, Dotson 1959, Shuster and Kisliak 1952) relied on visual

techniques for analysis and interpretation of these fluctuations. Later

on, with the development of improved techniques for the analysis of random

time signals, several statistical properties of these pressure fluctuations

could be calculated from a digital record of these disturbances. (Winter

1966, Taylor et al 1973).

Lirag (1969) appears to be the first investigator who attempted

to relate these statistical properties to events occurring in the fluid

bed. The two most important statistical properties of the pressure fluctu-

ations are the autocorrelation and the frequency spectrum. The first is

a measure of the degree of correlation between neighbouring values of the

pressure fluctuations. The autocorrelation is also used as an intermediate

step in the estimation of the spectrum, which measures the distribution of

energy with frequency.

The objective of this paper is to show how the behaviour of these statistical properties changes with different bed operating conditions, and secondly to present data from which dimensionless correlations have been obtained which allow the estimation of the peak frequency f and the magnitude Y of the frequency spectrum, from the gas and particle properties, and bed dimensions, for fluid beds operating in the slugging regime.

EXPERIMENTAL DETAILS

The basic apparatus consisted of four transparent lucite columns with diameters 2.5, 5, 10 and 21 centimetres, with 3 to 4 metres of vertical height available to accommodate deep beds, Figure 1. Various gases (helium, air and freon -12) were fed to the column through pressure regulators and a rotameter manifold. The rotameters were calibrated with each gas at a discharge gauge pressure of 173 kPa by an absolute water displacement method. The rotameters were operated at this high discharge pressure so that the readings would be unaffected by pressure oscillations

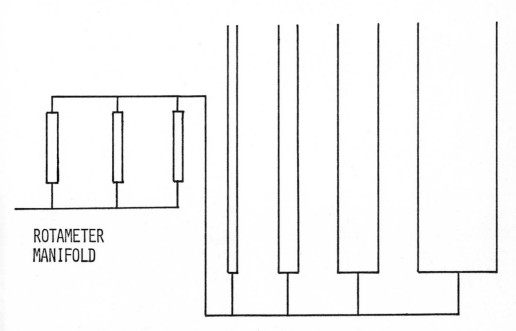

ROTAMETER
MANIFOLD

Figure 1. Flow diagram of the four columns used for the measurement of pressure fluctuations.

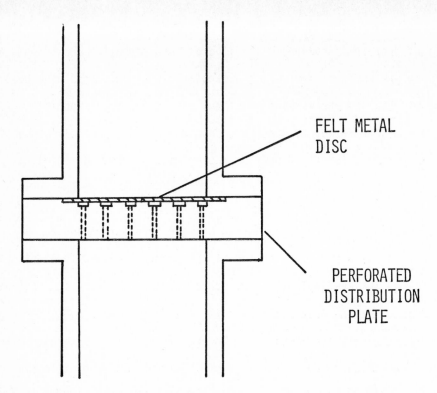

Figure 2. Arrangement used to distribute gas at the base of the fluidized bed.

in the plenum of the fluid bed. The gas was distributed at the base of
each column using a disc of Feltmetal (Huyck Metal Company) supported
on a perforated aluminum plate, Figure 2. The physical properties of the
particles used are shown in Table 1.

A pressure fitting was inserted in the lower flange of each of
the four columns, Figure 3. The purpose of this fitting was to measure
the pressure as close as possible to the base of the bed. A small piece
of 400 mesh bronze screen was mounted at the column wall to prevent sand
from entering the pressure transducer.

A Pace P7D pressure transducer was mounted on the pressure fitting.
A diaphragm giving a full scale deflection of ± 69 kPa was used. With this
pressure range, it was unnecessary to balance out the mean pressure at the
base of the fluid bed, and the negative port was left open to the atmosphere.

Table 1 Physical properties of particles

Material and Supplier	D_p mm	ρ_p kg/m^3	ψ
Sand; American Graded Sand Co., Paterson, New Jersey	0.343	2650	0.85
	0.213	2660	0.85
	0.183	2660	0.85
	0.150	2660	0.85
	0.106	2670	0.85
	0.071	2680	0.85
Glass beads; Minnesota Mining and Mfg. Company, St. Paul 6, Minn.	0.165	2450	1.0

The column was charged with the required amount of material, and the bed was fluidized with the quantity of gas required to create a slugging bed. Correlating equations for minimum slugging presented earlier by Broadhurst and Becker (1975) were used to ensure that the gas velocity was in excess of the minimum slugging velocity. The pressure signal was recorded on the VIDAR 5203 Digital Data Acquisition System using

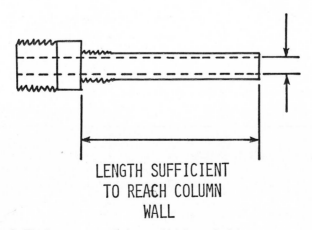

LENGTH SUFFICIENT
TO REACH COLUMN
WALL

Figure 3. The flange pressure fitting to which is attached the pressure transducer.

magnetic tape with a sampling interval of 22.5 milliseconds.   The autocorrelation and spectrum were computed for a record length containing
1000 data points, and plotted as shown in Figure 4.   The computation
procedure is described by Broadhurst (1972) and Jenkins and Watts (1968).

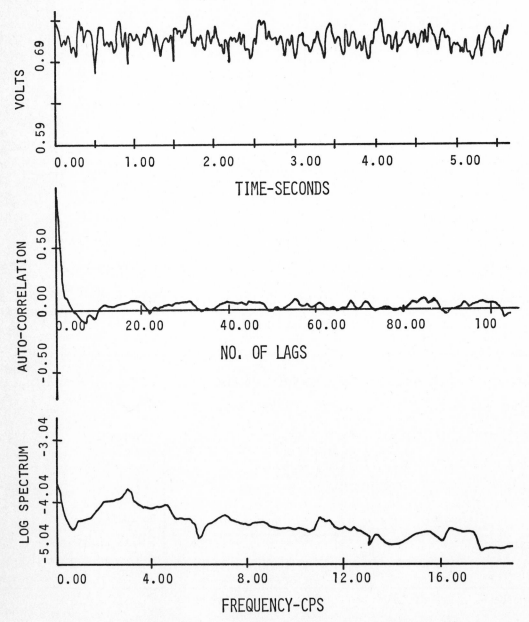

Figure 4. Spectral analysis of pressure fluctuations occurring in a 10 cm bed of sand for H/D = 1.

DISCUSSION OF RESULTS

Qualitative Behaviour of the Autocorrelation and Spectrum at Different Bed Conditions

The change in behaviour of the autocorrelation and spectrum with different bed operating conditions was illustrated by fluidizing a bed of sand (D_p = 0.0183 cm) with air at a moderate gas velocity of 10 cm/sec, in the 10 cm diameter column, using settled depths of sand ranging from 10 to 40 cm. Figure 4 shows the pressure fluctuation recorded with a bed depth of 10 cm. In this example, the autocorrelation function damps to zero almost immediately, indicating a low degree of correlation between neighbouring values of the pressure fluctuation. This behaviour is characteristic of a random time series, and the resulting spectrum is quite flat indicating a uniform distribution of energy with frequency. In this example, numerous bubbles are bursting in a completely random fashion at the bed surface. This pattern of pressure fluctuation would be characteristic of smooth fluidization.

Figure 5 shows the pressure fluctuations recorded at a bed depth of 20 cm. The autocorrelation now has a pronounced, slow oscil- lation, which is indicated by a peak in the spectrum at 2 cps. In this example, the small random bubbles have coalesced to form a small number of large bubbles which are sufficient to remove the gas in excess of that required for incipient fluidization. These large bubbles burst at the bed surface with a frequency of 2/sec.

The existence of a single peak frequency as shown in this example indicates that the bed surface is moving under the influence of single bubbles, and is approaching the slugging condition. Further increase in bed depth or gas velocity merely intensifies this affect. In this work, a thorough investigation of the statistical properties

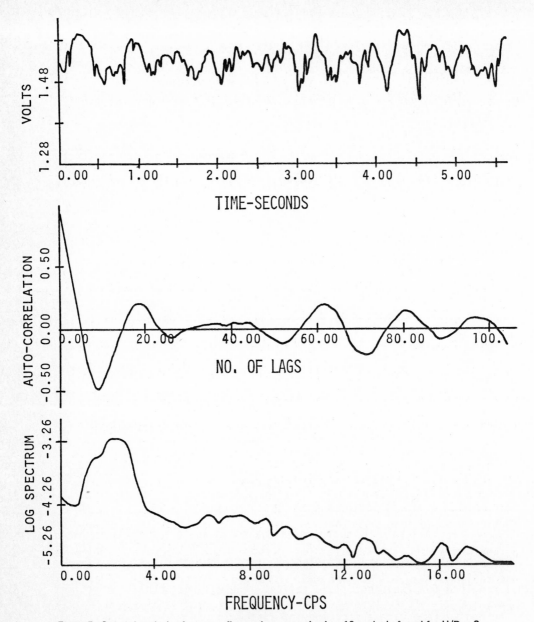

Figure 5. Spectral analysis of pressure fluctuations occurring in a 10 cm bed of sand for H/D = 2.

of pressure fluctuations in this last, or slugging region, was carried

out.

**Quantitative Behaviour of the Spectrum at
Different Bed Operating Conditions**

The frequency spectrum of the pressure fluctuation in a fluid

bed possesses three characteristic parameters, the peak frequency f, the peak height Y, and the peak bandwidth. The first two parameters were found useful in this investigation, where the periodic component of the pressure fluctuation is well defined. The peak bandwidth would probably be useful in an investigation of the pressure fluctuation occurring in the transition region between smooth fluidization and slugging, but the concept was of little use in this work. Knowledge of how the parameters f and Y depend on the gas and solid properties and bed operating conditions would allow the prediction of the frequency and magnitude of the pressure oscillations in slugging beds.

In common with many other phenomena occurring in fluidized beds, it is necessary to use the theory of dimensions in order to correlate the results of experimental investigation, and to arrive at some understanding of the phenomena. The general application of the theory of dimensions to fluid beds has been described in detail previously by Broadhurst and Becker (1973 and 1975). In these previous papers, it was shown that the behaviour of fluidized beds depends on the set of dimensionless groups

$$X_A \equiv \phi_A\left(\frac{D_p{}^3 \gamma \rho_f}{\mu^2}, \frac{v^2 \rho_f}{\gamma D_p}, \frac{\rho_p}{\rho_f}, \frac{D_p}{D}, \frac{H}{D}, \psi\right) \tag{1}$$

That is, any dynamic phenomena taking place in the fluid bed will be a function of the six dimensionless groups shown. The void fraction is not included in the above set because in fluid beds it is a dependent variable. It is possible that one or more of the above groups will have a negligible effect on the phenomena being studied, but each group must be considered as being potentially significant until experimental evidence indicates otherwise.

The two dynamic quantities of interest here are the peak frequency f, and peak height Y of the spectrum. The dimensionless expressions of these quantities are:

$$\frac{f^2 \rho_f D}{\gamma} \equiv \phi_f \left(\frac{D_p{}^3 \gamma \rho f}{\mu^2}, \frac{v^2 \rho_f}{\gamma D_p}, \frac{\rho_p}{\rho_f}, \frac{H}{D}, \frac{D_p}{D}, \psi \right) \qquad (2)$$

$$\frac{\gamma}{\gamma^{1.5} \rho_f{}^{0.5} D^{2.5}} \equiv \phi_y \left(\frac{D_p{}^3 \gamma \rho f}{\mu^2}, \frac{v^2 \rho_f}{\gamma D_p}, \frac{\rho_p}{\rho_f}, \frac{H}{D}, \frac{D_p}{D}, \psi \right) \qquad (3)$$

Each of the independent dimensionless groups on the right hand side of Equations (2) and (3) was varied systematically to find the effect on the dimensionless frequency and spectrum.

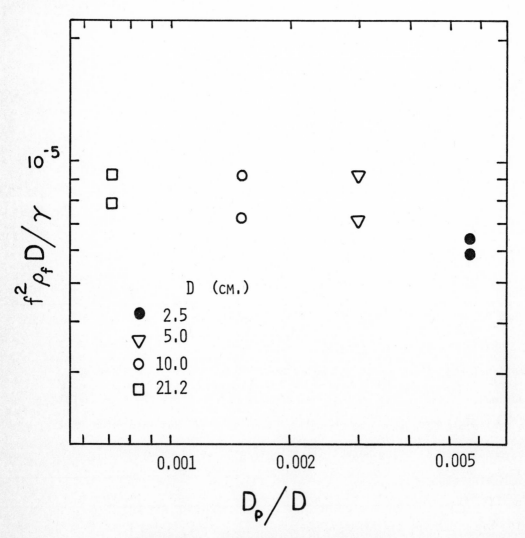

Figure 6. Effect of D_p/D on the dimensionless frequency.

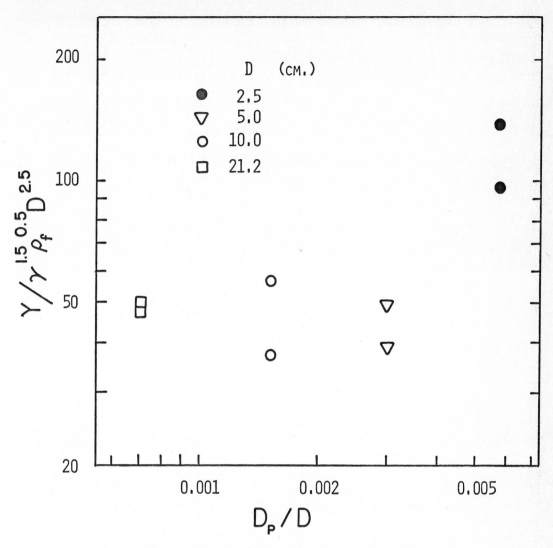

Figure 7. Effect of D_p/D on the dimensionless spectral peak.

       The effect of D_p/D was measured by observing the peak frequency
and spectrum using a single size of sand ($D_p = 0.015$ cm), a constant
velocity of air, and a settled bed depth of $H/D = 3$, on all four column
sizes. In this way, the remaining five groups remained constant while
D_p/D varied. The results are plotted in Figures 6 and 7, and show that
with the exception of the smallest column (2.5 cm dia), the dimensionless
frequency and spectrum are independent of D_p/D. It was concluded that

for columns with diameter 5 cm or greater, the effect of D_p/D was
negligible, and this group could be omitted from further consideration.
This conclusion permitted the use of a single column for the remainder
of the experimental program. The 10 cm column was used for all subsequent
runs. In a similar way, the effect of sphericity ψ for the range 0.85 to
1.0 was found to be negligible and was also omitted from further consideration.

The remaining groups $D_p{}^3\gamma\rho f/\mu^2$, ρ_p/ρ_f, $v^2\rho_f/\gamma D_p$, and H/D were
then varied in an orthogonal experimental design shown in three dimensional
form in Figure 8.

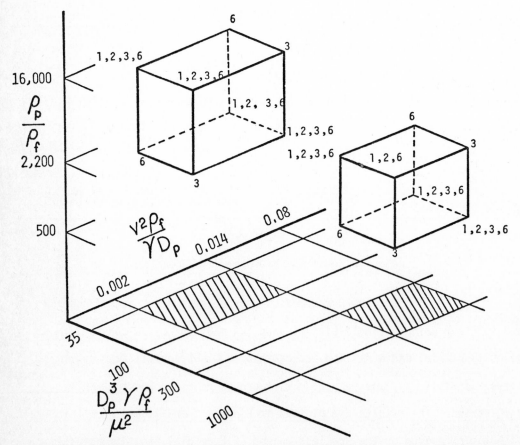

Figure 8. The experimental design used to measure and correlate the peak frequency and spectrum of pressure
fluctuations in a slugging bed. Values of H/D are shown at the corners of the cubes.

RESULTS

Effect of H/D, ρ_p/ρ_f, $v^2\rho_f/\gamma D_p$, **and** $D_p{}^3\gamma\rho_f/\mu^2$
on the Dimensionless Slugging Frequency

Graphical examination of the data indicated that a linear model in the logarithms of the dimensionless parameters is a reasonable approximation to (2). Regression analysis gave the following correlation for the slugging frequency.

$$\frac{f^2\rho_f D}{\gamma} = 0.117 \left(\frac{\rho_f}{\rho_p}\right)^{1.02}\left(\frac{D}{H}\right)^{1.71} \tag{4}$$

This equation estimates the slugging frequency with a 95% confidence interval of ±34%. The contribution of the groups $v^2\rho_f/\gamma D_p$ and $D_p{}^3\gamma\rho_f/\mu^2$ in the regression analysis was small as shown in Table 2 and these groups were omitted from the correlating equation.

A plot of Equation (4) with some data is shown in Figure 9. A comparison of the predictions of Equation (4) with some of Lirag's (1969) data is shown in Table 3. The agreement between predicted and measured frequencies is quite good.

For gas fluidized systems, usually $\rho_f \ll \rho_p$, and thus $\gamma \approx g\rho_p$. Equation (4) can then be written

$$f = 0.34\ g^{\frac{1}{2}}D^{-\frac{1}{2}}H^{-0.85} \tag{5}$$

The main factors determining slugging frequency are the bed diameter and dimensionless bed height, with gas flow and particle properties playing no part.

It is of interest here to examine the theoretical prediction of slug frequency. The rate of rise of a slug in an inviscid fluid contained in a column of infinite height is usually given as

$$v_s = 0.34\ g^{\frac{1}{2}}\ D^{\frac{1}{2}} \tag{6}$$

Table 2   Data on the stepwise regression analysis of $f^2\rho_f D/\gamma$ on ρ_p/ρ_f H/D, $D_p^3\gamma\rho_f/\mu^2$, and $v^2\rho_f/\gamma D_p$

Variable Entered	Standard Error	Multiple Correlation Coefficient	't' Values on the Parameter Estimates			
			$\dfrac{H}{D}$	$\dfrac{\rho_p}{\rho_f}$	$\dfrac{v^2\rho_f}{\gamma D_p}$	$\dfrac{D_p^3\gamma\rho_f}{\mu^2}$
H/D	.48	.735	8.1			
H/D, ρ_p/ρ_f	.13	.983	29	28		
H/D, ρ_p/ρ_f, $v^2\rho_f/\gamma D_p$	.11	.987	32	25	4	
H/D, ρ_p/ρ_f, $v^2\rho_f/\gamma D_p$, $D_p^3\gamma\rho_f/\mu^2$	.11	.987	33	25	3.4	1.2

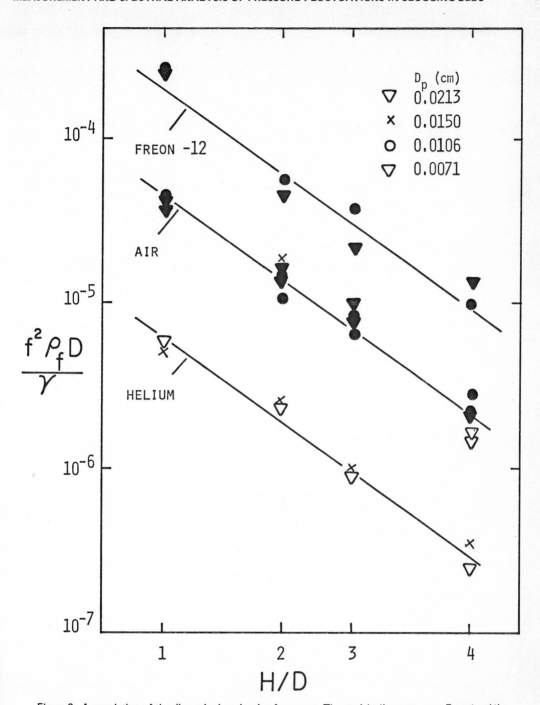

Figure 9. A correlation of the dimensionless slugging frequency. The straight lines represent Equation (4).

Table 3 Slugging frequency: a comparison of values predicted by Equation (4) with the results of Lirag (1969)

Lirag's Run No.	Material	$\frac{D_p}{cm}$	$\frac{D}{cm}$	$\frac{H}{D}$	$\frac{v^2 \rho f}{\gamma D_p}$	f (cps)	
						Measured	Predicted from Eq. 4
12	glass	0.05	6.35	0.7	0.017	5.8	5.4
47				5.0	0.0056	1.25	1.0
64	glass	0.022		3.3	0.0047	1.70	1.43
79	copper	0.013		3.2	0.002	1.3	1.46
84				4.3	0.002	1.10	1.14

Considerable evidence exists that slugs are spaced about three column diameters apart, therefore the frequency of slugs bursting at the bed surface is

$$\frac{V_s}{L_s} = \frac{0.34 \; g^{\frac{1}{2}}D^{\frac{1}{2}}}{3D} = 0.11 \; g^{\frac{1}{2}}D^{-\frac{1}{2}} \tag{7}$$

Equation (7) shows an equal dependence on g and D but does not indicate the effect of bed height shown in Equation (5). It is expected that the reason for this is that for the range of settled bed heights used in this work ($1 < H/D < 6$), bubble coalescence is still taking place, thus causing the decrease in slugging frequency with increased bed height. It is anticipated that for very deep beds, this dependence on bed height would cease, and the frequency would in fact become independent of H/D, thus satisfying Equation (7). This expectation is supported by slugging data presented elsewhere (Broadhurst and Becker, 1975) which showed that for deep beds ($H/D > 10$), the gas velocity required to produce slugging reached a constant value, the implication being that the slug volume and rise velocity had become independent of H/D.

Effect of H/D, ρ_p/ρ_f, $v^2\rho_f/\gamma D_p$, and $D_p{}^3\gamma\rho_f/\mu^2$ on the Dimensionless Spectrum

The dimensionless spectrum proved to be the most difficult property to correlate with the above groups. Again, graphical examination of the data suggested that a linear model in the logarithms of the dimensionless parameters would be suitable. Regression analysis gave the following correlation for the dimensionless spectrum

$$\frac{Y}{\gamma^{3/2}\rho_f{}^{1/2}D^{5/2}} = 0.177 \times 10^{-8} \left(\frac{D_p{}^3\gamma\rho_f}{\mu^2}\right)^{0.56} \left(\frac{\rho_p}{\rho_f}\right)^{3.11} \left(\frac{v^2\rho_f}{\gamma D_p}\right)^{1.46} \left(\frac{H}{D}\right)^{2.37} \tag{8}$$

In this regression, all groups contributed significantly as shown in Table 4. The 95% confidence interval for the estimate of Y provided by this equation

Table 4   Data on the stepwise regression analysis of $Y/\gamma^{1.5}\rho_f^{0.5}D^{2.5}$ on ρ_p/ρ_f, H/D, $D_p{}^3\gamma\rho_f/\mu^2$, and $v^2\rho_f/\gamma D_p$

Variable Entered	Standard Error	Multiple Correlation Coefficient	't' Values on the Parameter Estimates			
			$\dfrac{\rho_p}{\rho_f}$	H/D	$\dfrac{v^2\rho f}{\gamma D_p}$	$\dfrac{D_p{}^3\gamma\rho f}{\mu^2}$
ρ_p/ρ_f	0.955	0.607	59			
ρ_p/ρ_f, H/D	0.692	0.821	7.8	7.4		
H/D, ρ_p/ρ_f, $v^2\rho_f/\gamma D_p$	0.365	0.954	19.2	12.3	13.5	
H/D, ρ_p/ρ_f, $v^2\rho_f/\gamma D_p$, $D_p{}^3\gamma\rho_f/\mu^2$	0.243	0.957	35.4	24.5	15.9	12.3

is 3Y to 0.3Y. While this interval appears excessive, about half of this error is unavoidably contained in the computation of the magnitude of the spectral peak Y, which has a 95% confidence interval of 2Y to 0.6Y.

Equation (8) is plotted with some data in Figure 10. The data for freon-12 is quite linear, but some non linearity is apparent in the data for air, and this becomes more pronounced in the case of helium. The effect is for the dimensionless spectrum to level off as $H/D \simeq 1$ for light gases.

A non linear regression produced the following correlating equation giving a somewhat better fit to the data.

$$\frac{Y}{\gamma^{\frac{3}{2}}\rho_f^{\frac{1}{2}}D^{\frac{5}{2}}} = 0.117 \times 10^{-8}\left(\frac{D_p{}^3\gamma\rho_f}{\mu^2}\right)^{0.57}\left(\frac{\rho_p}{\rho_f}\right)^{3.10}\left(\frac{v^2\rho_f}{\gamma D_p}\right)^{1.61}\left(\frac{H}{D}\right)^{3.13} +$$

$$0.606 \times 10^{-8}\left(\frac{D_p{}^3\gamma\rho_f}{\mu^2}\right)^{0.61}\left(\frac{\rho_p}{\rho_f}\right)^{2.11} \tag{9}$$

Equation (9) gave a 25% reduction in the residual variance over Equation (8). However, this improved fit only occurs in the region $H/D \simeq 1$, and for deeper beds, the simpler model can be used.

Again using the simplification $\gamma \simeq g\rho_p$ for gas fluidized beds, Equation (8) can be rewritten.

$$Y = 0.177 \times 10^{-8}g^{0.6}\rho_p{}^{3.71}\rho_f{}^{-0.59}D^{0.13}\mu^{-1.12}D_p{}^{0.22}H^{2.37}v^{2.92} \tag{10}$$

It is clear that particle density, gas velocity, and bed height exert a very strong positive influence on the peak height of the pressure spectrum. The effects of particle and bed diameter are quite small, while Y is diminished by high gas density and viscosity.

These observations agree with the qualitative observations of other investigators. The predictions of Equation (9) are compared with

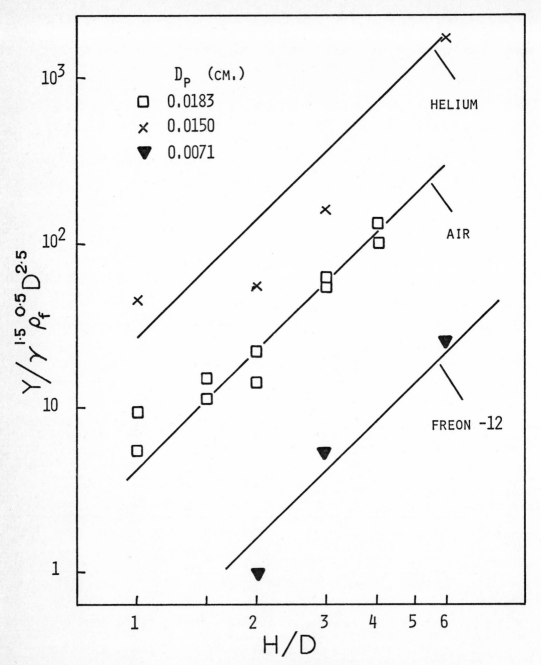

Figure 10. A correlation of the dimensionless spectral peak. The straight lines represent Equation (8).

some of Lirag's data (1969) in Table 5. There is reasonable agreement in the case of glass particles, but considerable discrepancy for the copper particles. The reason for this discrepancy is not clear, but is probably

Table 5 Magnitude of spectral peak: a comparison of values predicted by Equation (8) with the results of Lirag (1969)

Lirag's Run No.	Material	D_p cm	D cm	H/D	$\dfrac{v^2 \rho f}{\gamma D_p}$	rms Value of Pressure Peak (mm Hg)	
						Measured by Lirag	Predicted from Eq. 8
33	glass	0.05	6.35	3.4	0.0094	3.4	5.8
42				3.4	0.014	5.65	7.9
50				5.0	0.0094	5.88	9.3
65	glass	0.022		3.3	0.046	1.16	1.7
81	copper	0.013		3.2	0.0024	2.35	12.3
85				4.2	0.0019	3.22	14.7

due to differences in probe location and procedures for computing the spectrum. However, for most particles, the correlation provided by Equation (8) will provide a reasonable estimate of the intensity of the pressure oscillations that can be expected in slugging beds. Further work is required to improve the accuracy of this correlation, and to fill in the region between smooth fluidization and slugging.

CONCLUSIONS

The different modes of fluidization ranging from smooth to slugging were found to give pressure fluctuations with characteristically different autocorrelation and spectral density functions. For smoothly fluidized beds, the autocorrelation damped quickly to zero, and the spectrum was flat indicating a uniform distribution of energy with frequency.

For slugging beds, the autocorrelation had a pronounced oscillation which showed as a peak in the spectrum at the oscillation frequency. The slugging frequency was found to be dependent on bed diameter and dimensionless bed height, with negligible contribution from the gas and particle properties. The predictions of the dimensionless correlating equation agreed well with the results of Lirag (1969).

The intensity of the pressure fluctuation in a slugging bed, as measured by the peak in the spectrum, was very strongly dependent on the particle density, gas velocity, and bed height. Increased gas density and viscosity reduced this intensity, while particle and bed diameter had a small positive effect. The predictions of the dimensionless correlating equation were in fair agreement with Lirag's data for glass and sand particles. Further work is required to investigate non linear effects using light gases in shallow beds.

ACKNOWLEDGEMENTS

The work was supported by grants from the National Research Council of Canada.

LITERATURE CITED

Broadhurst, T. E., "Particle Motion, Slugging and Pressure Fluctuation Phenomena in Fluidized Beds", Ph.D. thesis, Queen's University, Kingston, Ontario (1972)

Broadhurst, T. E. and Becker, H. A., "The Application of the Theory of Dimensions to Fluidized Beds", Proceedings of the Toulouse, France International Congress on Fluidization and its Applications, Societé de Chimie Industrielle, 10-27 (1973)

Broadhurst, T. E. and Becker, H. A., "Onset of Fluidization and Slugging in Beds of Uniform Particles", AIChE J., 21, 238 (1975)

Dotson, J., "Factors Affecting Density Transients in a Fluidized Bed", AIChE J. 5, 169 (1959)

Jenkins, G. M. and Watts, D. G., "Spectral Analysis and its Applications", Holden Day Inc., San Francisco (1968)

Lirag, R. C., "Statistical Study of the Pressure Fluctuations in a
    Fluidized Bed:, Ph.D. thesis, Rennselaer Polytechnic Institute,
    Troy, New York (1969)

Morse, R.D . and Ballou, C.W., Chem. Eng. Progr., 47, 199 (1951)

Shuster, W. and Kisliak, P., Chem. Eng. Progr., 48, 455 (1952)

Taylor, P. A., Lorenz, M. H., and Sweet, M.R., "Spectral Analysis
    of Pressure Noise in a Fluidized Bed", Proceedings of the Toulouse
    France International Congress on Fluidization and its Applications,
    Societe de Chimie Industrielle, (1973)

Winter, O., "Density and Pressure Fluctuation in Gas Fluidized Beds",
    AIChE J., 14, 426 (1966)

DYNAMIC CHARACTERISTICS OF A GAS FLUIDIZED BED

W. R. A. GOOSSENS

NOMENCLATURE

A = bed cross-section
a = coefficient of the covariance function
b = apparent radiant frequency of the covariance function
D_p = particle diameter
F = force acting on a particle
f_n = eigenfrequency of the vibrations
f(t) = random signal
Ga = Galilei number = $[D_p{}^3 \, \rho_g(\rho_s - \rho_g)g/\mu^2$
g = acceleration of gravity
h = height in the bed
H = bed height
K = stiffness coefficient of the air spring
P = average value, defined in equation (2)
Δp = pressure drop
Re = Reynolds number = $D_p\rho_g U/U$
$R(\tau)$ = autocorrelation function
S = root mean square value, defined in equation (3)
T = observation time of the random signal
U = superficial gas velocity

Greek Symbols

ϵ = porosity
ζ = damping factor of the vibrations
μ = dynamic viscosity
ρ_g = gas density
ρ_s = solid density

1. INTRODUCTION

Fluctuations occurring in a gas-fluidized bed are characterized by either statistics means or dynamic system identification methods [2]. The dynamic analysis of a gas fluidized bed has often been limited to the determination of the average frequency of the pressure or density fluctuations. Complete dynamic analysis of pressure changes has been performed by Kang e.a. [4], Lirag [5] and Paul e.a. [6] applying random signal analysis techniques.

In the present study, the experimental autocorrelation function of the bed density changes within a microwave beam is discussed in order to derive the eigenfrequency and the damping factor of the
particle oscillations. The data obtained will be used to clarify the dynamic behavior of a fluidized bed.

2. EXPERIMENTAL PROCEDURE

The flow diagram of the experimental set-up is shown in Fig. 1. The microwaves with a nominal
frequency of 9100 Hz are transmitted through a square perspex fluidization column (10 cm in size)
along horn antennae having a rectangular window (4 X 8 cm). The power of the transmitted microwave
beam is detected by a diode the amplified outlet signal of which is analyzed with a specific purpose
computer, called autocorrelator. More details of this set-up have been described elsewhere [2].

The experimental runs were performed with 7 kg of glass spheres in narrow fractions having a
mean diameter of 95, 267 and 371 μm and a solid density of 2.78, 2.81 and 2.89 g.cm^{-3} respectively.

In each experimental run, the first hundred values of the autocorrelation function were calculated by sampling the outlet signal 136,192 times at time intervals of 10 ms. exceptionally 3.33 ms.
Mathematical analyses demonstrated that the experimental curve can be described in general by the
following equation:

$$R_{11}[\tau] = P^2 + S^2\, e^{-a[\tau]}\, \cos b\tau - \frac{a}{b}\, \sin b\, [\tau]. \tag{1}$$

This fitting is shown in Fig. 2 for a typical data set. The apparent radian frequency b was calculated for the arithmetic mean of the time differences between the zero points of the periodic term. The
coefficient a was taken to be equal to the intermediate of its values derived either at $\tau_7 = \pi b$, at τ_2 or
at the combined points 3 and 5. The periodic term could thus be approached within 10% in general.

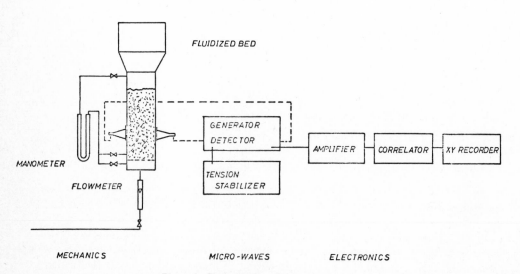

Figure 1. Flow diagram of the set up.

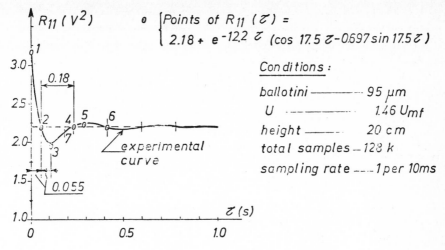

Figure 2. Example of autocorrelation function.

3. SIGNIFICANCE OF THE OBSERVED AUTOCORRELATION FUNCTION

The first term of the autocorrelation function equals the square of the average value, usually defined as

$$P = \frac{1}{T} \int_0^T f(t) \cdot dt \tag{2}$$

for a sufficiently large observation time T of the random signal f(t).

The second term is called covariance function. The coefficient S is the root mean square value of the fluctuations of the random signal as given by the following equation:

$$S^2 = \frac{1}{T} \int_0^T [f(t) - P]^2 \cdot dt \tag{3}$$

The coefficients a and b are related to the natural frequency f_n and the damping factor ζ of the investigated system according to the following equations:

$$(2\pi f_n)^2 = a^2 + b^2 \tag{4}$$

$$2\pi f_n \zeta = a \tag{5}$$

Mathematical manipulations as summarized in Table 1, allow to transform the covariance function into any characteristic function of a system [1]. This has been done in Table 1 for the displacement of a second-order system subjected either to velocity or to acceleration disturbances. Thus, it becomes

Table 1   Characteristic functions of second order system with transfer function $H(s) = 1/(Cs^2 + Ds + 1)$

Function	Relationship	Acceleration Pulses	Velocity Pulses
Time domain			
impuls response	$h(t) = \mathcal{L}^{-1}\{H(s)\}$	$\gamma e^{-at} \sin bt$	$b\gamma e^{-at}(\cos bt - \frac{a}{b}\sin bt)$
step response	$H(t) = \int_0^t h(\tau)\,d\tau$	$1 - e^{-at}(\cos bt + \frac{a}{b}\sin bt)$	$\gamma e^{-at} \sin bt$
covariance func.	$R_{11}(\tau) = F^{-1}\{P_{11}(w)\}$	$\frac{b\gamma}{4a} e^{-a\lvert\tau\rvert}(\cos b\tau + \frac{a}{b}\sin b\lvert\tau\rvert)$	$\frac{(b\gamma)^2}{4a} e^{-a\lvert\tau\rvert}(\cos b\tau - \frac{a}{b}\sin b\lvert\tau\rvert)$
Frequency domain			
transfer ratio	$P(jw) = H(s = jw)$	$\dfrac{1}{1 - Cw^2 + jDw}$	$\dfrac{jw}{1 - Cw^2 + jDw}$
power spectrum	$P_{11}(w) = P(jw)P(-jw)$	$\dfrac{(b\gamma)^2}{(b\gamma - w^2)^2 + 4a^2 w^2}$	$\dfrac{(b\gamma)^2 w^2}{(b\gamma - w^2)^2 + 4a^2 w^2}$

Relationship between parameters : $C = \frac{1}{b\gamma}$; $D = \frac{2a}{b\gamma}$; $b\gamma = a^2 + b^2 = w_n^2$; $a^2 = \zeta^2 w_n^2$

clear that the experimental covariance function, as expressed in equation (1), is typical for the displacement of a second-order system excited by random velocity pulses. The fluctuations of the bed density within the microwave beam being caused by random vertical displacements of the fluidized particles, these particles will oscillate as a second-order system the eigenfrequency, f_n, and damping factor, ζ of which being related to the experimental coefficients a and b by equations (4) and (5).

Under the experimental conditions of this study, only at the 10 cm height of the fluidized 95 μm beads, another type of covariance function was obtained expressing the displacement of a second-order system disturbed at random by a combination of velocity and acceleration pulses.

4. DISCUSSION OF THE DYNAMIC RESULTS

The derived damping factor of the oscillations does not show any perceptible dependency on the various working conditions used. This is illustrated in Fig. 3: the experimental points are the arithmetic

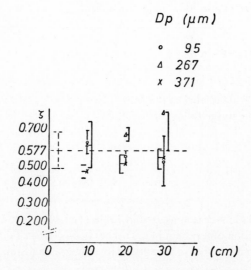

Figure 3. The damping factor as a function of height.

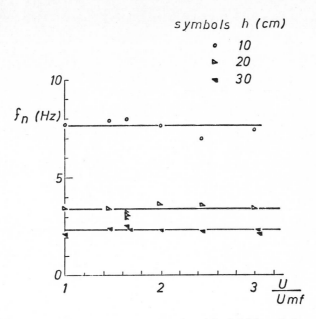

Figure 4. The eigenfrequency as a function of the fluidizing velocity.

means of the values obtained at various fluidizing gas velocities for a specific granular size and a specific height within the fluidized bed: the vertical lines show the related 68% confidence region. The overall mean value equals 0.577 with a standard deviation of 0.093. This value indicates that the oscillations are slightly underdamped as already observed by Paul e.a. [6] and Lirag [5] but for pressure fluctuations.

Figure 4 shows that the derived frequency does not depend on the fluidizing gas velocity in contrast to Hiby's predictions. In Fig. 5 is demonstrated that the eigenfrequencies of the 267 and 371 μm glass beads are inversely proportional to the square root of the measuring height. These data fit in the results obtained by Hiby [3], Lirag [5], Paul e.a. [6] and Verloop e.a. [7] for the eigenfrequency of pressure fluctuations if the total bed height is taken as abscissa. This means that the eigenfrequency of the pressure fluctuations anywhere in a gas-fluidized bed equals the eigenfrequency of the fluidized particles at the bed surface. This equality confirms Lirag's observation that the pressure waves are created at the bed surface by the escaping bubbles.

Furthermore, the full lines drawn on the figure illustrate that it is possible to predict a lot of the data on the basis of equation (9), which will be derived theoretically later on. Why so many literature values are somewhat out of range is not clear at the moment.

The eigenfrequencies of the 95 μm particles used in this study show to be inversely proportional to the measuring height. This different relationship can be attributed to their low Froude number indicating homogeneous fluidization even for this gas-solid fluidization. The same explanation holds also for a few of Paul's data.

5. CONSEQUENCES ON LITERATURE EQUATIONS FOR THE EIGENFREQUENCY

The foregoing observations indicate that it is not necessary to average the frequency of the oscillating particles all over the bed as was done by Hiby to predict the eigenfrequency of a shallow fluidized bed. If additionally Ergun's expression in Hiby's paper is substituted by the general fluidization law:

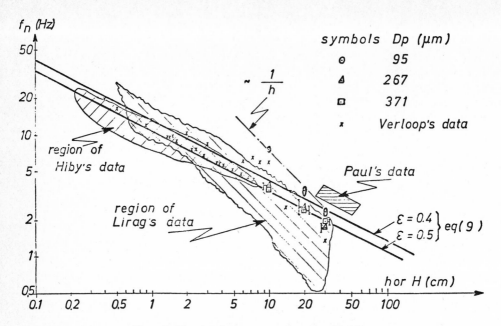

Figure 5. The eigenfrequency as a function of height.

$$\epsilon^{4.7}\,\mathrm{Ga} = 18\,\mathrm{Re} + 0.3\,\mathrm{Re}^2 \tag{6}$$

the pressure drop over a fluidized bed height can be expressed by:

$$\frac{\Delta p}{h} = \frac{1-\epsilon}{\epsilon^{4.7}} \cdot \frac{\mu}{\mathrm{Dp}^2}\,(18+0.3\,\mathrm{Re})\mathrm{U} \tag{7}$$

and the partial derivative of the force acting on a single particle by:

$$\frac{\mathrm{dF}}{\mathrm{d}\epsilon} = \frac{4.7\pi}{6} \cdot \frac{\epsilon}{\mu} \cdot \mathrm{D_p}^3\,(\rho_s - \rho_g)\mathrm{g} \tag{8}$$

At last, the following equation for the eigenfrequency of an oscillating fluidized particle is obtained:

$$(2\pi\mathrm{f_n})^2 = 4.7\,\frac{1-\epsilon}{\epsilon} \cdot \frac{1}{h}\,\frac{(\rho_s - \rho_g)\mathrm{g}}{\rho_s} \tag{9}$$

An identical relationship, but with a factor 4, can also be obtained by adapting Paul's lumped parameter model for the pressure fluctuations in fluidized beds. Therefore, one has to assume that the pressure drop over the bed acts as an air spring on the cross-section of the particles present in the bed area separating the oscillating half-masses of the bed load. In this way, the model's air spring with a height of ϵH has a stiffness coefficient K gives by:

$$\mathrm{K}\epsilon\mathrm{H} = \Delta\mathrm{p}(1-\epsilon)\mathrm{A} \tag{10}$$

Further, the pressure drop herein has to be put equal to the force of gravity acting on the suspended fluidized particles per unit bed cross-section. Additionally, for the displacement fluctuations under consideration the bed height of this model must be replaced by the measuring height.

Recently, Verloop-Heertjes [7] independently derived a similar equation for the frequency of pressure fluctuations in a gas-fluidized bed. Again, replacing the bed height in their equation by the measuring height for the displacement fluctuations of this study, their equation differs from equation (9) only by its factor which was found equal to 3.7 in the specific case of an exponent 4.7 in the Richardson-Zaki expression for the expansion of a fluidized bed. These authors demonstrated that the pressure vibrations in a fluidized bed should be harmonic. The stiffness coefficient K was derived for various cases by calculating the partial derivative of the pressure drop to the bed porosity.

Consequently, these three literature models converge to the same relationship, as expressed by equation (9), but with an uncertainty range of 12% for the eigenfrequency of the vibrations due to the different values obtained for the equation's factor. Taking the measuring height h in equation (9), it predicts the eigenfrequency of the displacement oscillations of the fluidized particles at this height. If the total bed height H is taken, equation (9) allows to predict the eigenfrequency of the pressure fluctuations anywhere within the bed.

Finally, it should be noted that equation (9) does not show any direct dependence of the eigenfrequency on the fluidizing gas velocity. Indirectly, the over-all bed porosity could take this dependence into account. However, Fig. 4 illustrated that experimentally the eigenfrequency of the particle oscillations does not sensitively depend on the fluidizing velocity and consequently not on the related over-all bed porosity. Hence, in the case of particle oscillations the porosity ϵ in equation (9) has to be set equal to the porosity of the emulsion phase which is generally considered to be constant and equal to the bed porosity under minimum fluidization conditions.

6. CONCLUSIONS

This dynamic analysis of the bed fluctuations has shown that the particles of a gas-fluidized bed oscillate due to random excitations. The eigenfrequency of these particle oscillations can in general be predicted by equation (9) with a theoretical uncertainty on the value of its factor. Considering the total bed height instead of the height in the bed, this equation seems also be be valid for the eigenfrequency of pressure fluctuations.

The damping factor has the experimental value of about 0.6 indicating that the particle oscillations have an underdamped nature.

REFERENCES

[1] J. S. Bendat, A. G. Piersol. Measurement and Analysis of Random Data, John Wiley, N.Y. (1966).
[2] W. R. A. Goossens. Dr. Appl. Sci. thesis, K.U.L., Leuven (1972).
[3] J. W. Hiby. Proc. Int. Symp. Fluidization, Netherlands, Univ. Press, Amsterdam (1967).
[4] W. K. Kang, J. P. Sutherland, S. L. Osberg. IEC Fundamentals, 6, 4, 499–504 (1967).
[5] R. C. Lirag. Ph.D. Thesis, Rensselaer Polytechn. Inst., N.Y. (1970).
[6] R. J. A. Paul, T. T. Al-Naimi, D. K. Das-Gupta. Int. J. Control, 12,5, 817–34 (1970).
[7] J. Verloop, P. M. Heertjes. Chem. Eng. Sci., 29, 1035–1042 (1974).

PARTICLE VELOCITY IN FLUIDIZED BED

KATSUYA OHKI AND TAKASHI SHIRAI

NOMENCLATURE

$C(\tau)$: cross correlation function $\quad\quad$ (V)

$\quad\quad$: distance between two detecting points of a fiber optic probe $\quad\quad$ (mm)

N_c : mean frequency of the output signals of bubble detecting circuit $\quad\quad$ (1/sec)

N_0 : flash frequency of stroboscope $\quad\quad$ (1/sec)

$n_{p,up}$: the number of particles moving upward $\quad\quad$ (-)

$n_{p,down}$: the number of particles moving downward $\quad\quad$ (-)

r : radial distance of measuring point from the central axis of a fluidized bed $\quad\quad$ (cm)

R : Radius of fluidizing column $\quad\quad$ (cm)

T : time $\quad\quad$ (sec)

t : time $\quad\quad$ (sec)

u_f : fluidizing air velocity $\quad\quad$ (1/sec)

u_{mf} : incipient fluidizing air velocity $\quad\quad$ (1/sec)

v_p : velocity of particle $\quad\quad$ (cm/sec)

$\bar{v}_p$: local mean particle velocity
$$\left(= \frac{1}{n_{p,up} + n_{p,down}} \sum_{i=-\tau_0/\Delta\tau}^{\tau_0/\Delta\tau} (\ell/i\Delta\tau)n_i\right) \text{ (cm/sec)}$$

$\bar{v}_{p,up}$: local upward mean particle velocity
$$\left(= \frac{1}{n_{p,up}} \sum_{i=1}^{\tau_0/\Delta\tau} (\ell/i\Delta\tau)n_i\right) \quad\quad \text{(cm/sec)}$$

$\sqrt{\overline{v_p^2}}$: local root mean square velocity

$$\left(= \sqrt{\dfrac{1}{n_{p,up} + n_{p,down}} \sum_{i=-\tau_0/\Delta\tau}^{\tau_0/\Delta\tau} (\ell/i\Delta\tau)^2 n_i}\right) \quad (cm/sec)$$

$y_A(t), y_B(t)$: signals detected by a fiber optic probe

(V)

$y_{p,A}(t), y_{p,B}(t)$: pulse trains   (V)

z : height from air distributor   (cm)

Greek Symbols

τ : time lag   (sec)

τ_i : time lag for particle i   (sec)

τ_m : time lag which gives the muximum of crosscorrelation

function   (sec)

$\Delta\tau$: sampling period   (sec)

1. INTRODUCTION

Little is known to date about the local behaviour of particles in freely bubbling fluidized bed due to the lack of suitable measuring methods.

In this report, a new method for measuring the local velocity distributions of fluidizing particles with a fiber optic probe is presented and the measured results in a gas fluidized bed of 15 cm in diameter and 15 cm in bed height are shown. The behaviour of particles are compared with those of bubbles also measured simultaniously in the bed.

2. EXPERIMENTAL APPARATUS AND PARTICLES USED

The fluidizing column is a steel cylinder having a inside diameter of 15 cm as shown in Fig. 1. The dimensions of the fluidized bed, details of the gas distributor and the properties of solid particles are all summarized in Table 1.

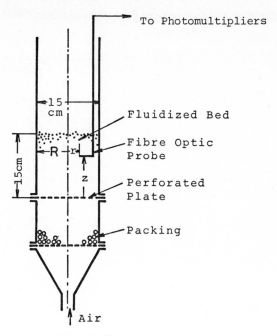

Figure 1. Experimental apparatus.

Table 1 Experimental conditions

Apparatus

  Fluidizing Column

    Inner Diameter D_T = 15 cm
    Particle Bed Height at
    Incipient Fluidization = ca. 15
                                   cm

  Distributor (Perforated Plate)

    Hole Diameter = 0.1 cm
    Pitch   = 1.15 cm
    Fraction of Hole Area =
              0.0067

Particle (Alumina Particles)   **I**   **II**

	I	II
Mean Diameter d_p (μ)	281	612
Standard Deviation (μ)	67	65
Particle Dencity ρ_p (g/cm^3)	1.29	1.29
Minimum Fluidizing Gas Velocity u_{mf} (cm/sec)	3.0	11.0

Fluid (Air at 1 atm., Room Temp.)

    Fluidizing Gas Velocity u_f/u_{mf} =
              2, 3, 4, 6

3. MEASURING INSTRUMENTS AND METHODS

Particle Velocity Distribution

The details of the fiber optic probe for measuring the
velocity of fluidizing particles are shown in Fig. 2.
A fiber in the center of three fibers is used to illuminate
the moving individual particles and the other two fibers, A
and B are used to catch the light reflected on the surfaces
of those particles.   The variations of the reflected lights
are converted to electric signals by photomultipliers and
are processed as shown in Fig. 3.

As presented in the previous papers,[1,2,3] when particles are
moving with a constant speed in one direction, the velocity
of those particles, v_p can be determined by computing the
crosscorelation function of the signals, $y_A(t)$ and $y_B(t)$
detected by two fibers, A and B as follows;

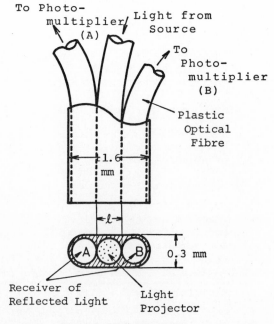

Figure 2. Fiber optic probe.

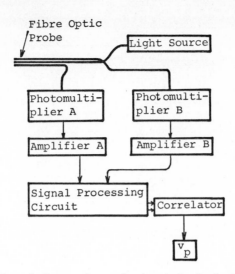

Figure 3. Schematic diagram of measuring the velocity distribution of particles.

$$v_p = \ell/\tau_m \qquad\qquad (1)$$

where ℓ is the distance between two detection fibers and τ_m is the time lag where crosscorrelation function takes the muximum.

In the fluidized bed, the particles form a certain circulation pattern in a large time scale. But, since both velocity and direction of particles at any local points in the bed change rapidly with time due to the rising bubbles, the correlation between two signals detected by two fibers is poor and thus, it is difficult to determine exactly the mean velocity of particles.

Figure 4-(a) and (b) show the examples of the signals detected by the upper fiber A and lower fiber B of the probe in 50 msec time duration at the bed height z = 7.5 cm and the radial position r/R = 0.62 in the bed where many bubbles are rising. In these relatively short time durations, signals, $y_A(t)$ and $y_B(t)$ show quite similar variations with

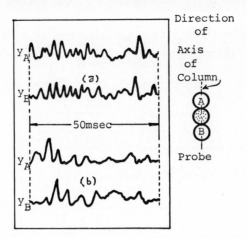

Figure 4. Examples of the signals detected by the probe.

each other, except that either of the signals are at different times. Therefore, it can be seen that the particles moved from B to A (upward) from Fig. 4-(a) and vice versa (downward) from Fig. 4-(b). In these cases, detected signals, $y_A(t)$ and $y_B(t)$, respectively, are passed through the signal processing circuits to be converted to the pulse trains, $y_{p,A}(t)$ and $y_{p,B}(t)$ as shown schematically in Fig. 5. Subsequently, the time lags of the pulses in $y_{p,A}(t)$ and $y_{p,B}(t)$ can be computed by a real-time correlator as briefly explained below.

The pulse trains, $y_{p,A}(t)$ and $y_{p,B}(t)$ as shown in Fig. 5 are assumed to be given by the following formula;

$$y_{p,A}(t) = a\delta(t - t_A) \qquad (2)$$

$$y_{p,B}(t) = a\delta(t - t_B) \qquad (3)$$

where $\delta(t)$ represents a Dirac's delta function and time, t_A and t_B, respectively, represent the times when a particle passes near the fibers, A and B of the probe. The cross-correlation function of $y_{p,A}(t)$ and $y_{p,B}(t)$ is ;

$$C(\tau) = \frac{1}{T}\int_0^T y_{p,A}(t)y_{p,B}(t + \tau)dt$$

$$= \frac{a^2}{T}\{\tau - (t_B - t_A)\} \qquad (4)$$

where $0 \leq t_A, t_B \leq T$. The crosscorrelation function of Eq. (4) have a certain value only when the value of τ is equal to $(t_B - t_A)$, which is just the time required for a particle to move from the position of fiber A to that of fiber B. Therfore, the velocity of the particleis given by ;

$$v_p = \ell/(t_B - t_A) = \ell/\tau_{AB} \qquad (5)$$

In the real cases, many pulses appear in the pulse trains, $y_{p,A}$ and $y_{p,B}$, so that the outputs of the real-time correlator will be as follows.

$C(\tau)$ = (Computed results for pulses generated

by the same particles) + (Computed

results for pulses generated by the

different particles) (6)

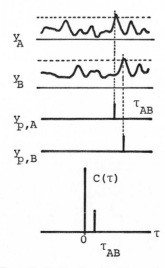

Figure 5. Outline for signal processing.

When the pulse separations in the pulse trains are taken large enough to neglect the second term in Eq. (6), the output of correlator might be as follows;

$$C(\tau) = \frac{a^2}{T} \sum_i \delta(\tau - \tau_i) \qquad (7)$$

where τ_i is the time required for the particle i to move from fiber A to B.

Equation (7) just gives the histogram of time lags which are required for particles to move from A to B. Therefore, the velocity distribution of particles can be easily calculated with Eq. (5) from the histogram of time lag.

Bubble Characteristics

The block diagram for measuring the characteristics of bubbles which allows the simultanious measurement of the velocity, fraction, frequency and diameter of the bubble in the bed and the optical probe used are, respectively, shown in Figs. 6 and 7. The bubbles can be detected by the probe based on the light transmission principle. When a bubble wrapps the probe, the light from stroboscope of flash frequency N_0 is transmitted to the light receiving fiber of the probe and converted to an electric signal with a photomultiplier. The local bubble fraction ε_B can be calculated by dividing the mean frequency, N_c of the photo-multiplier output signal by the stroboscope flash frequency, N_0 as follows.

$$\varepsilon_B = N_c/N_0 \qquad (8)$$

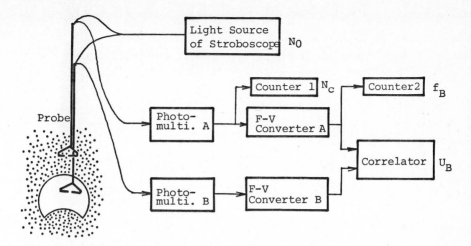

Figure 6. Block diagram of measuring the characteristics of bubbles.

The mean rising velocity of bubbles, u_B can be computed from the output signals of the frequency to voltage converters in terms of correlation method.

4. RESULTS AND DISCUSSIONS

Particle Behaviours

Figures 8, 9 and 10, respectively, show the cross-correlation functions, time lag distributions and velocity distributions, all computed from the same signals detected at the height z = 7.5 cm from the air distributor in the fluidized bed with an air velocity of $u_f = 3u_{mf}$. In Fig. 8 which is obtained from correlation method, it can be seen

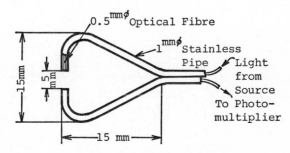

Figure 7. Optical bubble detector.

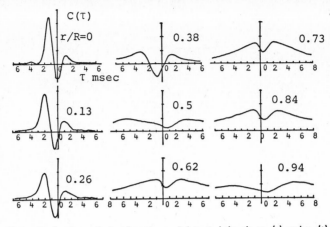

Figure 8. Crosscorrelation functions of detected signals, $y_A(t)$ and $y_B(t)$.

that the particles move downward with relatively constant speed only in the center region of the bed. On the other hand, the same signals, $y_A(t)$ and $y_B(t)$ can be processed as explained in section 3 as Fig. 9 (time lag distribution), which can be converted to the velocity distributions as Fig. 10. Figure 10 well shows the vertical velocity distributions of particles in all radial positions of the bed.

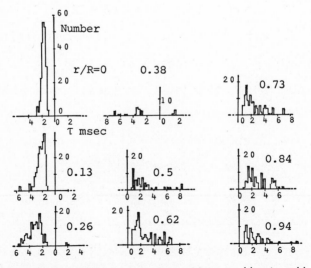

Figure 9. Time lag distributions computed from $y_{p,A}(t)$ and $y_{p,B}(t)$.

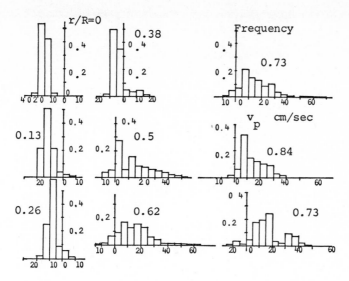

Figure 10. Vertical velocity distributions of particles in fluidized bed.

In the cases of $u_f = 3u_{mf}$ and $4u_{mf}$, various characteristics of particle motions are calculated from time lag distributions as shown in Fig. 11, where local mean velocity $\bar{v}_p$, local upward mean velocity $\bar{v}_{p,up}$ and local root mean square velocity $\sqrt{\overline{v_p^2}}$ are plotted as a function of radial position r/R from the center of the bed at three different bed heights z.

In the preliminaly experiments, the symmetry of those measurements about the radial distance of the bed was confirmed. Thus, from Fig. 11, a considerably intensive circulation of particles as schematically shown in Fig. 12 is found to be formed so that particles move downward in the central region of the bed. When the fluidizing air velocity is increased, those tendencies of particle circulations are almost similar to the cases described above.

Those circulation patterns are stable in the long time period of experiment, when the experimental conditions are fixed. Those circulation patterns are remarkably different

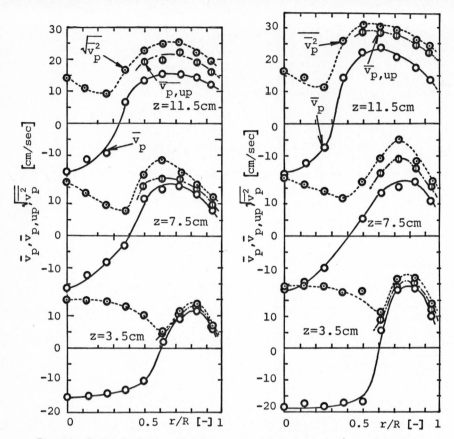

Figure 11. Radial distributions of various velocities in the cases of u_f/u_{mf} = 3 and 4.

from the results[4] reported so far in the case of deep bed height which state that particles move upward in the center region of the bed and downward in the outer region near the wall of the bed.

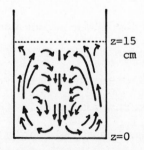

Figure 12. Possible flow pattern of particles in fluidized bed.

Comparisons of the Behaviours of Particles and Those of Bubbles

In order to explain such flow pattern of solid particles in the bed, the radial profile of bubble characteristics was also measured. Figure 13 shows the local velocity of bubbles, u_B and local bubble fraction, ε_B in the cases of $u_f = 3u_{mf}$ and $4u_{mf}$ with the same measuring positions of the probe as described in Fig. 11. It can be seen from Fig. 13 that bubbles are generated in an annulus near the wall at the bottom of the bed and then come to the center with their rise to the surface of the bed. It is suggested that the particle circulations in the bed are formed as the result of mutual interferences between particles and bubbles from the comparison of the movements of particles in Fig. 11 with those of bubbles in Fig. 13.

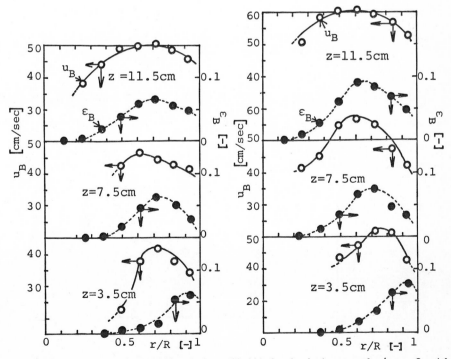

Figure 13. Radial distribution of bubble velocity and bubble fraction in the cases of $u_f/u_{mf} = 3$ and 4.

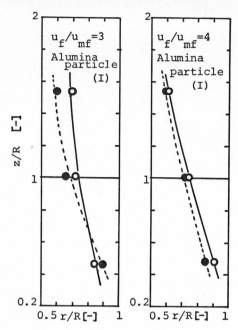

Figure 14. Comparisons of particle motions and bubble motions (● Particle, ○ Bubble).

In order to make the mutual interferences more clear, the radial positions where bubble fraction ε_B take the muximum values are compared in Fig. 14 with those where the number of upward particles, $n_{p,up}$ minus those of downward particles, $n_{p,down}$ take the muximums. It can be seen in Fig. 14 that those positions at different bed heights z/R agree with each other.

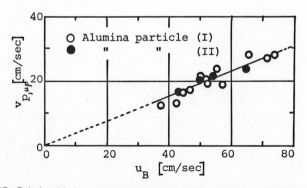

Figure 15. Relationship between upward mean particle velocity and bubble velocity.

The upward mean particle velocity, $v_{p,up}$ is correlated with the rising velocity of bubbles, u_B in Fig. 15, where linear relationship between those velocities are well indicated.   The rising velocity of alumina particles can be correlated as follows.

$$\bar{v}_{p,up} \doteqdot 0.37u_B \tag{9}$$

5. CONCLUDING REMARKS

By using a small fiber optic probe, the local behaviour of fluidizing particles in a free-bubbling fluidized bed can be made clear.

In the rather shallow fluidized bed as used in these experiments, intensive circulation of particles so that particles move downward in the central region of the bed and upward in the outer region of the bed, are formed.

From the simultaneous measurements of the behaviours of particles and those of   bubbles, mutual interferences of those behaviour are confirmed.   And it is found that the rising velocity of particles is proportional to the bubble velocity.

LITERATURE CITED

1) K. Ohki, T. Akehata and T. Shirai, A New Method for Measuring the Velocity of Solid Particles with Fiber Optic Probe, Kagaku Kogaku, 37 (1973) 965.

2) K. Ohki, T. Akehata and T. Shirai, A New Method for Evaluating the Size of Moving Particles with a Fiber Optic Probe, Powder Technology, 11 (1975) 51.

3) K. Ohki, T. Akehata and T. Shirai, On the Measurement
of Local Particle Velocity in the Bulk-Flow of Solid Particles,
J. of the Res. Ass. of Powder Technology, Japan, 12(1975)222.
4) R. M. Marsheck and A. Gomezplata, Particle Flow Patterns
in a Fluidized Bed, A.I.Ch.E. Journal, 11(1965)167.

LOCAL AND INSTANTANEOUS VARIATION OF THE HEAT TRANSFER COEFFICIENT IN THE VICINITY OF AN ARTIFICIAL BUBBLE RISING IN A THREE-DIMENSIONAL FLUIDIZED BED

A. BERNIS, F. COEURET, F. VERGNES, AND P. LE GOFF

The heat transfer properties of the gas solid fluidised bed are essentially determined by the renewal of the particles along the transfer surface. This vigourous agitation is led by the motion of the particles around the bubbles rising in the bed. With a view to analysing the elemental steps of this process, instantaneous heat transfer coefficients have been measured as function of the time in the neighbourhood of the track of artificial bubbles.

EXPERIMENTAL SYSTEM

In a cylindrical sand bed ($d_p \approx 154 \mu$I.D.=0.35m H/D=1), maintained by an air stream at incipient fluidisation, an injection device periodically feeds bubbles of known volume at an adjustable frequency. A small platinum film (2x1mm) is set on an isolating cylinder with diameter of 5mm immersed in the bed at a location which is adjustable. An electronic device controls the electrical current in the film in a such a way that its resistance V/I and therefore its temperature is kept constant, at about 30°C above the temperature of the bed. The power input VI is digitally recorded on a multichannel analyser, of which the sweeping is synchronised with the injection of the bubbles. Such an accumulation procedure increases the signal to noise ratio. The time constant of the whole measuring linkage is about one millisecond. The heat transfer coefficient between the probe and the bed is deduced from the power input and the temperature difference between the probe and the bed.

RESULTS AND DISCUSSION

With the probe kept in the axis of the bubble track, a typical record is given in Fig. 1a. Between A and B, h increases due to the

111

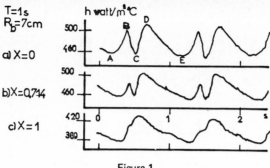

Figure 1.

renewal of the grains pushed upwards by the bubble, between B and C
the probe is immersed in the air of the bubble and obviously h decreases but after reincreases during the sweeping of the probe by
the bubble wake and finally tends towards the value corresponding to
incipient fluidisation After the time T : the period of the injection,

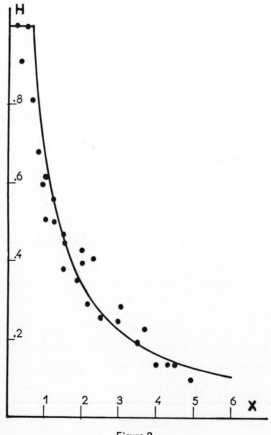

Figure 2.

the phenomenon begins again. Similar results in a two dimensional bed have been obtained by TUOT and CLIFT (1). The variation of the instantaneous and mean values of h have been studied as a function of the two parameters : bubble diameter and injection period, respectively in the range 2.6 to 14cm and 0.3 to 12 sec. No influence of the diameter of the bubble on any value of h is detected. But obviously the value of h decreases when T increases. The different experimental values of h (Watt/m^2.°C) are correlated with T (sec.) by the relations :

$$h_A = 470 \ T^{-0.15} \qquad \bar{h}_{AB} = 500 \ T^{-0.15} \qquad \bar{h}_T = 480 \ T^{-0.15} \ ; \ \dots$$

If the probe is kept at a distance x=XR$_b$ away from the bubble axis, the point A occurs progressively later and later after the injection, the point D sooner and the decreasing part BC vanishes (Fig. 1b, 1c).

The mean heat transfer coefficient $\bar{h}_T(X)$ is never higher than $\bar{h}_T(0)$ and decreases with X. On Fig. 2, $\bar{h}_T(X)$ is plotted in the dimensionless form H(X) relative to its value on the axis and to its value at the onset of fluidisation h_{mf}=260 Watt/m^2.°C

$$H(X) = (\bar{h}_T(X) - h_{mf})/(\bar{h}_T(0) - h_{mf})$$ A logarithmic regression leads to the relations :

H(X) = 1 for X < 0.66 and H(X) = 0.66/X for X > 0.66

Fig. 3 shows the heat transfer coefficient to the probe immersed axially in the free bubbling fluidised bed as a function of

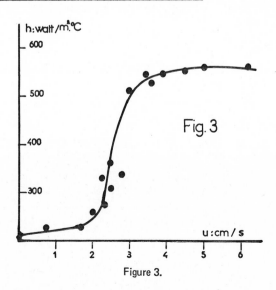

Figure 3.

the superficial air velocity. A very sharp increase of h starts from
the onset of fluidisation. If the flowrate of air at incipient fluidisation is deducted from the total flowrate, the flowrate rising
through the bed in the form of bubbles is obtained, giving a relation between the mean size and frequency of the bubbles. By averaging our results on artificial bubbles against the time and on a cylindrical space arbitrarily bounded to a radius of $2R_b$, the heat
transfer coefficient in the free bubbling bed is found again. The
possible values of the size and the frequency of the bubbles agree
well with the work of GELDART (2).

REFERENCES

(1) TUOT, CLIFT, Chem. Eng. Progress, Symp. Series 128, <u>69</u>, 1973,p.78
(2) GELDART, Powder Technology, I (1967/1968), p. 355

THE PHYSICAL AND MASS TRANSFER PROPERTIES OF BUBBLES IN FLUIDIZED BEDS OF ELECTRICALLY CONDUCTING PARTICLES

P. H. CALDERBANK, J. PEREIRA AND J. M. BURGESS

NOMENCLATURE

A tube cross-sectional area, m^2

A_j projected area of bubbles of size class j.

Cb concentration of tracer in bubble (cm^3 He m^{-3})

C_p concentration of tracer in dense phase (cm^3 He m^{-3})

D diameter of the column (m)

De diameter of a sphere with same volume as the bubble (m)

Dcl diameter of bubble cloud (m)

Dg diffusion coefficient of tracer in fluidizing gas (m^2 sec^{-1})

d_c bubble column diameter, m

d_p probe separation distance, m

$\hat{d}_e$ truncated value of d_e

f_j number fraction of bubbles of size class j per unit volume of dispersion

f_k number fraction of bubbles of size k in the dispersion

g acceleration due to gravity, $m\ sec^{-2}$

h height (cm)

Kg mass transfer coefficient ($m\ sec^{-1}$)

Ko gas exchange coefficient ($m^3\ sec^{-1}$)

Kv gas exchange coefficient per unit of volume of bubble (sec^{-1})

Lb bubble vertical axis length (m)

L_c bubble central axis length, m

L_d distance to bubble frontal surface at radial distance x_p, m

L_o bubble vertical chord length at radial distance x_p, m

N bubble injection frequency

n_j number of bubbles of size class j recorded

n integer

Q gas exchange flux for bulk flow $(m^3 \ sec^{-1})$

Q_B bubble phase total gas flow, $m^3 \ s^{-1}$

Rb Radius of curvature of the bubble nose (m)

r_p probe upper element radius, m

r radial position in column

Sf bubble frontal area (m)

S Surface area of bubble (m^2)

T bubble injection period (sec)

tc exposure time

t_c response correction time Eq.(17), sec

t_i delay time on channel i, sec

t_m linearisation time interval, sec

Δt_m step time interval for curve examination, sec

t_s start time for compensation process, sec

t^* dimensionless delay time ratio, Eq.(13)

t_L^* limit time ratio for discrimination, Eq.(15)

T_i pulse duration time on channel i, sec

T_{1c} corrected pulse duration time on channel 1, sec

T^* dimensionless pulse duration time ratio, Eq.(14)

T_L^* limit time ratio for discrimination, Eq.(15)

U_B bubble vertical rise velocity, m

Ub bubble velocity $(m \ sec^{-1})$

Umf superficial velocity of gas at minimum fluidization $(n \ sec^{-1})$

U_{mf} minimum fluidisation velocity, ms^{-1}

U_s total gas flow per unit area, ms^{-1}

U_i velocity of bubble in size class i, ms^{-1}

U_j velocity of bubbles of size class j.

V_i instantaneous voltage on channel i, volt

V_o transition voltage between gas and liquid, volt.

V_{min} minimum voltage, volt.

V_{max} maximum voltage, volt.

V_i volume of bubble i , m^3

V_b bubble phase flow per unit area in dispersion, ms^{-1}

V bubble volume

x_p probe shape element radial distance, m

z_a axial distance above gas distributor, m

z_r radial distance from dispersion centreline, m

σ standard deviation of normal size distribution

σ_{de}^2 variance of bubble size distribution

γ probe voltage pulsing frequency, s^{-1}

$\Gamma(x)$ gamma - function of x

α gamma - function size distribution parameter,
β gamma - function size distribution parameter,

ϵ_b bubble phase voidage

ε_{mf} bed porosity at minimum fluidization

μ . mean of normal bubble size distribution function

THE DEVELOPMENT OF AN IMPROVED PROBE FOR THE MEASUREMENT OF BUBBLE SIZE, SHAPE AND VELOCITIES

Abstract

An improved probe technique for the measurement of bubble size and velocity in turbulent bubble dispersions has been developed.   The technique

makes use of fast real-time data processing methods to identify the bubble
orientation during measurement and to accept readily interpretable data for
only those bubbles whose central axes are coincident with the probe axis.
The method is evaluated experimentally using well defined single bubbles
and shown to be accurate as well as rapid.

Introduction

The measurement of bubble size and velocity in bubble dispersions is
an important problem in both gas-liquid and gas-fluidised particle systems.
Several optical techniques using light scattering [1,2] and high speed flash
photography [3,4,5,6] allied to statistical bubble counting procedures [7]
have been used to evaluate bubble sizes in sieve tray froths. Similarly,
X-ray photography [8,9] has been applied to single bubbles in three dimensional
gas-fluidised beds and, in addition, a number of probe techniques have been
developed to evaluate gas bubble properties in freely bubbling gas-fluidised
beds [10].

Park et al.[12] and Rigby et al.[13] developed a two-element resistivity prob
to examine the fluidisation state of two- and three-phase fluidised beds and
Werther and Molerus [10] used a similarly designed capacitance probe for the
examination of non-conducting particle beds. In these cases uncertainty exists
in deducing the bubble size and velocity distribution at a point in the bed
because of the varying and unknown positions at which the probe contacts the
bubble frontal surfaces.

Probe Design

To eliminate some of the uncertainties associated with previous
techniques and produce a device which gives accurate bubble sizes and
velocities in dispersions the following conditions must be satisfied:

1.   The probe must be capable of resolving the position at which it is struck by the bubble relative to the bubble centre-line and thus select for measurement only those encounters where the bubble and probe axes of symmetry are coincident within fine limits.

2.   The instant when the probe tip enters the bubble must be accurately determined, bearing in mind that the probe may deform the interface and this could result in a significant departure from a step voltage change.

In order to achieve these aims, a three dimensional resistivity probe with five channels was designed in order to sence the bubble local interface approach angle as well as measure bubble size and velocity.   This device has been coupled to a high speed digital computer* with facilities for rapid, accurate conversion of analog voltage signals to discrete binary numbers and with software to undertake logical decisions consequent on the spatial orientation of the bubble with respect to the probe axis.   The logic selects only those bubbles whose central axes are coincident with the probe axis and calculates their size, shape and velocity by proper correction of the resulting non-square pulses.   It is assumed in deriving the characteristic bubble-size distribution function of a dispersion that bubbles rise into the probe in a random fashion and that the measurements are therefore characteristic of unbiased sampling.

Probe discrimination

The probe discrimination hardware was contrived by symmetrically disposing three contacts around and above the first contact so that all three exist in a horizontal plane a known distance above the central contact and radically spaced from it.   A fifth contact was placed in the same

* Digital Equipment Corporation PDP/8e

horizontal plane as the central contact but somewhat distant from it.
The arrangement is shown schematically in Fig.1.

The probe array was connected so that each contact formed part of
an electrical resistivity circuit whereby current flows from an external
D.C. power supply whilst that contact is resident in the conducting phase
and thus develops a voltage across a load resistor which is measured by
the computer in digital form simultaneously for each channel.

The method of operation may be illustrated by considering the voltage
pulse sequence which is generated when a single bubble approaches the
assembly of contacts.   A schematic diagram of such a typical sequence is
shown in Fig.2, where the voltage channels are numbered with reference

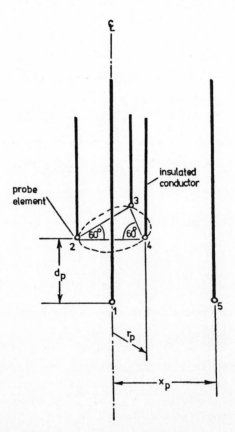

Figure 1. Isometric projection of the spatial orientation of the probe contact elements.

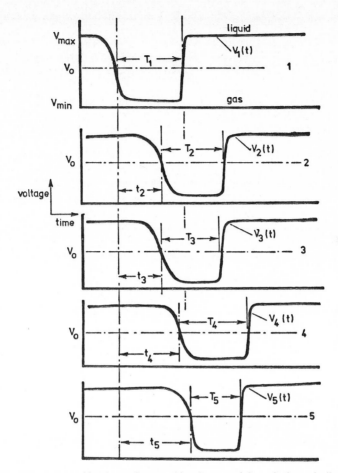

Figure 2. Ideal pulse sequence generated by the probe assembly when struck by a single vertically rising bubble. Channel numbers refer to the code in figure 1.

to the code in Fig.1.   When the voltage is at a maximum, the probe tip resides in the conducting phase and <u>vice versa.</u>

Defining an arbitrary threshold voltage:

$$V_{min} < V_o < V_{max} \qquad (1)$$

and requiring that initially:

$$V_1 > V_o \qquad (2)$$

so that channel 1 resides in the conducting phase, the voltage fall curve,

$V(t) ; V_1 > V_o$, in continuously stored in computer memory so that the voltage fall function may be examined retrospectively in the event of a successfully co-incident bubble hit.   This curve is used to account for the effect of interface deformation in order to compute the correct bubble size.   The data analysis process is discussed in further detail below.

The above sample, decide and store process continues until:

$$V_1 \leqslant V_0 \qquad\qquad (3)$$

indicating that the approaching gas bubble interface has enveloped the probe tip of the central contact.

At the instant this occurs the system immediately examines the state of the other channels, since the arrival of a voltage pulse at the upper channels $(2,3,4)$ must be uniquely associated with the pulse at the lower contact (1).

Now if the condition

$$V_i \leqslant V_0 \quad ; \quad i = 2,3,4 \text{ or } 5 \qquad (4)$$

is satisfied at the same instant when inequality (3) above is satisfied, one or more of the upper channels is already in the gas phase when the lower, central channel strikes the bubble.   It is difficult under these conditions to unequivocally determine the velocity of the bubble since the sequence is generated by either a markedly off-centre bubble, a bubble moving in a direction not parallel with the probe axis or bubbles which are very close together.   For this case the analysis is abandoned, the bubble rejected, and the examination of the arriving pulse sequences recommenced from the start.

However, if

$$V_i > V_0 \quad ; \qquad i = 2,3,4 \text{ or } 5 \qquad (5)$$

is satisfied at the instant when inequality (3) above is satisfied, the bubble has struck the central channel first and may be travelling in the axially coincident fashion. The system logic enters a continuous cyclic count mode under these conditions, with incremental real time counting of the times t_i and T_i defined by Fig.2.

The conditions where

$$T_1 < t_i \quad ; \quad i = 2,3,4 \tag{6}$$

also presents an interpretation problem as the pulse sequence cannot be explicitly related to a unique bubble under these conditions, as two small bubbles may co-exist between the probe contacts. The sequence is therefore again abandoned when inequality (6) is detected. This imposes an important limiting constraint on the output of the system as bubbles in the dispersion whose vertical lengths are smaller than the probe separation distance d_p are rejected.

For the acceptable condition:

$$T_1 \geq t_i \quad ; \quad i = 2,3,4 \tag{7}$$

the logic process continues with incremental counting of

$$t_i \quad ; \quad i = 2,3,4,5 \tag{8}$$

and
$$T_i \quad ; \quad i = 1,2,3,4,5 \tag{9}$$

with the discrimination condition

$$V_i \leq V_0 \quad ; \quad i = 2,3,4,5 \tag{10}$$

terminating the counting of t_i and

$$V_i > V_0 \quad ; \quad i = 1,2,3,4,5 \tag{11}$$

terminating the counting of T_i

On complete termination defined by

$$V_i > V_0 \quad ; \quad \text{all } i = 1,---5 \tag{12}$$

the system leaves the real time sampling routine to check the criteria for a successful bubble hit.

Since the probe array has a three dimensional shape and the upper contacts 2,3 and 4 are equispaced around the central contact 1, the delay times t_i; i = 2,3,4 provide a useful measure of the degree of local curvature of the bubble interface when expressed in dimensionless form:

$$t^* = \frac{t_{i_{min}}}{t_{i_{max}}} \tag{13}$$

and

$$T^* = \frac{T_{i_{min}}}{T_{i_{max}}} \tag{14}$$

For a perfectly coincident bubble-to-probe centreline encounter both t* and T* approach unity.   However, in a bubble dispersion a perfect hit is rare and it is better to define limit conditions for the time ratios to achieve a satisfactory rate of bubble sampling, acknowledging that a small area surrounding the bubble centreline will be acceptable to the probe.   Thus, the limit ratios t_L^* and T_L^* lie in practice between 0.8 and 1.0 and for all cases where either

or

$$t^* < t_L^* \tag{15}$$

$$T^* < T_L^* \tag{16}$$

the system returns to the start to await the arrival of another bubble interface at the central probe channel.

Corrections for the probe response-time

It is well known [10,12,13] that a resistivity or capacitance element when struck by a bubble, produces a voltage pulse which is not square.

As noted previously, the system stores the analog voltages in descrete form in sequence whilst the central contact is resident in the conducting phase until the instant inequality (3) is satisfied.   Thus, whilst the examination outlined above is being undertaken in real time, a curve which represents the fall of the voltage is retained in digital form in computer memory as shown schematically in Fig.3.   For a successful bubble hit, the system examines the curve slope over a known linearised small time interval.

The curve examination process proceeds backwards in time from the voltage level V_0 (time t_s), incrementing small time steps of length Δt_m and linearising the curve over the time interval t_m.   When the desired conditions are met for the definition of pulse commencement considered below, the additional pulse duration time is given by:

$$t_c = (n+1)\Delta t_m - \frac{t_m}{2} \qquad (17)$$

The corrected pulse time duration then becomes

$$T_{1_c} = T_1 + t_c \qquad (18)$$

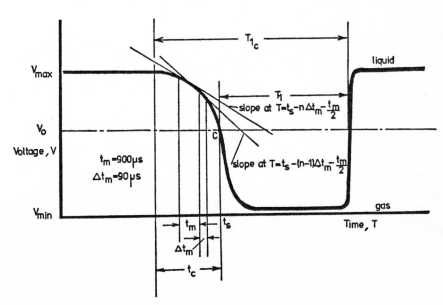

Figure 3. Ideal voltage pulse on a single probe channel showing compensations for the time response of the wetted element.

Bubble velocity and size

The delay times t_i ; i = 2,3,4, defined on Fig.2, are approximately equal for a central, co-axial, bubble-to-probe encounter since the bubble interface has taken the same time to travel between the central lower contact and the upper contacts.   For a defined and accurately known probe separation d_p, the bubble velocity is given by;

$$U_B = \frac{3\ d_p}{\sum\limits_{i=2}^{4} t_i} \qquad\qquad (19)$$

The size of the bubble, defined in terms of its central vertical chord length, is then given by:

$$L_c = U_B\ T_{1c} \qquad\qquad (20)$$

Bubble shape

The function of the probe contact designated 5 on Fig.1 is to measure approximate bubble shapes for those bubbles whose horizontal axis length is greater than x_p, as is shown in Fig.4. The probe array selects bubbles very close to their centrelines,

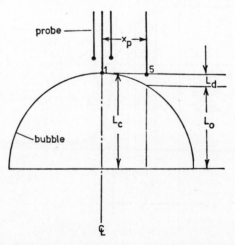

Figure 4. Definition of the bubble shape measurement parameters.

so the outer contact 5 measures the vertical distance, L_d, between the
bubble leading surface at the centreline and at the radial distance, x_p.
It also measures the bubble vertical length at this position, L_0.
Accordingly,

$$L_d = U_B t_5 \tag{21}$$

and

$$L_0 = U_B (T_5 + t_c) \tag{22}$$

Complete details of the system logic for the above data processing
method may be found elsewhere[15], together with a discussion of timing
methods.   The developed probe system can examine all five channels of analog
signals at a rate in excess of ?kHz which is sufficient to resolve bubble
velocities up to approximately 2 m.s^{-1} with a probe separation dimension
of around 3×10^{-3} m.

Probe Fabrication

The probe is depicted in Fig.5.   Polytetrafluoroethylene (PTFE)
coated nickel wire of diameter $0.?? \times 10^{-3}$ m was used as the insulated
conductor to the probe element comprising the bared tip of the wire.
Each insulated wire was passed through hypodermic stainless steel tubing
so that the geometric relationships outlined in figure 1 were rigidly
conformed to.   An optical technique was used to align the probe elements
initially and periodic checks were made during use with a travelling
cathetometer[15].   The probe separation distances for the device utilised
here were given by

$$d_p = (3.57 \pm 0.04) \times 10^{-3} \text{ m}$$

and

$$r_p = (1.52 \pm 0.02) \times 10^{-3} \text{ m}$$

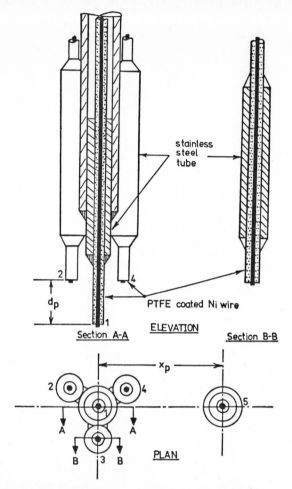

Figure 5. Engineering details of the probe used for gas liquid systems.

Probe Evaluation

The system described in this work was designed to measure bubble sizes and velocities in bubble swarms but was tested with single bubbles rising vertically in water before use in this connection.

Figure 6 shows the experimental single bubble apparatus used; full details may be found elsewhere[15,16]. The probe A was constructed as above and clamped securely in place with an adjustable support bracket B being positioned both axially and radially with respect to the axis of the 0.13m diameter glass column C of length 2m. The probe tip was normally sited

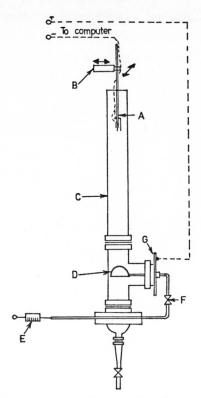

Figure 6. Schematic diagram of the experimental equipment used for the measurement of the properties of single
isolated air bubbles in water.

0.3m below the level of water in the column and the column end plate G

served as a positive terminal for current flow through the liquid to the

probe and thence to the data processing system.   Release of the bubble

was achieved through rotation of the plastic dump cup D after the required

volume of gas was injected from syringe E through the non-return valve F.

Voltage pulse shape and sequence

By rapid sampling of the analog voltages associated with the probe

output and sequential storage of the digital form of the voltages, an

accurate trace of the five output channels from the probe may be obtained.

A typical trace for a single spherical cap bubble is shown in Fig.7, with

channel numbers corresponding to the code in Fig.1.

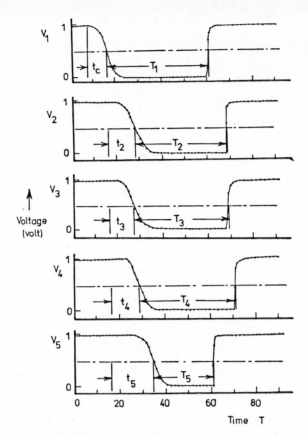

Figure 7. Typical pulse sequences generated by the probe when a single bubble in water rises into the probe assembly. Each point represents an instantaneous voltage sampled by the computer. The channel numbers are given by the code in figure 1.

The pulses shown in the figure exhibit the following characteristics. The voltage fall from fully conducting to non-conducting follows a smooth relatively slowly responding "s" shape curve caused by liquid film thinning around the wire tip and on the PTFE wire coating, clearly illustrating the importance of the correction process outlined above. Examination of probe outputs in fluidised beds (vide infra) indicates that this effect is of some significance in all submersed probe techniques. By contrast, the voltage rise at the end of the pulse is extremely rapid.

The sequence of pulses which exist on all channels of the probe assembly when struck by a vertically moving single bubble follows the system logic imposed by the data processing softwave.   The delay and pulse duration times are clearly in evidence and of the correct sequence, enabling computation of the bubble parameters.

Bubble rise velocity

The measured bubble rise velocity and its variation with bubble equivalent spherical diameter is shown in Fig.3 for individual experiments, where the bubble velocity is calculated from equation (19).   The relationship clearly reflects the influence of the column walls and the subsequent approach of the larger bubbles to the Dumitrescu[17] slug flow velocity relation

$$U_B = 0.35 \ (g \, d_c)^{1/2} \tag{23}$$

The high accuracy available with the technique is indicated by the approach of the data to the best-fit curve reported by Johnson[16]

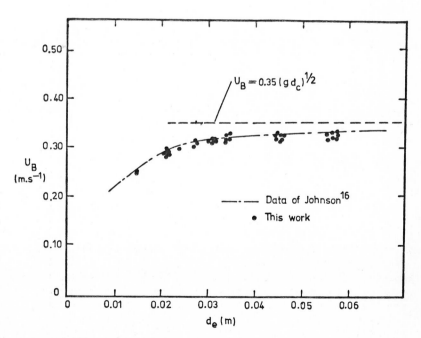

Figure 8. Instantaneous bubble velocities reported by the probe for single bubbles, where bubble velocity is plotted against equivalent spherical diameter.

to detect the bubbles.   Reference to the figure also indicates that the reproducibility of the device is satisfactory, with a deviation of less than ten percent from the mean.

Bubble axis length

Consider Fig.9 which shows an accepted pulse from the leading probe. Taking three slope transition criteria, defined by A,B and C. on Fig.9, and plotting the central axis length – equivalent spherical diameter relationship for each criterion leads to Fig.10 where the central axis length has been calculated from equation (20), and the probe and bubble axes are coincident.   Also shown thereon is the accurately determined photographic data of Johnson[16] for spherical cap bubbles in the same apparatus.

It is obvious that transition criterion A (arbitrary selection of the mid-voltage point V_o) fails to accurately record the bubble axis length. Indeed, at pierced lengths associated with small bubbles the reported parameter is more than fifty per cent in error.   When criterion C (movement

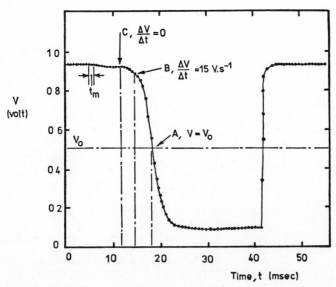

Figure 9. Typical voltage pulse on a single probe element. Criteria A, B and C are arbitrary definitions for the transition from gas to liquid.

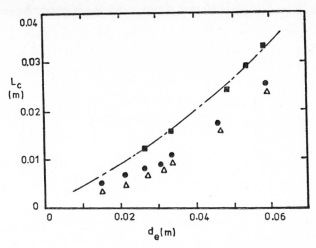

Figure 10. Variation of bubble central axis length with equivalent spherical diameter and the criteria defined in figure 9.

and obtained on the same apparatus using two vertically separated photocells along the voltage fall curve to the point where the voltage is constant over the time interval t_m) is used, the data confirm the accurately determined photographic bubble axis lengths. Previous techniques using resistivity probes [12,13] have not employed this rather important correction.

The detection of bubble orientation may be tested by forcing a bubble interface of known geometry to strike the probe array. This was achieved by siting the probe vertically and moving it radially away from the column centreline so that the approach angle became progressively greater. For each radial position the local bubble interface angle was calculated from the bubble eccentricity and spherical cap geometry [16] and the ratio t^* defined by equation (15) determined. The variation of t^* with the calculated interface angle is shown in Fig.11.

Application to Bubble Dispersions

Channels 2,3 and 4 together constitute an AND gate which opens for a small adjustable time so that only the characteristics of those bubbles which enter and leave the measurement field 'squarely' are stored. The further constraints that all channels must initially and finally be in the

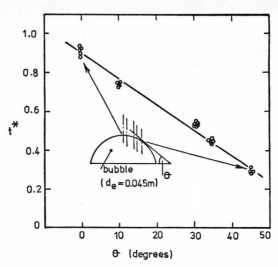

Figure 11. Variation of the probe discrimination ratio t^* with the computed local bubble interface angle seen by the probe.

continuous phase (conducting), that all channels must at some intermediate time be simultaneously non-conducting and that the channels must change logic state in a preordained sequence, ensures that samples of single well-defined bubbles are alone selected from dispersions for measurement.

The compound probe thus selects from a bubble dispersion rather rare events, the variable descrimination being adjusted to secure a satisfactory rate of sampling with adequate accuracy (typically 1000/hr).   The delay times, from which bubble velocity and height are calculated were punched onto paper tape after each successful sample, the measurement process being temporarily halted while this took place.

In a dispersion, it may be necessary to measure the abundance of bubbles in each of a number of specific narrow size ranges.   The probability of a bubble passing the AND gate, which is of fixed size, is obviously greater, the larger the bubble.   It was argued that the number abundance per unit volume, was related to the bubble cross-sectional area and rise velocity as,

$$f_j = \left[\frac{\dfrac{n_j}{U_j \, A_j}}{\displaystyle\sum_{j=1}^{m} \dfrac{n_j}{U_j \, A_j}} \right] \qquad j = 1, \text{- - -} \ m \qquad\qquad (24)$$

The above relationship was confirmed by computer simulation in which "normal" distributions of bubbles of various mean size, randomly positioned in space, rose into a simulated probe array.   It was from this simulation that the optimum probe geometry was determined ($r_p = \dfrac{d_p}{2}$), d_p being the smallest height of bubble deemed to be significant in a particular bubble size distribution.

APPLICATION TO A FREELY BUBBLING GAS-FLUIDIZED BED

Although there has been some considerable effort devoted to the study of freely bubbling gas-fluidised beds, there is still doubt concerning the bubble velocity and size distributions in these dispersions. Goddard and Richardson (14) concluded that there was no relation between bubble velocity and size while Park et al. (12) assumed that there was a unique relationship between the two.

The description of a freely bubbling bed which specifies one bubble size with a mean velocity is obviously incomplete. However, this is the only information to be derived from the most recent work using a probe technique (10).   Again, the influence of macro bed circulation on bubble velocities has not been evaluated with any certainty.

The probe developed in this work has been used to reveal some of the above properties.

Experimental

The vessel containing the gas fluidised bed consisted of a 1 m (3ft) length of 0.155m (6 inch) diameter glass tube. The air flow through the bed was metered by a calibrated rotameter and the bed was supported by a 0.625 cm ($\frac{1}{4}$ in.) thick sintered brass plate.

The particle material was petroleum carbon* which had been treated at a high temperature to make it electrically conducting. The screen size range was -105 + 90 μm. The bed static height was set at 0.48 m (19 ins) and during operation the expanded bed height averaged 0.55 m (22 ins).

The operating gas rate was maintained at around three times the minimum fluidisation velocity.

$$U_{mf} = 0.011 \text{ ms}^{-1} \text{ (Minimum fluidisation velocity)}$$
$$U_{s} = 0.031 \text{ ms}^{-1} \text{ (Operating velocity)}$$

Figure 12 gives details of all the important probe dimensions.

The probe was moved both vertically along the bed axial centreline and horizontally along the radius of the bed to provide a complete traverse of the bubble dispersion. Each position in the bed took from four to ten hours of computer time to collect the required large number of bubbles (usually set at 500 to 1000) and the results presented here therefore consist of around four hundred hours of on-line computer time.

Pulses generated when a bubble encounters the probe show a high level of noise when the initial voltage falls from the fully conducting level, as demonstrated in Figure 13a   . This noise was filtered with 4μF

* Kindly supplied by Conoco Ltd.

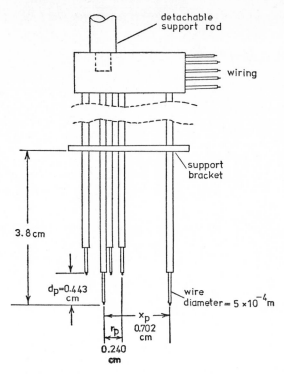

Figure 12. Engineering details of the fluidized bed bubble detection probe.

capacitors. Figure 13b demonstrates the effect of this on a typical pulse showing that the frequency response of the system remains adequate.

Figure 14 shows typical pulses for two bubbles striking the multi-element probe sequentially. Clearly shown on the figure are both the imperfect and the acceptable bubble hits.

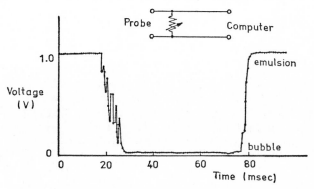

Figure 13(a). Fluidized bed bubble pulse before electronic filtering.

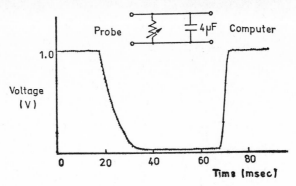

Figure 13(b). Fluidized bed bubble pulse after electronic filtering.

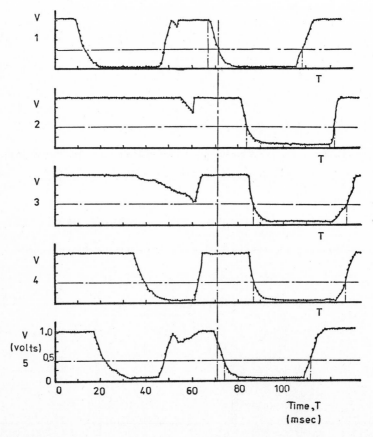

Figure 14. Typical fluidized bed probe sequence generated by sequential bubbles striking the array. Each point represents a voltage sampled by the computer in digital form. Channel numbers are as given in figure 3.2.

Results

Bubble shapes

Figure   15   shows typical measured bubble shapes.   These are very
similar to those reported by Rowe and Partridge (8) for single bubbles
using X-ray photography.   Considerable deviation from the mean shape occurs,
some bubbles having a flat back shape whilst others have marked concave
indentation of the rear surface.

In order to process the data further, a bubble shape-fitting procedure
was developed to select between four shapes to secure the best fit to the
measured coordinates Lc, Ld and Lo.   These were: hemispherical cap,
spherical cap with eccentricity greater than 2, spherical cap with eccentricity

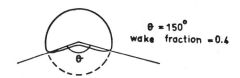

Figure 15(a). Mean bubble shape in a fluidized bed of 100 m particulate material (after Rowe and Partridge(143)).

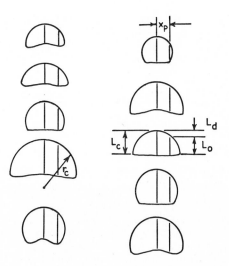

Figure 15(b). Typical sequential bubble shapes reported by the probe for this work.

less than 2 and paraboloid of revolution. A hemispherical cap was chosen to represent those small bubbles which failed to activate the outer shape probe.

The data interface between the PDP/8e machine and the ICL 4/75 processor was provided by ASC III coded punched paper tape and the shape selection and data processing software was written in IMP language.

Bubble velocities

Figure 16   shows the velocity of individual bubbles in the bed and their computed equivalent spherical diameter obtained from the shape fitting process above. The data represent a sequential sample of about 150 bubbles at the top of the bed on the bed centreline.

The bubble velocities exhibit considerable deviation from the mean in agreement with the photographically determined data of Goddard and Richardson (14) and X-ray work of Rowe and Matsuno (9   ). However, the mean bubble

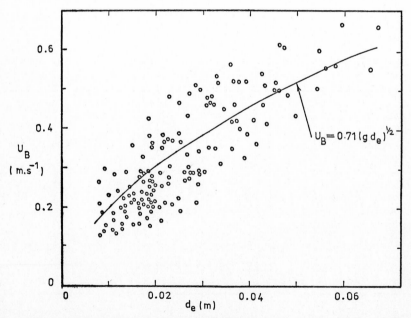

Figure 16. Typical variation of fluidized bed bubble velocity with equivalent spherical diameter for individual bubbles in the bed.

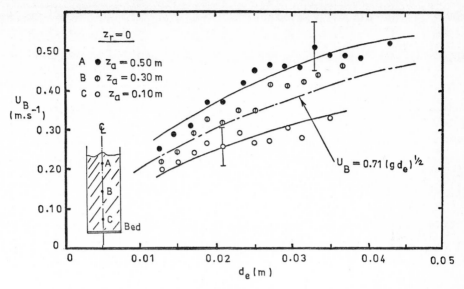

Figure 17. Variation of mean bubble velocity with bubble size and axial level in the bed.

velocity for a given size shows a definite trend with change in size. Figure 16 also shows the Davies-Taylor (18) relationship for spherical-cap bubbles which indicates its probable applicability.

Figure 17 shows the mean bubble velocity-size relationships at three vertical positions in the bed on the bed centreline. Each mean bubble velocity has a large standard deviation. However, there is clearly a difference in the mean bubble velocity of a given size of bubble with position in the bed the bubbles increasing their velocity with height above the distributor. Bubbles near the top of the bed have velocities which are above those predicted by the free rise relation (18) whilst those near the distributor have velocities below this. Figure 18 shows a traverse in the radial direction at the top of the bed (z_a = 0.50 m). The bubble velocity decreases with radial distance from the centreline near the top of the bed, which leads one to suppose that macroscopic bed mixing and circulation are influential.

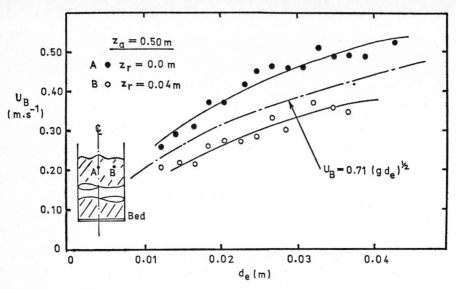

Figure 18. Variation of bubble velocity with size and radial position in the fluidized bed at an axial position at the top of the bed.

Figure 19 shows the variation of the mean velocity – equivalent diameter relationship with radial position at a vertical distance 0.10 m above the distributor. Here, the velocities on the bed centreline are lower than those near the wall, indicating that there is downward movement of emulsion at the bed centre and an upwards movement of bubbles and emulsion

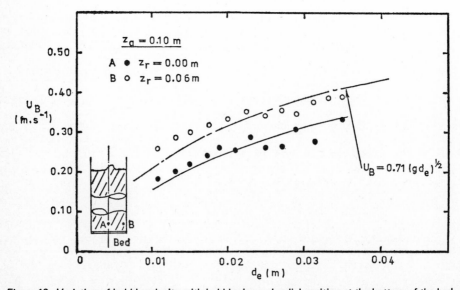

Figure 19. Variation of bubble velocity with bubble size and radial position at the bottom of the bed.

adjacent to the walls. Thus, it appears that there are two circulation cells, one close to the distributor and one in the upper levels of the bed. This behaviour is implied in previous work (19) (20) (21) and will be further considered later.

Bubble size distributions

Typical bubble size distribution functions are shown in figures 20 and   21   where the characteristic bubble size is the equivalent spherical diameter computed by the shape fitting process.

Figure   20   shows the size distribution (z_a = 0.10 m) on the bed centreline; Most bubbles at this position are small. Figure   21   shows the bubble size distribution at a greater vertical level above the distributor (z_a = 0.50 m) and we see here that coalescence has increased the number of large bubbles and decreasing the number of small bubbles in the distribution.

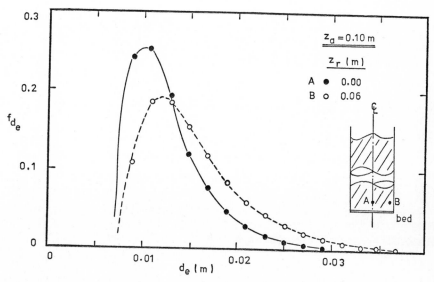

Figure 20. Variation of the gamma-function bubble size distributions with radial position in the fluidized bed at the bottom of the bed.

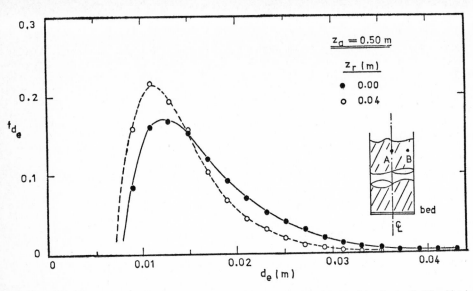

Figure 21. Variation of the gamma-function bubble size distributions with radial position in the fluidized bed at the top of the bed.

The truncated gamma function distribution with the same mean parameters as the experimental distribution,

$$p(\hat{d}_e) = \frac{\beta^{\alpha+1}}{\Gamma'(\alpha+1)} \, e^{-\beta\, \hat{d}_e} \, (\hat{d}_e)^{\alpha} \qquad (25)$$

$$\alpha = \frac{\mu\hat{d}_e^{\,2}}{\sigma_{\hat{d}_e}^2} - 1 \qquad \beta = \frac{\mu\hat{d}_e}{\sigma_{\hat{d}_e}^2}$$

provides an excellent fit to the data, as is shown in figures 22 and 23

Figure 24 shown the variation of mean bubble size on the bed centreline with vertical level above the gas distributor. The influence of bubble coalescence is clearly seen.

Figure 25 shows the changes in the size distribution variance with position in the bed. Larger bubbles exist at the bed centreline

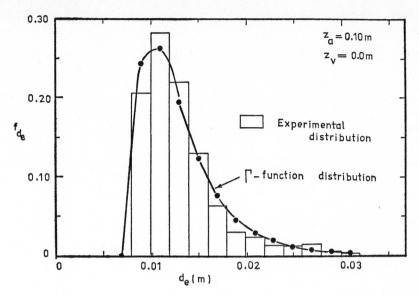

Figure 22. Typical bubble size distribution function in the fluidized bed together with the gamma-function analytical fit to the experimental data.

being the position of maximum upward emulsion phase velocity and smaller

bubbles exist in the region of downward emulsion movement near the vessel

walls.   At the bottom of the bed, the bubbles next to the wall are larger

than at the centre.   We may conclude that regions in which the bubble

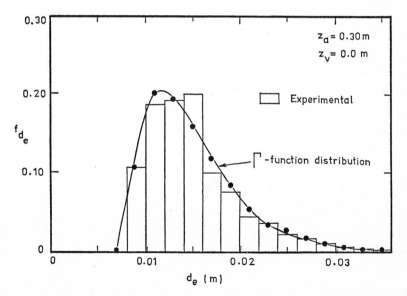

Figure 23. Typical bubble size distribution reported by the probe for the fluidized bed together with the gamma-function analytical fit.

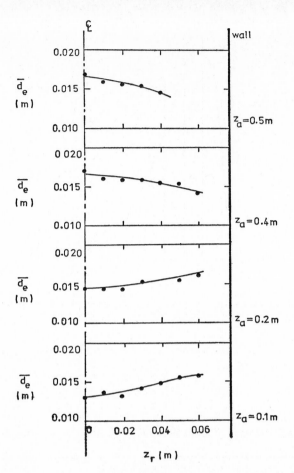

Figure 24. Variation of mean bubble size with axial and radial position in the fluidized bed.

velocity is enhanced relative to free rise are also regions containing the larger bubbles and vice versa.

Bubble frequency, gas flow rate and voidage

The frequency of pulses at the measurement point is not a measure of the bubble spatial density at this point as shown by Park et al. (12).

However, the local gas flow rate in the bubble phase is available from the pulsing frequency as below provided bubbles rise randomly towards the probe. Radial variation of both bubble size and velocity tends to invalidate the assumption of randomness but the equations below are relatively insensitive to the slight changes in velocity and size which exist between adjacent positions in the bed.

$$V_b = \gamma \left[\frac{\sum_{i=1}^{m} f_i U_i V_i}{\sum_{i=1}^{m} f_i U_i A_i} \right] \qquad (\ 26 \ )$$

$$\epsilon_b = \gamma \left[\frac{\sum_{i=1}^{m} f_i V_i}{\sum_{i=1}^{m} f_i U_i A_i} \right] \qquad (\ 27 \ )$$

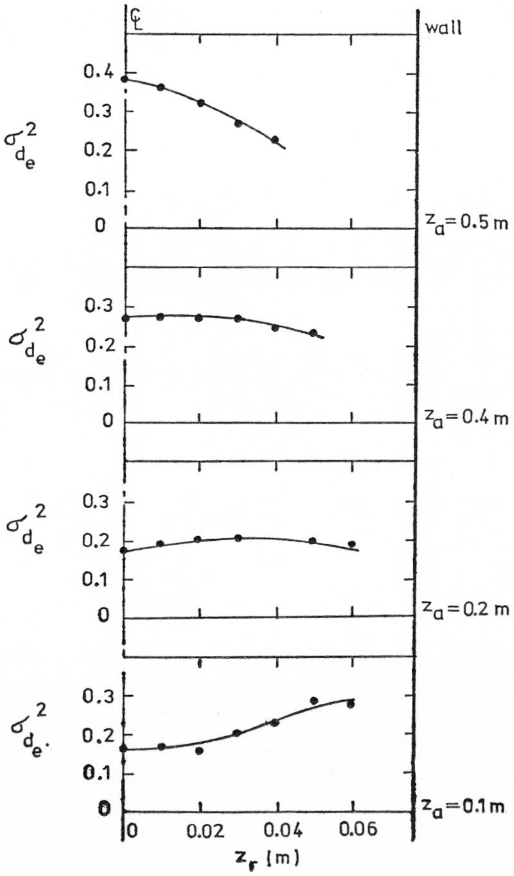

Figure 25. Variation of bubble size distribution variance with axial and radial position in the fluidized bed.

where V_b is the point bubble phase gas flow per unit area

   ϵ_b is the point bubble phase voidage

Figure 26   shows the variation of bubble phase voidage with position in the bed, calculated from equation 27   . The regions of high bubble phase voidage can be seen to be associated with regions of upward bed circulation, as expected.

Figure 27   shows the variation of point bubble gas flow per unit area (the "visible" bubble flow) with bed position.   The same trends as

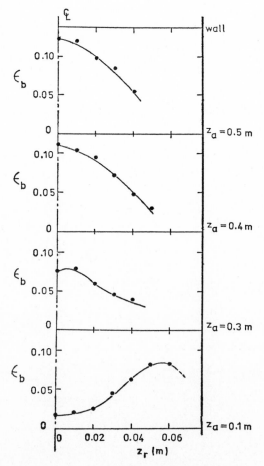

Figure 26. Variation of bubble phase voidage with axial and radial position in the fluidized bed.

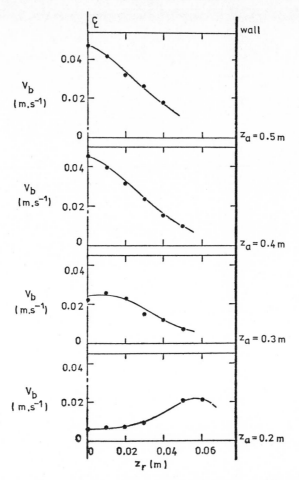

Figure 27. Variation of bubble phase gas flow per unit area with axial and radial position in the fluidized bed.

the voidage profile may be seen, with the position of maximum flow moving towards the bed centre with increase in height.

Many authors have found that the visible bubble flow is below that given by the simple two phase theory:

$$\frac{Q_B}{A} = \overline{V_b} = U_s - U_{mf} \qquad (\ 28\ )$$

as reviewed by Clift and Grace (22). The problem of accurate measurement of the flow in regions very close to the wall means that integration of the profiles is inaccurate. However, an estimate of the mean flow can be made if it is assumed that the bubble characteristics near the wall are the same as those at an adjacent radial measuring position. Taking the maximum level in the bed (z_a = 0.50 m) and integrating:

$$\overline{V_b} = \frac{2 \pi \sum (r\, V_b\, \Delta r)}{\pi\, d_c^2} \tag{29}$$

from which: $\overline{V_b}$ = 0.012 ± 0.002 ms^{-1}

which is significantly less than the two-phase theory prediction of 0.020 ms^{-1}.

Bed macro-circulation

From the above, a qualitative description of bed circulation becomes possible and figure   28   shows a schematic diagram of both emulsion and bubble phase flow patterns.

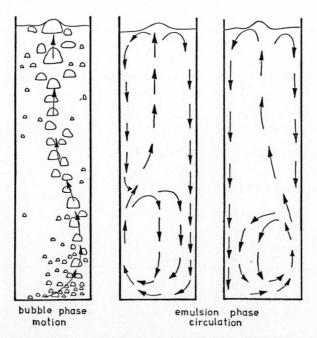

bubble phase   emulsion phase
motion   circulation

Figure 28. Schematic diagram of the observed bubble phase and emulsion phase circulation patterns in the fluidized bed.

The bed consists of two circulation cells: a relatively stable cell at the top of the bed analogous to a gas-liquid bubble column, and an unstable cell immediately above the gas distributor. The former cell provides a stable emulsion phase circulating up the column centre and down the column walls. In regions of upward circulation at the top of the column the emulsion phase has interstitial upward velocities of about 0.10 m^{-1} whilst near the walls the downward velocity is about 0.05 ms^{-1}. The bubble size spatial distribution is modified by this cell, in that the larger bubbles travel preferentially with the upward flow.

The circulation cell close to the distributor is unstable, with switching and rotation in three dimensions and random bubble flow near the vessel walls.

RATES OF GAS TRANSFER BETWEEN BUBBLES AND DENSE PHASE

Davidson and Harrison (23) give an interpretation of bubble behaviour in fluidized beds and a model for rates of gas-exchange between the two phases.   From their analysis, two mechanisms are apparent due to percolation of the fluidizing gas through the bubble void and to convective diffusion across the frontal surface of the bubble.   In a later model Kunii and Levenspiel (24) point out that the above two contributions to gas transfer only account for gas-exchange between bubble and cloud.   They estimate a rate constant for gas transfer between cloud and dense phase using a surface renewal model with the cloud dimensions obtained from Davidson's theory. Chiba and Kobayashi (25) assume that gas transfer is limited by convective diffusion across an equivalent sphere with the diameter of the cloud; their analysis is similar to above applied to Murray's stream function and bubble

geometry (26).   Partridge and Rowe (27) propose convective diffusion by analogy
with mass transfer from a solid sphere in an extensive flowing fluid.

The evaluation of such models through direct measurement of the rate of gas
transfer between the phases requires knowledge of bubble volume, surface area
and velocity as well as the concentration of gas in both phases and the time
derivative of the latter.

Experimental measurements of gas transfer rates have
previously been carried out by Szekely (28) Davies et al. (29)
Stephens et al. (30) and Chiba et al. (25) but without clear
definition of the bubble properties pertaining thereto.

In the present work a new probe for the simultaneous measurement of
all the above mentioned va iables is described and the rates of gas exchange
derived for a case where transport limitation is mainly due to convective
diffusion from the bubble interface, the cloud thickness being negligible.

Experimental

A probe for the simultaneous measurement of bubble velocity, size and
shape in fluidized beds of electrically conducting particles has already been
described.   In the present work the technique was extended to measure in
addition a tracer gas concentration within the bubble and in the dense phase
nearby as the bubble undergoes mass transfer with its surroundings.

The shapes, sizes and velocities of bubbles were determined as previously
described.   Two bubble shapes were fitted to each bubble as determined from
the relations between L_b, L_d, L_o and R_b.   Thus,

Low eccentricity cap $L_b > R_b$ and $L_b < 2 R_b$

Paraboloid $L_b > 2.R_b$

The concentration of a helium tracer in the bubble and in its surroundings
was measured by placing coaxially and at the same level between the three upper

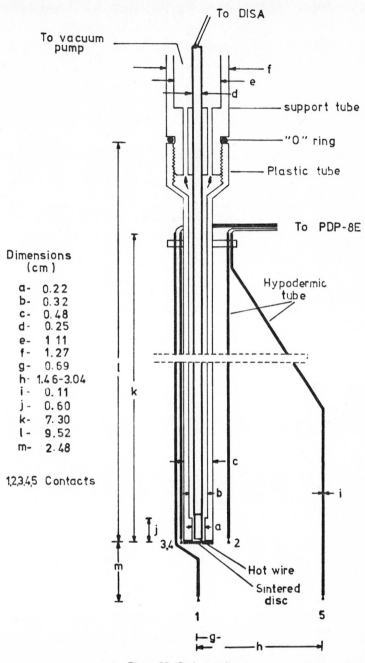

Figure 29. Prob   etails.

contacts a hot wire katharometer detector contained in a plastic tube (see Fig29) and over which gas was drawn through a sintered plate 0.02 cm thick with 25 μm pore diameter the hot wire being 0.5 - 1mm above the sintered plate.   The

analytical gas sampling rate was 3 cm^3 sec^{-1} (superficial velocity around the
hot wire 79 cm sec^{-1}).

The output of a 'DISA' constant temperature hot wire bridge (CTHWB) was stored
in computer memory when the three upper contacts were in the bubble or in the dense
phase just ahead of and behind the bubble and the relevant compositions were thereby
determined for only those cases where the bubble encountered the probe "squarely".
The CTHWB gave an output voltage linearly dependent on the helium concentration.

A hibrid computer (8DP-8E DIGITAL CORP.) used, on-line, analysed all probe
outputs using the necessary assembler software written to extract the relevant
bubble size and concentration functions from the five probe contacts and the
analytical channel.

The voltage pulses on the five channels were adjusted to between 0 and
one volt to suit the analogue-digital converter characteristics of the computer,
and the output of the CTHWB was conditioned through an off-set computer line
matching circuit with variable gain in order to obtain 0 volts for pure air
and a voltage between 0.5 and 1 for any concentration of tracer between 0.5
and 1 cm^{3}He/cm^3. Before an experiment the analytical channel was calibrated
automatically by the computer operating a magnetic valve and introducing
a known mixture of helium-air in order to determine the proportionality constant
between helium concentration and CTHWA output.

The helium concentration associated with the bubble and its near
environment is given by the CTHWB at particular times determined by the pulse
mode and delay times of the three upper channels. The output of the CTHWB
can only be associated with the bubble location if the rise time of the analytical
system response is very short. A test was carried out to measure the
analytical frequency at the gas sampling rate to be used in practice.   A
step change in composition from air to helium, was effected by puncturing
balloons of helium gas with the probe and using a microphone-actuated time

switch connected to the computer.   The output of the CTHWB was recorded in memory during the transient state and the rise time verified to be 3.2 $\pm$ 0.5 milliseconds.   Thus the CTHWB output was seen to be adequately synchronized with the pulsing mode of the three upper channels and hence with the bubble position.

A detailed diagram of the probe is presented in Fig. 29

Experiments were carried out by injecting helium bubbles at various frequencies and of different sizes in an incipiently air-fluidized bed of petroleum coke 15.4 cm i.d. and 90 cm high, supported on a sintered plate gas distributor 0.635 cm thick (pore size 25 µcm).   The coke particles had been partially graphitised and were screened to give 11.5 Kg with diameters between 80 µ m and 105 µ m.   The experimentally determined gas superficial velocity at incipient fluidization was 0.575 cm sec^{-1} the density of the coke particles 1.53 g cm^{-3}.   The porosity of the bed at incipient fluidization was 0.557.

The helium bubbles were injected into the bed through an expansion nozzle on the axis of the column, immediately above the distributor and connected to a solenoid valve which by means of a variable time-delay switch could be made to stay open for o.10 - 0.13 seconds.   Helium tracer was supplied from a pressurized tank connected to the valve.   In order to obtain bubbles of different sizes the pressure in the tank was regulated in the range .2 - 2.5 Kg cm^{-2} gauge and two different sizes of tank were used (590 and 139 cm^3).   Bubble sizes from 30 to 900 cm^3 were obtained.

The frequency of operation of the valve was controlled by the time-delay system to give 20, 40 and 60 bubbles per minute.   Bubble properties and tracer concentrations were measured between 10 cm and 65 cm above the injector at 5 cm intervals.   In each experimental run the time delay valve circuit was automatically switched on by the computer and then bubbles introduced.   The

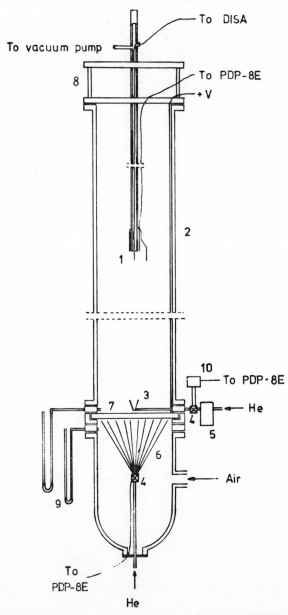

Figure 30. 1. Probe and supporting tube. 2. Column. 3. Expansion nozzle. 4. Solenoid valve. 5. Pressurized tank. 6. Hypodermic tube distributor. 7. Sintered plate distributor. 8. Supporting bracket. 9. Manometers. 10. Variable time delay switch.

analysis by the computer began after the introduction of 60 bubbles.   For each

bubble encounter, pulse durations and delay times between contacts were stored

in memory as was the tracer concentrations in the dense phase, just ahead of, behind and within the bubble as it passed a given level in the bed.　If the descrimination function described above was satisfied and the bubble had passed the probe assembly "squarely" the data was kept in computer memory.　After six successful bubble encounters, the analysis and the bubble injection stopped; the data were then punched out on paper tape for further encounter processing. The system automatically repeated these experiments, the probe level being normally changed during the print-out of the data.　Fig. 30 shows schematically the apparatus used.

Results

Bubble properties

Table I shows the experimentally determined bubble velocity, Ub, central height, Lb, radius of curvature of nose, Rb, shape, volume frontal area and the diameter of an equivalent sphere, De.

Bubble velocities are compared in Fig.31 with Davidson's equation for bubble and slug velocities (33).

Table 1

Experiment	U_b (cm sec^{-1})	L_b (cm)	R_b (cm)	Shape	V (cm^3)	S_f (cm^2)	D_e (cm)
1	58.83	10.03	3.475	PAR.	810.5	904.0	11.6
2	56.3	8.50	3.55	PAR.	612.6	723.78	10.53
3	53.6	5.76	4.67	LEC.	287.0	172.0	8.18
4	49.4	5.06	4.01	LEC.	187.0	130.21	7.10
5	47.7	4.13	3.51	LEC.	114.0	91.92	6.01
6	46.5	3.37	2.95	LEC.	65.3	63.00	4.99
7	43.8	3.06	2.47	LEC.	42.8	48.42	4.33
8	39.4	2.66	2.25	LEC.	30.3	38.04	3.86

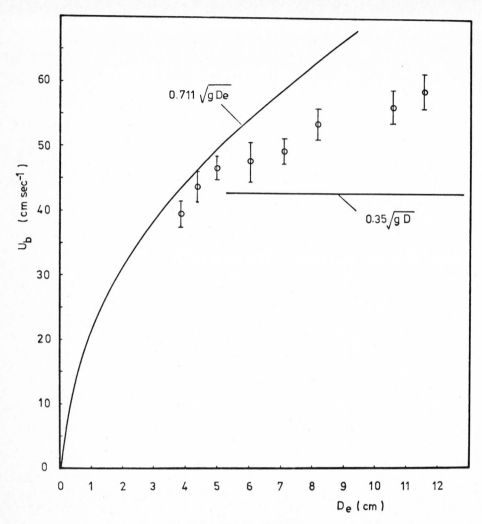

Figure 31. Bubble velocities.

Bubbles $U_b = 0.711 \sqrt{g\,D_e}$ (30)

Slugs $U_b = 0.35 \sqrt{g\,D_e}$ (31)

It is seen that the bubbles examined follow a relationship intermediate between the above.

Mean bubble shapes are given in each series of experiments in Fig. (32). From the mean shape fitted to each of the different bubble sizes, a transition from a low eccentricity cap to a parabaloid occurs as the bubble volume increases.

This agrees with the observations of Goldsmith et al (32) No systematic variation of shape with height in the column was noticed.

Both the bubble volume and frontal area were calculated using the equations for each of the solids of revolution corresponding to the fitted shape. The volume of the mean bubbles was compared with the bubble volume obtained from a mass balance upon the tracer concentration and its value shown to be lower by 7 to 10%.

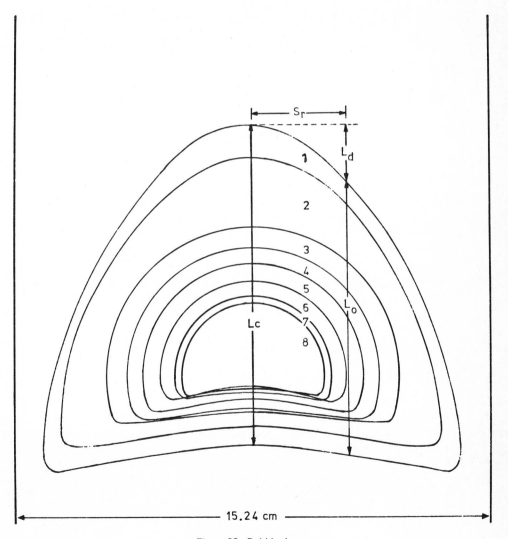

Figure 32. Bubble shapes.

Tracer concentration

Tracer concentrations were recorded as previously described.   Fig. 33 shows four typical tracer concentration profiles at four different levels in the bed corresponding to bubbles of the same size;   they distinguish three distinct regions;   A – the region ahead of the bubble corresponding to the sampling carried out between the pulse appearance on the leading channel and its appearance on the upper channels.   B – The bubble compositions corresponding to the pulse duration on the upper channels.   C – The region behind the bubble, analysed when the three upper channels had returned to the dense-phase mode.

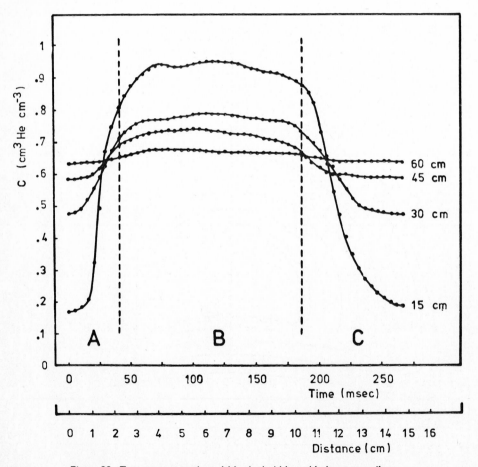

Figure 33. Tracer concentration within the bubble and in its surroundings.

Axial profiles of concentration within the bubble and dense phases were recorded as previously described over the height 10 to 65 cm at 5 cm intervals, where the bubble concentration was taken as the concentration measured in the centre of the bubble and the dense phase concentration as determined just ad of the bubble.   The value of dense phase concentration differs by less than 8% from the value at the later time T/2 so that a quasi-steady-state can be assumed to be reasonably valid where T is the bubble injection period.

Mass transfer

Assuming that the regular and rapid injection of bubbles will be conducive to an almost steady state condition the following equations will describe transfer between the bubble and dense phases.

$$V \quad Ub \quad \frac{dCb}{dh} \quad = \quad - Ko \; (Cb - Cp) \qquad (32)$$

and

$$N \; V \quad \frac{d \; Cb}{d \; h} \quad + \quad A \quad Umf \; \frac{d \; Cp}{d \; h} \quad = \quad 0 \qquad (33)$$

from (32) and (33)

$$\frac{d \; \log \; (Cb - Cp)}{d \; h} \quad = \quad - \; \frac{Ko}{2.306} \left[\frac{A. \quad Umf + N. \; V}{A. \quad Umf \; V. \; Ub} \right] \qquad (34)$$

Values of Cb - Cp at each level were plotted against height on semi-log coordinates to give the overall mass transfer coefficient from equation 34. This plot is shown in Fig.34 - for all the experiments, the best straight lines being determined by the least squares method.   In order to compare the present results with models proposed in the literature, values of Kv = Ko/V were calculated and plotted versus De in Fig.35 .

Assuming a two-component model for gas transfer as in (23) the lumped mass transfer flux is

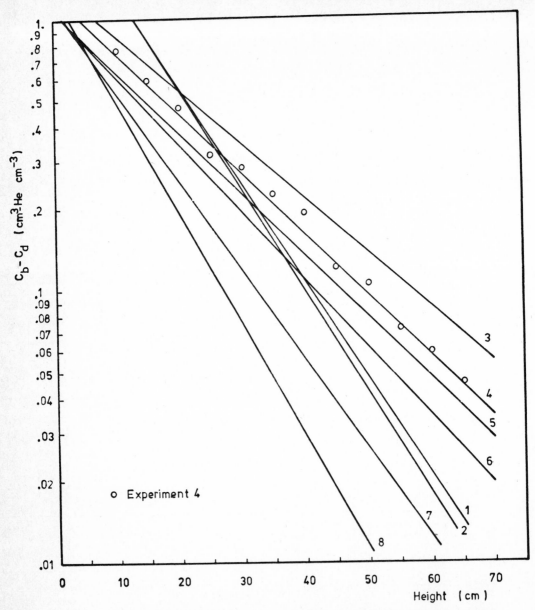

Figure 34. Difference between bubble and dense phase concentrations at various heights in the bed.

$$K_o = Q + K_g \, S \qquad\qquad (35)$$

where Q is the ascribable to bulk flow and $K_g \, S$ due to convective diffusion.

Davidson proposed (23) that Q for spheres is given by

$$Q = 3\pi \, U_m \, D_e^2 \qquad\qquad (36)$$

and for slugs

$$Q = \pi \ Um \ D^2 \qquad (37)$$

The surface area for gas transfer S is considered to be the unimpeded frontal surface area of the bubble

$$S = Sf \ . \ \varepsilon_{mf} \qquad (38)$$

where ε_{mf} is the porosity of the bed at minimum fluidization.

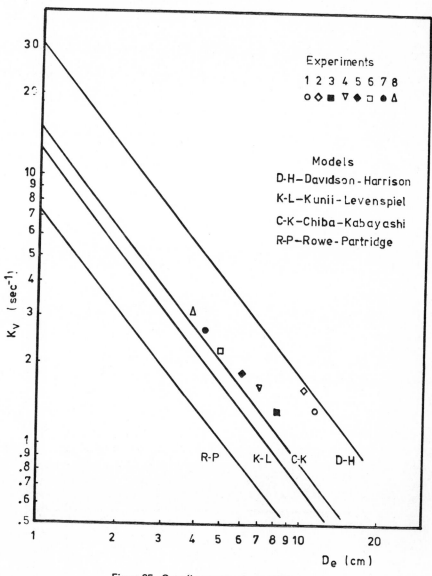

Figure 35. Overall mass transfer coefficients.

Further, following Higbie (31),

$$Kg = \sqrt{\frac{4 \, Dg}{\pi \, tc}}$$

where tc is the exposure time of elements of gas flowing over the bubble frontal envelope.

$$tc = \frac{Lb}{Ub} \qquad\qquad (39\,)$$

which is the pulse duration of contact 1 registered by the compound probe.

Fig. 36 shows the mass transfer coefficients calculated as above compared with those experimentally observed.   It appears that the assumption of a simple 'surface-renewal' mass-transfer model is adequate in this instance.   The

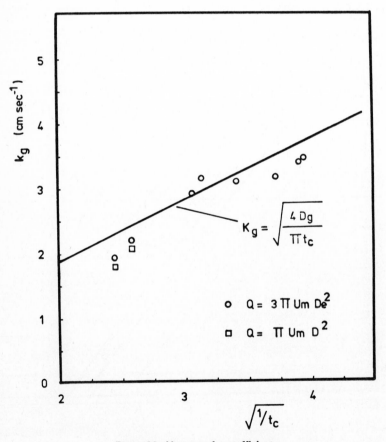

Figure 36. Mass-transfer coefficients.

absence of significant concentration gradients within the bubble (see Fig. 33)

shows that the principal resistance to mass transfer lies at the bubble-cloud

interface.   The concentration in the bubble wake does not greatly differ from

that in the adjoining dense phase so that this region is well mixed but mass

transfer between the rear surface of the bubble and the wake is slow since it is

not apparently necessary to make allowance for it in the prediction.

The thickness of the bubble 'cloud' in all this work, calculated from

Ref. (23) was such that

$$\frac{Dcl}{De} < 1.025$$

Thus it is seen that bubble mass-exchange rates agree well with a model for

surface-renewal at the cloud interface over the frontal surface of the bubble.

This model correctly accounts for changes in mass-transfer observed as due to

changes in the frontal areas of bubbles as well as their 'contact-times'.

The work is being extended to a study of gas-exchange when coalescence

between two bubbles occurs and the rates of exchange are largely enhanced.

REFERENCES

[1]   Calderbank P.H.
       Trans, Inst.Chem.Engrs 1958 36 443;
       Ibid 1959 37 173.

[2]   Calderbank P.H. and Moo-Young M.B.
       Chem.Eng.Sci. 1961 16 39; Ibid 1961 16 337.

[3]   Calderbank P.H. and Rennie J.
       Trans.Inst.Chem.Engrs 1962 40 3.

[4]   Calderbank P.H. and Evans F.
       Int.Symp.Distillation, Brighton 1960 pp 51,
       (London: Inst.Chem.Engrs.)

[5]   Porter K.E., Davies B.T. and Wong P.F.Y.
       Trans.Inst.Chem.Engrs 1967 45 T265

[6]   Ashley M.J. and Haselden G.G.,
       Trans.Inst.Chem.Engrs 1972 50 119.

[7]   Chalkley H.W., Cornfield J and Park H.,
       Science 1949 110 295.

[8]   Rowe P.N. and Partridge B.A.
       Trans.Inst.Chem.Engrs 1965 43 T157.

[9]   Rowe P.N and Matsuno R.,
       Chem.Eng.Sci. 1971 26 923.

[10]   Werther J and Molerus O.,
       Int.J.Multiphase Flow 1973 1 103;
       Ibid 1973 1 123.

[11]   Garner F.H. and Porter K.E.
       Int.Symp.Distillation, Brighton 1960 pp43
       (London: Inst.Chem.Engrs.)

[12]   Park W.H.,Kang W.K.,Capes C.E. and Osberg, G.L.,
       Chem.Eng.Sci.1969 24 851

[13]   Rigby G.R.,Van Blockland G.P.,Park W.H. and Capes C.E.,
       Chem.Eng.Sci. 1970 25 1729.

[14]   Godard K and Richardson J.F.,
       Chem.Eng.Sci. 1969 24 663.

[15]   Burgess J.M.
       Ph.D. Thesis, Univ.Edin. 1974.

[16]   Johnson D.S.L.
       Ph.D. Thesis, Univ.Edin.1970.

[17]   Dumitrescu D.T.,
       Z. Agnew.Math.Mech. 1943 23 139.

[18]   Davies, R.M. and Sir G.I. Taylor
       Proc.Roy.Soc., 1950, A200, 375

[19]   Geldard D., J.R. Kelsey
       I.Chem.E.Symp.Ser.,No.30, 1968, 114.
       (Lond: Inst.Chem.Engrs)

[20]   Merry J.M.D.,J.F.Davidson
       Trans.Inst.Chem.Engrs., 1973, 51, 361

[21]   Whitehead A.B.
       In: Davidson J.F.,D.Harrison
            "Fluidisation", Academic Press, London, 1971, 781.

[22]   Grace J.R. & Clift R.
       Chem.Eng.Sci.,1974, 29, 327

[23]   Davidson J.F. and Harrison, D.,1963, "Fluidized Particles"
       Cambridge University Press.

[24]   Kunnii, D. and Levenspiel, O., 1968,
       Ind.Eng.Chem.Fundls.,7, 446.

[25]   Chiba, T. and Kobayashi, H., 1970,
       Chem.Eng.Sci.,25, 1375

[26]   Murray, J.D.,1965,
       J.Fluid Mech., 22 (part 1), 57

[27]   Partridge, B.A.,Rowe,P.N.,1966
       Trans.Inst.Chem.Engrs.,44, T335

[28]   Szekely, J.,1962, "The interaction between fluids and particles"
       pg 197. Inst.Chem.Engrs.,London.

[29]   Davies, L. and Richardson, I.F.,1966
       Trans.Inst.Chem.Engrs.44, T293.

[30]   Stephens, G.K.,Sinclair, R.J.,Potter, O.E.,1967,
       Powder Technol.,1,157

[31]   Higbie, R.,1935,
       Trans.Am.Inst.Chem.Engrs.,31, 365.

[32]   Goldsmith, J.A.,Rowe,P.N.,1975,
       Chem.Eng.Sci.,30, 439

[33]   Davidson, J.F., Hovmand S.,1971,
       in"Fluidization" pg. 193, Academic Press, London.

PART II

Gas Exchange
and Fluid Bed Modeling

FLUIDIZED BED REACTORS

COLIN FRYER AND OWEN E. POTTER

INTRODUCTION

Davidson and Harrison (1963) have presented solutions of
a two-phase model for fluidized bed catalytic reactors, assuming
either plug flow or well-mixed behaviour of the dense phase gas. A
modification of their equation for gas exchange between phases has
been suggested by Kunii and Levenspiel (1968).

The backmixing model of Latham *et al* (1968) and Kunii and
Levenspiel (1968) aims to provide a more satisfactory description
of gas mixing, and predicts downflow of gas with solids in the
dense phase at gas flow-rates above some critical value. It is in
accord with experimental studies of tracer gas mixing (Hamilton
et al, 1970; Latham and Potter, 1970). For a catalytic reactor,
this model gives rise to the surprising prediction that axial
profiles of reactant concentration may pass through a minimum
within the bed, and also predicts dependence of conversion on rate
constant and on gas velocity somewhat different from that arising
from a two-phase model (Fryer and Potter, 1972a).

Fryer and Potter (1972b, 1974) have incorporated bubble size
variation into the two-phase and the backmixing reactor models, to
allow valid comparison with experiment. This paper reports on an
experimental study designed to distinguish between these models.

EXPERIMENTAL PROGRAMME

Equipment. The 200 cm x 22.9 cm diameter reactor, described by
Fryer and Potter (1974), has been used for catalytic decomposition
of ozone in air on a sand of mean particle size 117 μm. Probes
allow withdrawal of gas samples from within the bed so that concentration profiles can be recorded.

Bubble sizes. Photography of the bed surface at various heights has
shown that bubble diameter varies linearly with height, and has
provided data on bubble size for use in model computations.

Catalyst activity. The decomposition reaction is close to first
order; the rate constant has been determined on catalyst sampled
from the fluidized reactor.

The fluidized reactor. Blank runs, with no catalyst, ensured that
gas samples from all positions exhibit equal ozone concentration.
Calibration of a slight loss of ozone in the probe system has been
checked during every reaction run.

Reaction runs have been conducted with seven levels of catalyst
activity (0.049 to 7.75 s^{-1}) in beds of height from 11 to 66 cm,
using gas velocities from 2.4 to 15 cm s^{-1} (c.f. incipient
1.70 cm s^{-1}).

COMPARISON OF EXPERIMENTAL RESULTS AND MODEL PREDICTIONS

Bed expansion. Agreement between predicted and observed expansion
is good, especially for lower bed heights (see appended data).
Small discrepancies for higher, vigorously bubbling beds are not
surprising. Overall, it is clear that the model description of
bubble rise velocity is adequate, leading to good prediction of
expanded bed height.

Axial concentration profiles. Little radial variation of ozone
concentration occurs at any height within the fluidized bed,
indicating that no large-scale circulation pattern is established.
Data from four positions along one radius have been averaged to
give an axial profile. All experimental results display one
feature of major importance, as in the data of Figure 1. At gas
velocities below 5 cm s^{-1}, all profiles show gradual decrease in
concentration from distributor to surface, but at higher velocities
all profiles pass through a minimum at some position within the bed.
Only the backmixing model predicts such behaviour. Agreement
between measured and predicted profiles is quite good, and the
variations of the position of the minimum with changes in gas rate,
bed height and rate constant are well accounted for by the model,

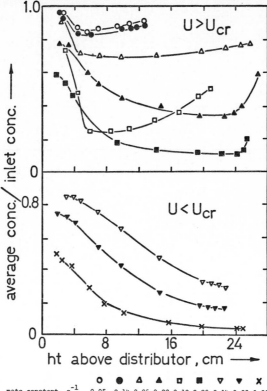

rate constant, s^{-1}	○	●	△	▲	□	■	▽	▼	✕
rate constant, s^{-1}	0.05	0.14	0.05	0.33	0.10	0.86	0.14	0.33	0.86
incip. bed ht., cm	11.2	10.6	23.7	23.1	20.2	24.0	22.8	23.1	24.0
gas velocity, cm s^{-1}	15.0	12.7	10.4	10.4	8.5	5.8	2.4	2.9	3.5

Figure 1. Experimental reactant concentration profiles.

(see appended data). In some cases, the minima occur quite low in
the bed; this fact, and the appearance of the minima only at high
gas velocities, show that the observed minima are a true indication
of bed behaviour, and are not caused by some sampling defect. The data
indicate a critical velocity for backmixing of about 5 cm s^{-1},
corresponding to a ratio of wake to bubble volume of 0.8.
Conversions. Experimental data on conversion, and its variation
with gas rate, are well predicted by the backmixing model. In
particular, the data do follow the model in a marked change of
slope at the critical fluidizing velocity. The two-phase models
predict a smooth curve on such a graph. The variation of conversion with reaction rate constant is also in accord with the backmixing model. In this regard, the two-phase models predict in-
correctly that conversion will "flatten out" at relatively low

values of rate constant. These points are illustrated by the
appended diagrams, which are typical of results and comparisons
made over the whole range of reactor conditions (Fryer, 1974).

CONCLUSIONS

The experimental conversions and concentration profiles provide
further strong support for the backmixing model. Conditions have
been chosen to minimise effects of radial non-uniformity and
adsorption, which the current model ignores. Further research is
needed in these areas and into prediction of wake size and bubble
size variation.

ACKNOWLEDGEMENT

This work has been assisted by the Australian Research Grants
Committee.

APPENDIX

The summary paper includes only one diagram, Figure 1,
illustrating the nature of reactant profiles at selected
conditions. Additional data, which illustrate several of the
points discussed in the paper, are given in the following
diagrams :

Figure 2 compares the observed bed expansion with that
predicted by the models which incorporate bubble size variation
(Fryer and Potter, 1972b, 1974).

Figure 3 shows part of a set of experimental reactant
concentration profiles and, for the same conditions, profiles
predicted by the backmixing model and by Davidson's two-
phase models. Only the backmixing model predicts the change
in shape of the profiles at gas velocities above 5 cm s^{-1}.

Figures 4 and 5 illustrate that the backmixing model is
more precise than the two-phase models in predicting the dependence of conversion on gas velocity and on reaction rate
constant.

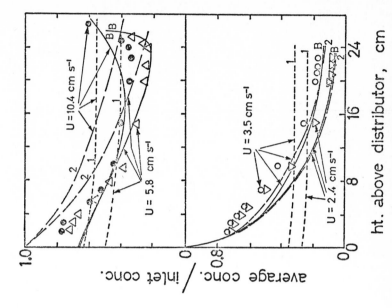

Figure 3. Experimental and predicted profiles (H_{mf} = 23.1 cm, k = 0.33 s^{-1}).

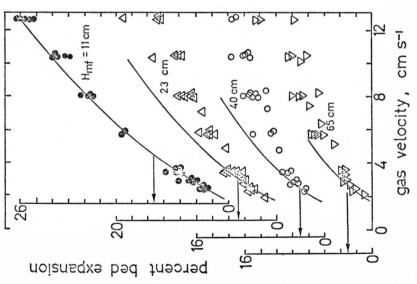

Figure 2. Bed expansion compared with model predictions.

175

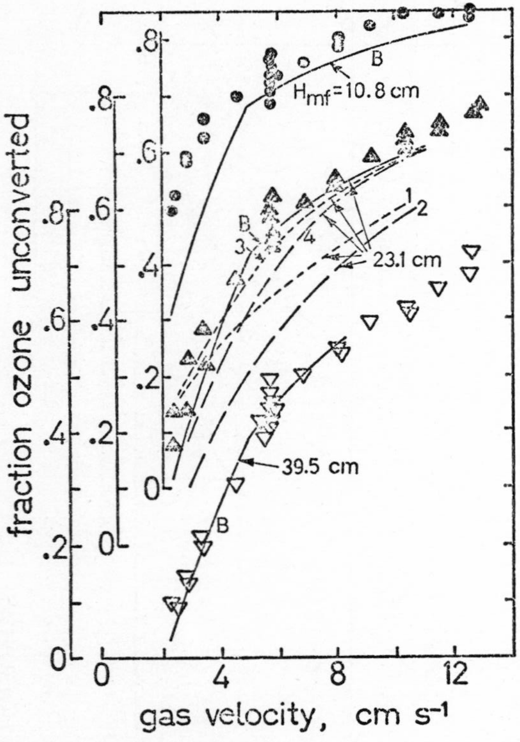

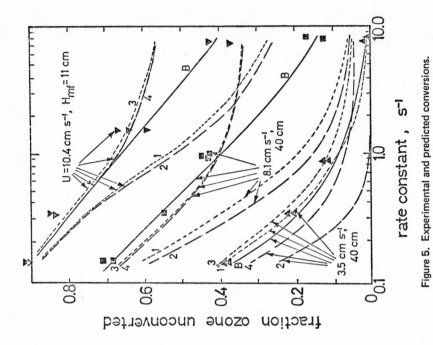

Figure 5. Experimental and predicted conversions.

Figure 4. Experimental and predicted conversions ($k = 0.33$ s^{-1}).

176

In Figures 3, 4 and 5, the model predictions are shown thus :

B ————————————— backmixing model.

1 — — — — — — — — Davidson model, mixed dense phase gas, Davidson exchange coefficient.

2 —————— —————— —————— Davidson model, plug flow in dense phase, Davidson exchange coefficient.

3 — — — — — — — — Davidson model, mixed dense phase gas, Kunii exchange coefficient.

4 ————— ————— ————— Davidson model, plug flow in dense phase, Kunii exchange coefficient.

REFERENCES

Davidson, J.F. and Harrison, D. "Fluidised Particles", C.U.P.(1963).

Fryer, C. "Fluidized Bed Reactors", Ph.D. Thesis, Monash Uni.(1974).

Fryer, C. and Potter, O.E. *Ind. Eng. Chem. Fund.* *11*, 338 (1972a).

Fryer, C. and Potter, O.E. *Powder Tech.* *6*, 317 (1972b)

Fryer, C. and Potter, O.E. Proc. Int. Symp. "Fluidization and its Applications", Toulouse 1973, Ste. Chimie Industrielle, p.440(1974)

Hamilton, C.J., Fryer, C. and Potter, O.E. Proc. "Chemeca 70", Melb. 1970, Session 1, p.1. *Aust. Acad. Sci. and Instn. Chem. Engrs.* (1970).

Kunii, D. and Levenspiel, O. *Ind. Eng. Chem. Fund,* *7*, 446, (1968).

Latham, R.L., Hamilton, C.J. and Potter O.E. *Brit. Chem. Eng.,* *13*, 666, (1968.)

Latham, R.L. and Potter, O.E. *Chem. Eng. Jnl.* *1*, 152, (1970).

SIMULATION OF FLUIDIZED BED REACTOR PERFORMANCE BY MODIFIED BUBBLE ASSEMBLAGE MODEL

S. MORI* AND C. Y. WEN

NOMENCLATURE

A_t	cross-sectional area of the fluidized bed, cm^2.
C_{Bn}, C_{en}	nondimensional concentration of reactant in the bubble and emulsion phase at the nth compartment, respectively.
C_{An}	nondimensional average concentration in the nth compartment.
D_B	bubble diameter, cm.
D_{Bc}	critical bubble diameter at the condition of $\varepsilon_c = \varepsilon_c^* = 0.6$, cm.
D_{Bm}	maximum bubble diameter determined by the coalescence of bubbles, cm.
D_{Bn}'	bubble diameter at the boundary between the $n-1^{st}$ and n^{th} compartments, cm.
D_{Bo}	initial bubble diameter, cm.
D_d	diameter of an orifice on the distributor plate, cm.
D_s	diffusion coefficient, cm^2/sec.
D_t	diameter of the fluidized bed, cm.
d_p	diameter of the fluidized solid particles, cm.
h	vertical height in the bed measured from the top of the jets, cm.
h_c	critical height from the top of the jets where the value of ε_c becomes equal to $\varepsilon_c^* = 0.6$, cm.
h_j	vertical jet length, cm.

* present address Nagoya University, Nagoya, Japan.

$h_n^{\prime}$ height of the boundary between $n-1^{st}$ and n^{th} compartments measured from the top of the jets, cm.

Δh_n height of the nth compartment, cm.

k first order reaction rate constant, sec^{-1}.

L_f fluidized bed height, cm.

L_{mf} bed height at the minimum fluidization velocity, cm.

n_d total number of holes in the perforated plate.

u_o superficial gas velocity, cm/sec.

u_B bubble rising velocity, cm/sec.

u_{mf} minimum fluidization velocity, cm/sec.

u_j gas velocity in the orifice holes, cm/sec.

V_{bn}, V_{cn}, V_{en} volume of bubble phase including cloud, cloud without bubble, emulsion phase in the nth compartment, respectively, cm^3.

Z vertical height measured from the bottom of the bed, cm.

ε_{Bn} volume fraction of the bubbles in the nth compartment.

ε_c volume fraction of the bubble phase including the cloud associated with the bubble phase.

ε_c^* volume fraction of the bubble phase when the adjacent clouds just touch each other.

ε_{mf} void fraction in the bed at u_{mf}.

$\overline{\varepsilon}_n$ average void fraction in the bed at the nth compartment.

ρ_f gas density, g/cm^3.

ρ_p particle density, g/cm^3.

δ_c ratio of the cloud volume (excluding bubble) to the volume of the bubble.

ABSTRACT

The bubble assemblage model previously reported has been modified to incorporate a new bubble size correlation which takes into account the effect of bed diameter, bubble coalescence and initial bubble size.

Data obtained from fluidized bed experiments for fast chemical reactions are used to test the modified model. The model is found to satisfactorily represent the experimental performance. This model, therefore, may be useful in scale-up and design of a group of fluidized bed reaction system.

INTRODUCTION

A number of fluidized bed reactor models that take into account the behavior of bubbles have been proposed for predicting the extent of catalytic reactions in a fluidized bed. These models include those of Orcutt and Davidson (1962), Kunii and Levenspiel (1968), Kobayashi et al (1969), Partridge and Rowe (1966), Toor and Calderbank (1967), Mori and Muchi (1972), Fryer and Potter (1972) and Kato and Wen (1969).

The models presented by Partridge and Rowe (1966), Toor and Calderbank (1967), Kato and Wen (1969), Mori and Muchi (1972) and Fryer and Potter (1972) consider bubble coalescence. Kato and Wen (1969) and later Mori and Muchi (1972) used the following equation to predict the bubble diameter in the bed.

$$D_B = 1.4 d_p \, \rho_p \, (u_o/u_{mf}) Z + D_{Bo} \qquad (1)$$

Toor and Calderbank (1967), in their model, utilized a relation between bubble frequency, f, and bed height of the form $f = \alpha_1 \exp(-\beta Z)$, where α_1 and β are constants which must be determined experimentally. Partridge and Rowe (1966) and Fryer and Potter (1972) considered bubble growth by directly measuring the bubble sizes in the fluidized bed reactors operating without chemical reaction.

Other models such as those presented by Orcutt and Davidson (1962), Kunii and Levenspiel (1968) and Kobayashi et al (1969), utilize a single

valued equivalent bubble diameter throughout the bed. Kobayashi et al
(1969) suggested that the bubble size at $Z = L_f/2$ calculated from
Equation (1) may be used as the value of equivalent diameter of bubble.
In the models presented by Orcutt and Davidson (1962) and Kunii and
Levenspiel (1968), the equivalent bubble diameter has been treated
as an adjustable parameter.

Experimental results of Botton (1968) and Cooke et al (1968) indicate
that the grid geometry affects considerably the extent of chemical
reactions taking place in the fluidized bed. Behie and Kehoe (1973)
and Zenz (1968) have suggested that the jets formed just above a
perforated plate distributor may have an important effect on the overall
behavior of the bed. Also Cooke et al (1968) and Kobayashi et al (1969)
have demonstrated experimentally that the extent of chemical reaction
in a fluidized bed is affected by the diameter of the bed considerably.

At the present time, a model of fluidized beds which can predict the
effects of bed diameter and jet behavior on the extent of chemical
reaction, has not yet appeared in the literature. In this paper, the
Bubble Assemblage Model reported by Kato and Wen (1969) is modified to
incorporate jetting region at the distributor as well as a new bubble
size correlation (Mori and Wen (1975)) which takes into account the
effects of the bed diameter. This new bubble size correlation makes use
of a bubble coalescence relation and an initial bubble size equation.

Experimental data available in the literature are used to test the
validity of the proposed model. The data are obtained from first order
catalytic reactions in different sizes of fluidized beds having various
types of distributors.

MODIFICATION OF BUBBLE ASSEMBLAGE MODEL

The Bubble Assemblage Model of Kato and Wen (1969) will be modified, as discussed below, so that this model may be used to predict the effect of bed diameter and jet behavior on the extent of chemical reactions occurring in a fluidized bed.

(i) Jetting Above the Distributor

When a perforated or a multinozzled plate is used as the distributor in a fluidized bed, gas jets are observed to form just above the distributor.

Zenz (1968) proposed an empirical correlation for the jet heights obtained from a two dimensional bed:

$$\frac{h_j}{D_d} = \frac{1}{0.0144} \left(\log \frac{u_j \sqrt{\rho_f}}{3.86} - 1.3 \right) \tag{2}$$

Basov et al (1969) presented an empirical correlation, which relates the jet height based on their experimental data obtained from a three dimensional bed as follows:

$$h_j = \frac{d_p}{0.0007 + 0.556 \, d_p} \left\{ \frac{A_t}{n_d} (u_o - u_{mf}) \right\}^{0.35} \tag{3}$$

Behie and Kehoe (1973) showed that the values of h_j calculated from Equation (3) are in good agreement with their measurements for a three dimensional bed.

Since information regarding the jet height formed above different type distributors (bubble capes, tuyere caps, fixed bed packed with large particles, etc.) is inadequate, Equation (3) will be applied in this paper in examining the effects of the jet region and distributor geometry on the chemical conversion in a fluidized bed reactor.

(ii) Bubble Diameter

Recently a semiempirical bubble coalescence equation was proposed by Mori and Wen (1975) which showed improvement in predicting bubble diameters over the previous correlations in the literature. This correlation relates bubble diameter, bed diameter and initial bubble size as follows:

$$D_B = D_{Bm} - (D_{Bm} - D_{Bo}) \exp (-0.3 \ h/D_t) \tag{4}$$

where h (= Z - h_j) is the height of the bed measured from the top of the jets and D_{Bm} is the maximum diameter of the bubble determined by bubble coalescence. The diameter D_{Bm} is given by:

$$D_{Bm} = 0.652 \ \{A_t (u_o - u_{mf})\}^{2/5} \tag{5}$$

The initial bubble diameter, D_{Bo}, is calculated either from Equation (6) or (7) shown below.

$$D_{Bo} = 0.00376 \ (u_o - u_{mf})^2 \tag{6}$$

$$D_{Bo} = 0.347 \ \{A_t (u_o - u_{mf})/n_d\}^{2/5} \tag{7}$$

Equations (6) and (7) are due to Miwa et al. (1972) and predict the initial bubble diameter for porous and perforated plates, respectively.

Equation (4) has been shown (Mori and Wen (1975)) to predict bubble size agreeing fairly closely with many experimental data over a wide range of operating conditions and bed sizes.

(iii) Size of Compartment

In the Bubble Assemblage Model, the bed is divided into n compartments along the axial distance of the bed. It is assumed that the nth compartment has a height, Δh_n, equal to the diameter of the bubble evaluated from Equation (1) at the height h_n. Unlike the original bubble growth equation used by Kato and Wen (1969), the bubble size computed from Equation (4) is not directly proportional to the bed height, h; and therefore, the height of nth compartment, Δh_n, can be determined by expanding D_B given by Equation (4) and calculated at h_n in a Taylor series about $D_{Bn}{}'$. Thus

$$\Delta h_n = \frac{D_{Bn}{}'}{1 + 0.15 \ (D_{Bn}{}' - D_{Bm})/D_t} \tag{8}$$

where $D_{Bn}{}'$ is the bubble diameter at the boundary between $(n-1)^{st}$ and n^{th} compartment or at $h = h_n{}'$. Here h_n is given by

$$h_n{}' = \sum_{i=1}^{n-1} \Delta h_i \tag{9}$$

Equation (8) is used to calculate the size of all compartments in a bubbling fluidized bed except the first compartment where the bubbles are generated at the distributor plate and the final compartment where the bubbles erupt as they leave the bed. The method of computing the height of the first compartment is presented later. The bubble diameter at the nth compartment is the diameter of the bubble at the middle of the nth compartment or at a height, h_n, given by

$$h_n = h_n{}' + \frac{\Delta h_n}{2} = \sum_{i=1}^{n-1} \Delta h_i + \frac{\Delta h_n}{2} \tag{10}$$

The height of final compartment Δh_N is calculated by the following equation as,

$$\Delta h_N = L_f - h_N{}' \tag{11}$$

where N is determined by the condition at which $h_n{}' + \Delta h_n$ becomes greater than L_f. At this position $n = N$.

(iv)   Volume of Bubble Phase and Emulsion Phase

In the original Bubble Assemblage Model, the volume fraction occupied by bubbles, ε_{Bn}, below the incipient fluidized bed height, L_{mf}, was regarded as constant.

In an actual fluidized bed, the voidage near the porous plate distributor is considerably higher than that in upper part of the bed as shown in Figure 1 presented by Bakker and Heertjes (1960).

If the voidage in the emulsion phase is assumed to be constant and is equal to ε_{mf}, the average voidage in the $n\underline{^{th}}$ compartment, $\bar{\varepsilon}_n$ can be expressed as a function of ε_{Bn}, as follows:

$$\bar{\varepsilon}_n = \varepsilon_{Bn} + (1 - \varepsilon_{Bn}) \cdot \varepsilon_{mf} \tag{12}$$

The volume fraction of bubbles in the bed can be written as follows (for example Kunii and Levenspiel (1968)).

$$\varepsilon_{Bn} = \frac{u_o - u_{mf}}{u_B} \tag{13}$$

In this paper, in order to consider the variation of voidage in the vicinity of the distributor, the bubble rising velocity, u_B,

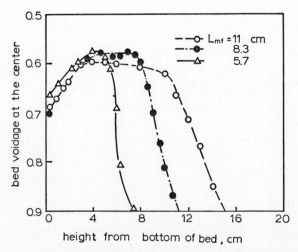

Figure 1. Profile of the bed voidage observed by Bakker and Heertjes (1960).

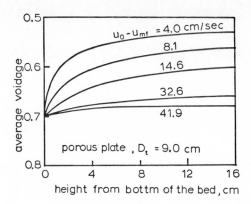

Figure 2. Calculated average voidage along the bed height.

in Equation (13) is treated as a function of bubble diameter. The

value of ε_{Bn} is also allowed to vary as a function of D_{Bn} (or h).

As an example, the calculated variation of $\overline{\varepsilon}_n$ in the bed obtained

from Equation (13) is shown in Figure 2. It can be seen from this figure

that the calculated value of $\overline{\varepsilon}_n$ at the bottom of the bed is considerably

different from the rest of the bed where $\overline{\varepsilon}_n$ is almost constant.

In this paper, Equation (13) is also assumed to be applicable to the

top part of the bed, $L_{mf} < Z \leq L_f$. This is because the method of

calculating the conversion near the top of the bed, where the voidage

tails-off rapidly as seen in Figure 1, does not affect greatly on the

estimated value of conversion except when rapid solid reaction of fine

particles takes place with an excess of reactant gas. For example,

the difference in calculated conversions of a catalytic first order

reaction obtained from the method described above and that obtained

using the method of original Bubble Assemblage Model is usually less

than 5%. Also, the length of the tail-off region is not considerably

long except under slug-flow conditions.

Since the volume of bubbles in the nth compartment can be written

as $\varepsilon_{Bn} \cdot A_t \cdot \Delta h_n$, the volumes of the bubble phase consisting of bubbles and clouds, V_{bn}, and the cloud volume excluding the bubble, V_{cn}, and the volume of emulsion phase, V_{en}, in the nth compartment can be expressed, respectively as:

$$V_{bn} = \varepsilon_{Bn} A_t \Delta h_n (1 + \delta_{cn}) \qquad (14)$$

$$V_{cn} = \varepsilon_{Bn} A_t \Delta h_n \delta_{cn} \qquad (15)$$

$$V_{en} = A_t \Delta h_n \{1 - \varepsilon_{Bn} (1 + \delta_{cn})\} \qquad (V_{en} \geq 0) \qquad (16)$$

where δ_{cn} is the ratio of the volume of the clouds excluding bubbles to the volume of the bubbles and is given by

$$\delta_{cn} = \frac{V_{cn}}{V_{bn} - V_{cn}} = \frac{3u_{mf}/\varepsilon_{mf}}{u_b - u_{mf}/\varepsilon_{mf}} \qquad (17)$$

where $u_b = 0.711 \sqrt{g\, D_{Bn}}$.

(v) First Compartment

From Equations (13), (14) and (17), the volume fraction of bubble phase in n^{th} compartment, ε_{cn}, can be written as,

$$\varepsilon_{cn} \equiv \frac{V_{bn}}{A_t \Delta h_n} = \frac{u_o - u_{mf}}{u_B} \{1 + \frac{3u_{mf}/\varepsilon_{mf}}{u_b - u_{mf}/\varepsilon_{mf}}\} \qquad (18)$$

When the value of u_{mf} is large and the initial bubble diameter is small, the value of ε_{cn} could become nearly one at the bottom of the bed. In such a case, the clouds around the bubbles will touch and join with the neighboring clouds as shown by Shichi et al (1968), Toei et al (1968) and Gabor and Koppel (1970). No quantitative correlation is available for the estimation of the cloud volume and the gas exchange rate under such conditions.

If an assumption is made that the shape of the cloud around a bubble is spherical, it is possible to estimate the critical volume fraction of clouds, $\varepsilon_c{}^*$, when the adjacent spherical clouds are just touching each other. It can be further assumed that if the calculated value of ε_c from Equation (18) becomes larger than ε_c^*, the clouds just above the distributor would join each other so that no emulsion phase could exist. Here, the value of ε_c^*, is assumed to be the volume fraction of spherical particles randomly packed or $\varepsilon_c^* = 0.6$.

Critical height from the top of the jets, h_c, is defined as the height where the calculated ε_c becomes equal to ε_c^*. The bubble diameter at h_c is also defined as the critical bubble diameter, D_{Bc}. The value of D_{Bc} can be determined from Equation (18) by substituting $\varepsilon_{cn} = \varepsilon_c^* = 0.6$. Then h_c is calculated from the following equation obtained from Equation (4)

$$h_c = \frac{D_t}{0.3} \cdot \ln \frac{D_{Bm} - D_{Bo}}{D_{Bm} - D_{Bc}} \qquad (D_{Bc} \geq D_{Bo}) \tag{19}$$

When the value of D_{Bc} becomes less than D_{Bo} the value of h_c is taken as zero, since the maximum value of ε_{cn} obtained from Equation (18) is the value at $D_B = D_{Bo}$ or $h = 0$.

Since the region from the distributor plate to the level at the top of the jets (jet region: $0 \leq Z \leq h_j$) is quite different from the remainder of the bubbling region of the bed, it should be treated separately. Behie and Kehoe (1973) proposed a model based on an assumption that the gas flow in the jets is plug-flow and gas in the emulsion phase is perfectly mixed. They showed an example of calculated results based on their model using experimentally obtained values of the mass transfer coefficient between the jets and emulsion phase. Toei (1973) calculated the gas flow pattern of a single jet based on a model assuming potential flow of the gas.

However, no quantitative correlation is available for estimation of the gas exchange coefficient between jets and emulsion phase.

In this paper in order to examine the effect of jet region on the chemical reaction, as far as the gas exchange is concerned, it is simply assumed that the jet region ($0 \leq Z \leq h_j$) can be treated the same as the bubbling region just above the jets. When h_c is larger than zero, the region, $0 \leq Z \leq h_j + h_c$, is treated as the first compartment where gas is assumed to be perfectly mixed. When h_c is equal to zero, the region $0 \leq Z \leq h_j + D_{Bo}$, is treated as the first compartment where the bubble diameter at $h = D_{Bo}/2$ is used to calculate the reactant gas concentration in each bubble and the emulsion phase.

RESULTS AND DISCUSSIONS

In order to test the validity of the modified Bubble-Assemblage Model, a comparison of the calculated conversions based on this model with the experimental conversions reported in the literature (see Table 1) is presented in Figure 3. As can be seen from the figure, when the values of kL_{mf}/u_0 are less than about 1.0, even a simple homogeneous model, either plug-flow or complete-mixing, can represent the experimental results quite closely. On the other hand, when the values of kL_{mf}/u_0 are larger than 1.0, experimental conversions are seen to deviate considerably from the conversions calculated based on the simple models. The conversions calculated from the Bubble-Assemblage Model are seen to agree with the experimental conversions sufficiently closely even for large values of kL_{mf}/u_0. Thus, examination of Figure 3 indicates that for slow reactions even simple models are good enough while for fast reactions more sophisticated model is needed to simulate the performance of fluidized bed reactors.

In addition to kL_{mf}/u_0, the extent of deviation between the calculated conversions based on bubbling models such as the Bubble-

Table 1 Summary of experimental conditions for fluidized bed reactor studies used in simulation

Authors	Reaction	D_t (cm)	Distributor nd	d_p (cm)	u_{mf} (cm/sec)	k (sec^{-1})	u_o (cm/sec)	L_{mf} (cm)
Kobayashi (1969)	decomp. of ozone	8.4 20	Pe 36 241	0.0194	2.1	0.7	4~18	34, 67
Cooke (1968)	carbonization of char	120	57 Pe 283 396	0.0375	5	1.0*	9~48	54~100
Ogasawara (1959)	synthesis of acetonitrile	20.0	PB	0.0100	0.54	12.1	6.6~18.2	8~30
Shen (1955)	decomp. of nitrous oxide	11.4	Po	0.0103 0.0139	0.42 0.71	0.012~0.02	0.9~5.7	27.9, 35.5
Massimilla (1961)	oxidation of ammonia	11.4	Po	0.0105	0.44***	0.081	1.4~17.4	19.4~58.3
Lewis (1959)	hydrogenation of ethylene	5.2	S	0.0122	0.73	1.41~8.7	4.2~35	24~47
Fryer (1974)	decomp. of ozone	22.9	BC 61**	0.0117	1.7	7.75	2~13	11.5, 39

Pe: perforated plate
BC: bubble cap
PB: packed bed
Po: porous plate
S:: screen

*   adjusted value
**   number of bubble cap
***   estimated value

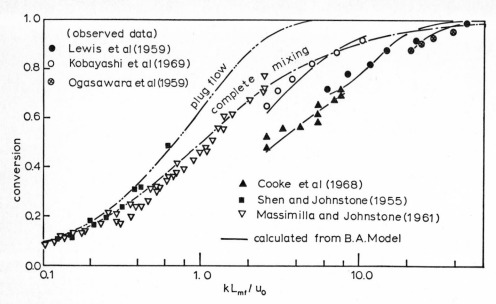

Figure 3. Comparison of observed conversion with calculated conversion based on plug-flow and complete mixing and bubble assemblage models.

Assemblage Model and that based on the simple homogeneous models (such as plug-flow or complete-mixing) is also affected considerably by the gas exchange between the bubble phase and the emulsion phase. This point was also discussed by Grace (1974). To clarify this point more the conversion calculated by simplified Bubble Assemblage Model,

$$f_{MS} = \frac{(kL_{mf}/u_o)}{(kL_{mf}/u_o)(x_B + 1) + 1} \qquad (20)$$

is compared with that computed from plug flow model,

$$f_p = 1 - \exp(-kL_{mf}/u_o).$$

Here, the simplified Bubble Assemblage Model is defined as that having a single compartment with $\delta_c = 0$ and an average value of gas exchange coefficient, $\overline{F}_o$, and an average value of volume fraction of the bubbles $\overline{\varepsilon}_B$. These values are assumed to be constant for a given fluidized bed reactor.

The difference between the two conversions, $\Delta f \equiv f_p - f_{MS}$ is shown
in Figure 4 as a function of kL_{mf}/u_0 with gas exchange number,
$x_B \equiv u_0/\overline{F}_0 \; \bar{\epsilon}_B \; L_f$, as the parameter. The broken line shown in the figure
at $x_B = 0.0$ represents the difference in conversion, Δf, for plug-flow
and complete-mixing and indicates the effect of gas mixing on conversion.
As can be seen from the figure, so long as kL_{mf}/u_0 is smaller than
about $1 \sim 2$, the values of Δf are rather small but are affected
significantly by kL_{mf}/u_0. However, when kL_{mf}/u_0 is larger than about
2, Δf is affected strongly by the values of x_B, the gas exchange. Thus,
for $kL_{mf}/u_0 > 2$, it is necessary to employ models such as the Bubble-
Assemblage Model which takes into consideration the bubble behavior,
the gas exchange and the effect of distributor plate to evaluate the
extent of conversion.

In Figure 5, the experimental data for a very fast reaction
$(k=7.75 \text{ sec.}^{-1})$ obtained by Fryer (1974) are compared to the calculated

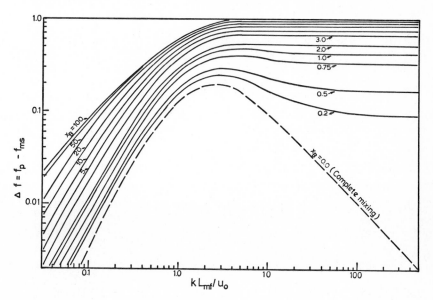

Figure 4. Differences in conversions between plug flow model and simplified bubble assemblage model indicating effect of gas exchange rates.

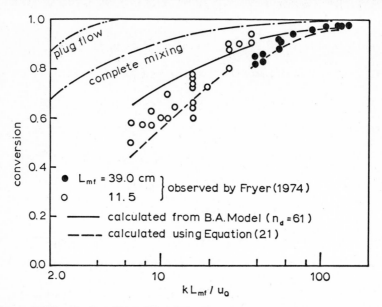

Figure 5. Comparison between observed data and calculated results from homogeneous plug-flow complete mixing and the bubble assemblage model as a function of kL_{mf}/u_0.

result based on the Bubble-Assemblage Model. Fryer used a distributor composed of sixty-one bubble-caps each of which had four downward holes. Two equations were employed to calculate the initial bubble diameter used in Figure 5. Equation (7) for perforated plates was used by setting $n_d = 61$. The result was compared to that obtained from an empirical equation, Equation (21), of Fryer (1974) who correlated the initial bubble diameter using the distributor employed in the chemical reaction experimentation.

$$D_{Bo} = 1.5 \ (u_0 - u_{mf})^{0.26} \quad \text{for } u_0 < 4.0 \text{ cm/sec}$$
$$D_{Bo} = 0.7 \ (u_0 - u_{mf})^{0.83} \quad \text{for } u_0 > 4.0 \text{ cm/sec} \qquad \} \qquad (21)$$

The results are shown in the figure with a solid line using Equation (7) and a broken line using Equation (21). As in such a case with fast chemical reaction, (large kL_{mf}/u_0), Figure 5 indicates that the actual experimental data seem to scatter between the two lines and are affected considerably by the initial bubble size.

In the following, therefore, we shall examine the validity of the
Bubble-Assemblage Model particularly for those cases where the behavior
of the bubbles and the gas exchange affects considerably on the extent
of conversion.   In other words, data from fluidized bed reaction
experiments of Cooke et al (1968) and Kobayashi et al (1969) are
selected for testing the effect of distributor design and the characteristics of model behavior.

A typical concentration profile of the reactant in the bubble
phase, C_{Bn}, in the emulsion phase, C_{en}, as well as the average concentration, C_{An}, calculated from the following equation, Equation (22),
are shown in Figure 6.

$$C_{An} = \varepsilon_{Bn} (1 + \delta_{cn}) C_{Bn} + \{1 - \varepsilon_{Bn} (1 + \delta_{cn})\} C_{en} \qquad (22)$$

Since for a fast reaction, a great deal of the reaction takes place
in the first compartment, the accuracy of calculation in the bottom

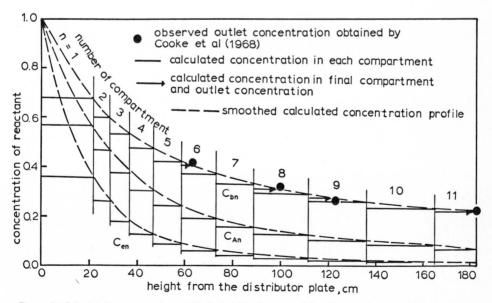

Figure 6. Calculated concentration profile in bubble phase, emulsion phase and average concentration.

region of the bed affects significantly the overall conversion of the

reactant in the fluidized bed as shown in Figure 6.

In Figure 7, the observed oxygen consumptions in the carbonization

of char in a fluidized bed using different perforated plate distributors

reported by Cooke et al (1968) are compared with that calculated from

the Bubble Assemblage Model. These distributors have different number

of holes as indicated in the figures. A reasonable value of the

reaction rate constant, k, is assumed in the calculation of the oxygen

consumptions. It can be seen from Figure 7 that the Bubble Assemblage

Model can predict the oxygen consumption in fluidized beds with

different perforated plate geometries and that the number of holes

on the perforated plate affects the conversion considerably.

Comparisons of observed conversions obtained by Kobayashi et al

(1969) using 8.4 cm and 20 cm diameter beds with the calculated con-

version obtained from the proposed model are shown in Figure 8.

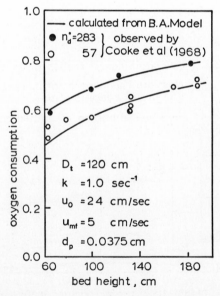

Figure 7. Comparison of calculated and observed oxygen consumption for a bed with two different perforated plates as a function of bed height.

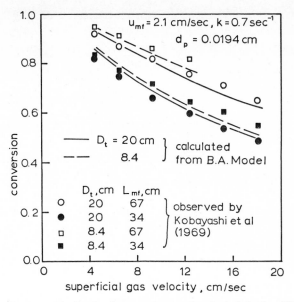

Figure 8. Comparison between calculated and observed conversion for beds having different bed diameters.

Although the effect of the bed diameter on the conversions is not as significant in comparison to that of distributor geometry as indicated in Figure 7, conversions are nevertheless affected substantially by the bed diameter. The effect of the bed diameter on the conversion, which is needed in scale-up can be predicted by the proposed Bubble Assemblage Model.

In Figure 9, the effect of distributor geometry on extent of conversion is shown for a wide range of conversion. As is evident from the figure, if the same status of fluidization could be maintained, that is, no channeling or maldistribution occurs with variation in pressure drops across the distributor, the conversion would increase in general with an increase in the number of holes, n_d, for the perforated plate and conversion eventually approaches that using a porous plate.

However, even in this case, if kL_{mf}/u_0 is smaller than 1.0, the effect of n_d on conversion is small. This is particularly true, when

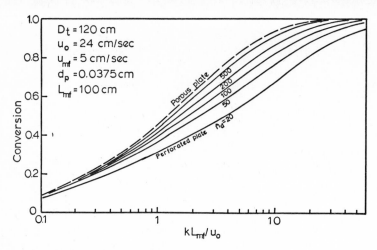

Figure 9. Effect of number of holes on the distributor plate on chemical conversions as a function of reaction constants.

the number of holes is increased beyond a certain value. If n_d is, say 50 or larger, no significant effect is seen, by further increases in the number of holes.

An example of effect of jet height correlations above the distributor on calculated conversions is shown in Figure 10. The

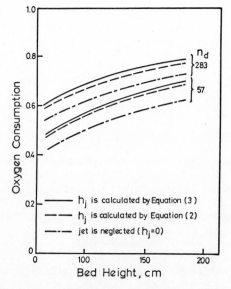

Figure 10. Effect of the jet heights by different correlation equations on the calculated conversion.

conditions of operation are the same as those used in Figure 7. The
solid line shown in Figure 10 represents oxygen consumption calculated
using Equation (3) proposed by Basov et al (1969) and is the same line
appeared in Figure 7. The broken line represents the calculated result
using Equation (2) proposed by Zenz (1968) and is seen to differ little
from that using Equation (3). The line with a dot and dash represents
the result of no jets formation by letting $h_j = 0$. As can be seen, the
oxygen consumption for the case of no jetting is considerably lower
than the cases with jetting indicating that the effect of the jet
formation cannot be neglected.

The sensitivity of key parameters in the Bubble-Assemblage Model
is tested in Figures 11 and 12. Two sets of experimental conditions
are selected to calculate the volume fraction of cloud δ_c, and the
gas exchange rate. For gas exchange coefficient, instead of Equation (23)

$$F_o = 11/D_B \qquad\qquad (23)$$

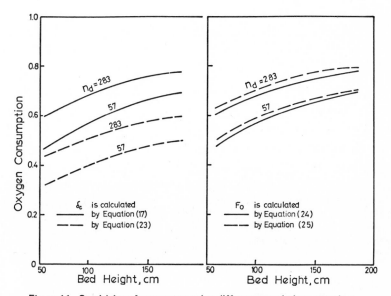

Figure 11. Sensitivity of parameters using different correlations equations.

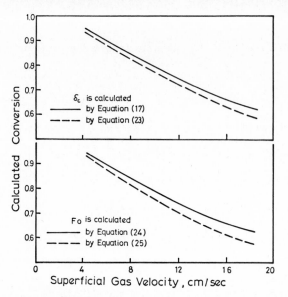

Figure 12. Sensitivity of parameters using different correlation equations.

the following equation proposed by Kunii and Levenspiel (1968) is used
to calculate the conversion.

$$F_0 = 6.78 \ (\frac{\epsilon_{mf} \ D_s \ u_B}{D_B^{\ 3}})^{1/2}$$

(24)

where D_s is the diffusion coefficient of gas in cm^2/sec.

The difference in gas exchange rate using Equations (23) and (24)
is relatively small.

For δ_c, Equation (25) proposed by Partridge and Rowe (1966) in
place of Equation (17) is used

$$\delta_c = 1.17/(u_B - u_{mf}/\epsilon_{mf})$$

(25)

The cloud volume calculated using Equation (25) is in general
substantially smaller than that from Equation (17). As the result,
the conversion calculated is also smaller. Particularly, as shown in
Figure 11, when kL_{mf}/u_0 is large and the gas exchange rate is small
(see Figure 3), the calculation of δ_c affects the conversion considerably.

Thus, it is necessary, under such condition, not only to evaluate jetting above the distributor and gas exchange rate carefully, but also to make an accurate estimation of δ_c.

However, δ_c estimated from Equation (17) and Equation (25) are both for the cloud around a single-bubble and application to the actual fluidized bed reactor in which the neighboring bubbles are affecting the shape and volume of cloud can be extended only with extreme caution.

It is hoped that future research will provide a better correlation for estimation of the cloud volume in a fluidized bed reactor.

Figure 13 shows a comparison of the calculated conversions for these selected models, namely the Bubble Assemblage Model, the model of Orcutt, Davidson and Pigford (1962) and the model of Kunii and Levenspiel (1968). An equivalent bubble diameter calculated at $Z = L_f/2$ from Equation (4) is used for the latter two models. It is seen from Figure 13 that when the value of k is large, the deviations in the conversions predicted by the three models become considerable.

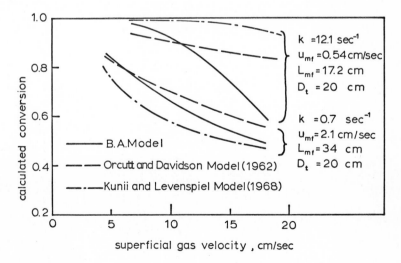

Figure 13. Comparison of calculated conversions from various models for fast and slow reactions.

CONCLUSIONS

Simulation of catalytic conversions and carbonization of char in a number of fluidized bed reactors is conducted based on the modified Bubble Assemblage model presented in this paper. The results of the simulation are compared with experimental data reported in the literature and found to be satisfactory using their operating conditions and kinetic data. The significant effect of distributor geometry on the extent of reaction, particularly for fast reactions, reported in experimental studies can be predicted by the proposed model. The effect of the bed diameter is found to affect the overall conversion particularly for a tall bed although this effect is not as significant in comparison to that of the distributor.

ACKNOWLEDGEMENTS

The authors express their gratitude to the Office of Coal Research, Department of the Interior, Washington, D.C., for financial support, and appreciation to Dr. W. J. McMichael for his suggestions and discussions during the preparation of this paper.

LITERATURE CITED

Baerns, M., Ind. Engng. Chem. Fundamentals, 5, 508 (1966).

Bakker, P. J. and Heertjes, P. M., Chem. Engng. Sci., 12, 260, (1960).

Basov, V. A., Markhevka, V. I., Melik-Akhnazarov, T. Kh. and Orachko, D.I., International Chem. Engng., 9, 263, (1969).

Behie, L. A. and Kehoe, P., A.I.Ch.E.J., 19, 1070 (1973).

Botton, R. J., Chem. Engng. Prg. Symp. Ser., 66, 101, 8, (1968).

Cooke, M. J., Harris, W., Highley, J. and Williams, D. F., Tripartite Chem. Engng. Conf., Symp. on Fluidization I, pp. 14-20, Montreal (1968).

Fryer, C., Ph.D. Thesis, Monash Univ. Australia (1974).

Fryer, C. and Potter, O. E., Powder Tech., 6, 317 (1972).

Gabor, J. D. and Koppel, L. B., Chem. Engng. Prg. Symp. Ser., 66, 105, 28, (1970).

Grace, J. R., A.I.Ch.E. Symp. Ser., 70, No. 141, 21 (1974).

Kato, K. and Wen, C. Y., Chem. Engng. Sci., 24, 1351 (1969).

Kobayashi, H., Arai, F., Chiba, T. and Tanaka, Y., Chem. Engng., Tokyo, 33, 274 (1969).

Kunii, D. and Levenspiel, O., Ind. Engng. Chem. Fundls., 7, 446 (1968).

Lewis, W. K., Gilliland, E. R. and Glass, E., A.I.Ch.E.J., 5, 419 (1959).

Massimilla, L. and Johnstone, H. F., Chem. Engng. Sci., 16, 105 (1961).

Mori, S. and Muchi, I., J. of Chem. Engng., Japan, 5, 251 (1972).

Mori, S. and Wen, C. Y., A.I.Ch.E. Journal, 11, 109 (1975).

Orcutt, J. C., Davidson, J. F. and Pigford, R. L., Chem. Engng. Prog. Symp. Ser., 58, 38, 1 (1962).

Ogasawara, S., Sasaki, A., Hojo, K., Shirai, T. and Morikawa, K., Chem. Engng., Tokyo, 23, 299 (1959).

Partridge, B. A. and Rowe, P. N., Trans. Instn. Chem. Engrs., 44, T335, T349 (1966).

Shichi, R., Mori, S. and Muchi, I., Chem. Engng Tokyo, 32, 343 (1968).

Shen, C. Y. And Johnstone, H. F., A.I.Ch.E.J., 1, 349 (1955).

Toor, F. D. and Calderbank, P. H., Proc. Int. Symp. on Fluidization, P. 248, Eindhoven, Netherland (1967).

Toei, R., Matsuno, R., Miyagawa, H., Nichitani, K. and Komagawa, Y., Chem. Engng., Tokyo, 32, 565 (1968).

Toei, R., A.I.Ch.E., Symp. Ser., 69, 128, 18 (1973).

Zenz, F. A., Tripartite Chem. Engng. Conf., Symp. on Fluidization II, pp. 36, Montreal (1968).

PREDICTION OF BUBBLE DISTRIBUTIONS IN FREELY-BUBBLING THREE-DIMENSIONAL FLUIDIZED BEDS

T. H. NGUYEN, J. E. JOHNSSON[1], R. CLIFT[2], AND J. R. GRACE

The importance of the frequency and size of gas bubbles in determining the behaviour of aggregatively fluidized beds is well-known. A number of empirical or semi-empirical correlations have been proposed (e.g. 1, 2) for predicting mean bubble diameters and frequencies. Aside from the uncertainty in applying such correlations to specific conditions other than those for which they were developed, previous methods have the disadvantage that they cannot predict the distribution of bubble sizes at a given level nor spatial non-uniformities across the bed.

The present work follows a more mechanistic approach in which the motion and coalescence of all bubbles in the bed are calculated to give full information on the temporal and spatial distributions of bubbles. The basis is a postulate developed by Clift and Grace (3) that the velocity of a bubble in a fluidized bed may be approximated by adding to its rise velocity in isolation the velocity which the particulate phase would have at the position of the nose if the bubble were absent. This approach has previously been shown to give a good description of the motion of

1 Present address: Danish Engineering Academy, Copenhagen
2 Present address: Imperial College, London

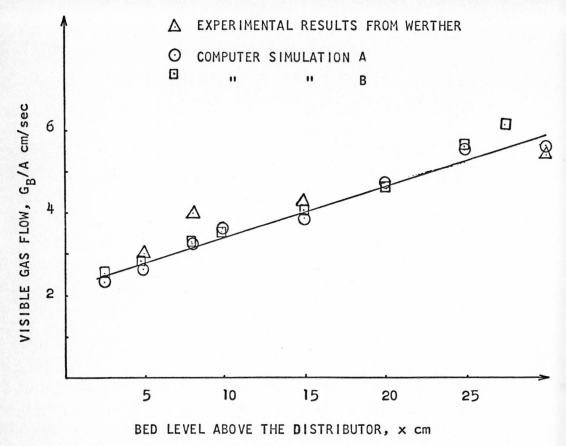

Figure 1. Visible gas flow at different levels above the distributor as measured by Werther and as predicted by the computer simulations.

bubble pairs (3, 4, 5) and vertical chains (6), and of the behaviour of two-dimensional freely-bubbling beds (7). The instantaneous velocity $\underset{\rightarrow}{U}_i \equiv (u_i, v_i, w_i)$ of bubble i in a three-dimensional bed containing N bubbles is then calculated as

$$\underset{\rightarrow}{U}_i = u_{Ai}\underset{\rightarrow}{i} + \sum_{\substack{j \neq i \\ j=1}}^{N} \underset{\rightarrow}{Q}_{ij}$$

where u_{Ai} is the isolated rise velocity of bubble i and $\underset{\rightarrow}{Q}_{ij}$ is the particle velocity at the position of the nose of bubble i

caused by bubble j moving at its instantaneous velocity $\underset{\rightarrow}{U}_j$. The
set of N equations is solved numerically to give the instantaneous
velocities of all bubbles in a freely-bubbling bed. Numerical
integration with respect to time then gives the positions of all

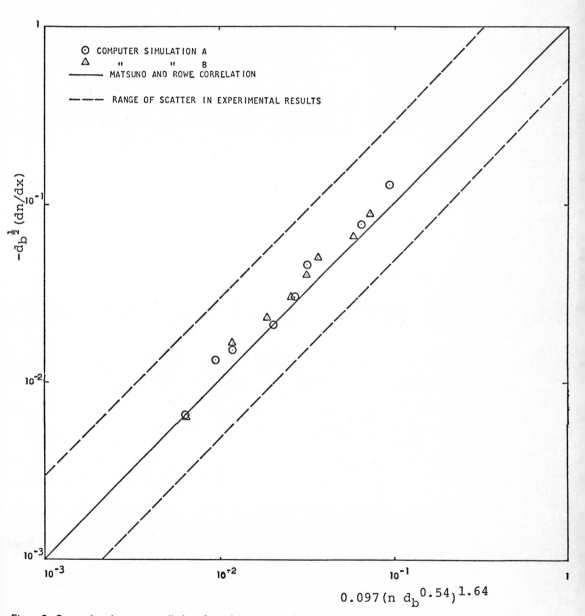

Figure 2. Comparison between predictions from the computer simulations, experimental results obtained by Werther, and an empirical correlation proposed by Matsuno and Rowe.

T. H. NGUYEN, J. E. JOHNSSON, R. CLIFT, AND J. R. GRACE

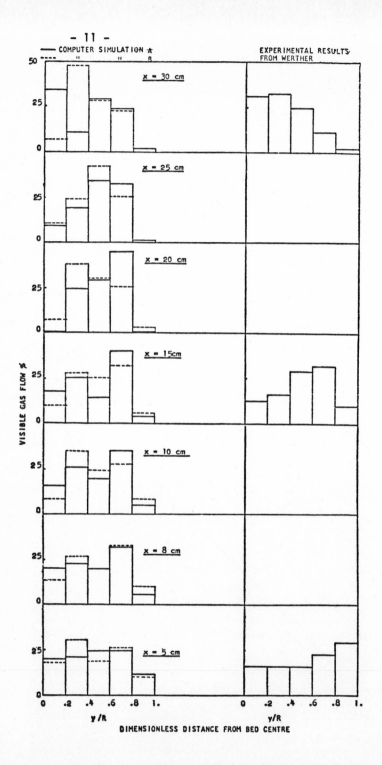

Figure 3. Spatial distribution of visible gas flow at different levels as measured by Werther and as predicted by the computer simulations.

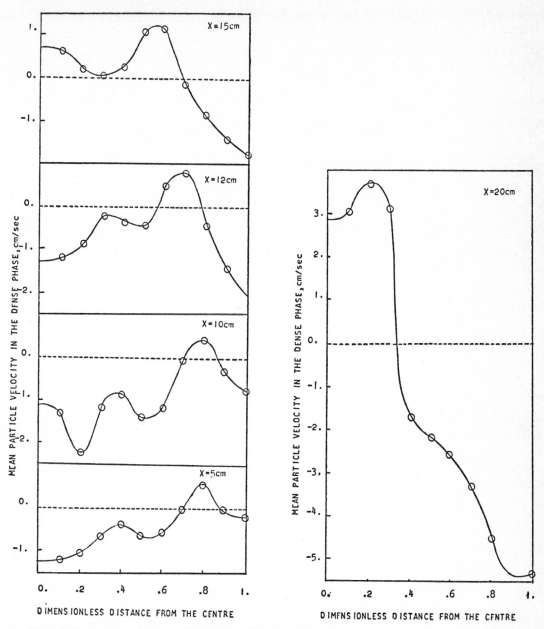

Figure 4. Dense phase velocity profiles (wakes excluded) at different levels above the distributor predicted by
simulation B.

bubbles at successive time intervals. Coalescence is assumed to

occur when bubbles come into close proximity and bubbles are taken

out of the simulation when they pass a level corresponding to the bed surface. The volume of a bubble formed by coalescence is assumed to be 15% greater than the sum of the primary bubble volumes, in accordance with measurements on coalescing two-dimensional bubbles (8).

This approach has no fitted parameters, and requires only the bubble distribution at a low level in the bed to allow predictions at higher levels. It has been applied to calculating the behaviour of a 0.2 m. I.D. bed of particles with minimum fluidizing velocity 0.018 m./sec. at a superficial velocity of 0.09 m./sec. These conditions correspond to a set of detailed measurements of bubble properties reported by Werther (9). Bubbles were assumed to enter the bed with constant radius of 0.01 m. and frequency 210 sec.$^{-1}$. The entering positions are taken to be randomly distributed across the distributor. The resulting simulation gives predictions of total bubble flow rate and distribution of bubbles by size and position which agree closely with the experimental measurements. In particular, it predicts that, at any level in the bed, there is an annular region in which the bubble frequency has a maximum. This maximum occurs close to the wall at low levels, moving towards the centre line of the column at higher bed levels.

The simulation has been extended to predict the velocities of the particles in the bed, by summing the effects of all bubbles moving at their instantaneous velocities. A small vertical velocity correction has been included to allow for the fact that the cal-culated velocities do not exactly satisfy particle continuity. The correction has some effect on the mean particle velocities, but the effect on the bubble population is negligible. In the region close to the distributor, the predicted mean particle velocity is upwards close to the walls and downwards near the axis. Such a flow pattern has been observed previously in shallow beds (10). At higher levels the more familiar "gulf stream" pattern is predicted, with the solids moving up near the axis and down at the walls. These solids circulation patterns result directly from the predicted non-uniform bubble distribution.

Summary of Experimental Conditions employed in a run by Werther (Doctoral Dissertation, Erlangen, 1972)

Bed shape:	Cylindrical
Bed diameter:	20 cm
Distributor type:	Sintered bronze
Particle size:	83 μm
Particle type:	Quartz sand
Particle shape:	Irregular
Minimum fluidization velocity:	1.8 cm/s
Static bed height:	50 cm
Superficial gas velocity:	8.9 cm/s

Boundary Condition Employed in Computer Simulation for above Experimental Run

Reference level:	2.5 cm
Initial bubble size:	Uniform, 0.97 cm diameter
Initial bubble frequency:	210 s^{-1}
Initial visible gas flow:	2.3 cm/s
Initial bubble positions:	Random, but with no bubble centre within one radius of the wall. Simulation A used RANDU subroutine while simulation B used RSTART subroutine.

It is concluded that this type of simulation provides a powerful tool for predicting local and overall behaviour of bubbling fluidized beds.

REFERENCES

1. H. Kobayashi, F. Arai, and T. Chiba, Kagaku Kogaku, $\underline{4}$, 147 (1966).

2. R. Matsuno and P.N. Rowe, Chem. Eng. Sci., $\underline{25}$, 1587 (1970).

3. R. Clift and J.R. Grace, Chem. Eng. Prog. Symp. Ser. No. 105, $\underline{66}$, 14 (1970).

4. R. Clift and J.R. Grace, AIChE Symp. Ser. No. 116, $\underline{67}$, 23 (1971).

5. J.R. Grace and R. Clift, Can. Jl. Chem. Eng., $\underline{52}$, 417 (1974).

6. R. Clift and J.R. Grace, Trans. Instn. Chem. Engrs., $\underline{50}$, 364 (1972).

7. J.E. Johnsson, R. Clift, and J.R. Grace, Instn. Chem. Engrs. Symp. Ser. No. 38, Paper B5 (1974).

8. J.R. Grace and G. Venta, Can. Jl. Chem. Eng., $\underline{51}$. 110 (1973).

9. J. Werther, Doctoral Dissertation, Erlangen (1972).

10. P.H. Calderbank, F.D. Toor, and F.H. Lancaster, Proc. Int. Symp. on Fluidn., p. 652 (Netherlands Univ. Press, 1967).

APPENDIX

Basis of the Model

The instantaneous velocity components for bubble i are calculated by summation:

$$u_i = u_{Ai} + \sum_{\substack{j=1 \\ j \neq i}}^{N} q_{ij} \qquad v_i = \sum_{\substack{j=1 \\ j \neq i}}^{N} p_{ij} \qquad w_i = \sum_{\substack{j=1 \\ j \neq i}}^{N} \sigma_{ij}$$

where u_{Ai} is the rise velocity in isolation of bubble i, i.e.

$$u_{Ai} = 0.67 \sqrt{ga_i}$$

and the velocity components for the particulate phase at the nose of bubble i caused by bubble j are calculated from potential flow theory with one doublet for each bubble:

$$q_{ij} = r_{ij}u_j + t_{ij}v_j + \ell_{ij}w_j$$

$$p_{ij} = t_{ij}u_j + s_{ij}v_j + m_{ij}w_j$$

$$\sigma_{ij} = \ell_{ij}u_j + m_{ij}v_j + n_{ij}w_j$$

where

q_{ij} = vertical particulate phase velocity at the nose of bubble i caused by bubble j

p_{ij} = horizontal particulate phase velocity in the y-direction at the nose of bubble i caused by bubble j

σ_{ij} = horizontal particulate phase velocity in the z-direction at the nose of bubble i caused by bubble j

u_j = vertical instantaneous velocity of bubble j

v_j = horizontal instantaneous velocity of bubble j in the y-direction

w_i = horizontal instantaneous velocity of bubble j in the z-direction

x, y and z are co-ordinates of the bubble centre (Cartesian), x vertical upwards

r_{ij}, s_{ij}, n_{ij}, t_{ij}, ℓ_{ij}, m_{ij} are interaction coefficients depending on the size and position of bubbles i and j given by

$$r_{ij} = a_j^3 \, (2X^2 - Y^2 - Z^2)/AB$$

$$s_{ij} = a_j^3 \, (2Y^2 - X^2 - Z^2)/AB$$

$$n_{ij} = a_j^3 \, (2Z^2 - X^2 - Y^2)/AB$$

$$t_{ij} = 3a_j^3 \, XY/AB$$

$$\ell_{ij} = 3a_j^3 \, XZ/AB$$

$$m_{ij} = 3a_j^3 \, YZ/AB$$

where $X = x_i + a_i - x_j$

$Y = y_i - y_j$

$Z = z_i - z_j$

$AB = 2(X^2 + Y^2 + Z^2)^{5/2}$

The bubble trajectories can be found by integration of the differential equations:

$$\frac{dx_i}{dt} = u_i; \quad \frac{dy_i}{dt} = v_i; \quad \frac{dz_i}{dt} = w_i$$

The above equations cover the general case where no wake entry has occurred. If the nose of bubble i has entered the wake of bubble j, we have simply:

$$u_i = u_{Ai} + u_j$$

$$v_i = v_j$$

$$w_i = w_j$$

BUBBLE GROWTH IN LARGE
DIAMETER FLUIDIZED BEDS

JOACHIM WERTHER

NOMENCLATURE

D = diameter of a spherical cap bubble, cm.

D_B = bed diameter, cm.

d_v = diameter of a sphere with a volume equal to the local average bubble volume, def. by equ. (17), cm.

$E[...]$ = expectation value of ...

f = local bubble frequency in a freely bubbling bed, $cm^{-2}s^{-1}$.

h = height above the distributor (in the case of the bubble chain injected into an incipiently fluidized bed: height above the nozzle), cm.

k = number of bubbles striking the probe per unit time, s^{-1}.

L = average distance between bubble centers, cm.

n = bubble frequency in the case of the bubble chain injected into an incipiently fluidized bed (number of bubbles passing the bed's cross-sectional area per unit time), s^{-1}.

$Q(s)$ = cumulative number density distribution of the nose-to-nose separations s, dimensionless.

s = nose-to-nose separation between successive bubble, cm.

T_c = time required for complete coalescence, s.

u = superficial fluidizing velocity, $cm\ s^{-1}$.

u_{mf} = superficial minimum fluidizing velocity, $cm\ s^{-1}$.

u_A = rise velocity of a bubble in isolation, $cm\ s^{-1}$.

v = local average rise velocity of single bubbles and of leading bubbles among the coalescing ones in the freely bubbling bed, $cm\ s^{-1}$.

$\bar{v}_b$ = local average rise velocity of bubbles in the freely bubbling bed, $cm\ s^{-1}$.

$\bar{V}_b$ = local average bubble volume, cm^{-3}.

$\dot{V}_b$ =   local visible bubble flow, $cm^3 \, cm^{-2} \, s^{-1}$.

$\bar{\dot{V}}_b$ =   average visible bubble flow obtained by integration over the bed cross-section, $cm^3 \, cm^{-2} \, s^{-1}$.

V_c =   net inflow of gas during an individual coalescence process, cm^3.

α, ν =   parameters, def. by equ. (6).

γ =   ratio of net inflow V_c to bubble Volume $\bar{V}_b$, dimensionless.

SCOPE

Although many investigations of bubble growth in small diameter laboratory fluidized beds are available in the literature, sufficiently accurate information on local hydrodynamic properties of large diameter fluidized beds is still lacking. On the other hand this information is needed for the design of industrial fluid beds, for the results of small-scale experiments cannot simply be used for scale-up as has been shown in a previous investigation of the author [1].

In this paper results of measurements of the bubble development in beds of 45 and 100 cm diameter for the fluidization of quartz sand and spent cracking catalyst, respectively, are presented. A statistical model for bubble coalescence is developed, the basic ideas of which are tested by comparing the predictions of the model to the results of X-ray measurements of Botterill et al. [2]. Based on this model an empirical correlation for bubble growth is derived which is shown to be in agreement not only with the measurements of the present investigation but also with X-ray observations by Rowe and Everett [3].

EXPERIMENTAL

The measurements have been carried out in fluid beds of 45 and 100 cm diameter fitted with porous plate distributors. The experimental set-up for the 100 cm diameter unit shown in Fig. 1 basically consists of a cylindrical fluidized bed of 100 cm diameter with a freeboard section of 190 cm diameter. The elutriated solid particles are collected in cyclones and

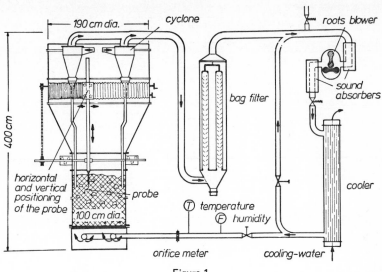

Figure 1.

are continuously fed back into the bed via standpipes. The
fluidizing air is recycled to the roots blower after passing
a bag filter where the fines which have passed the cyclones
are held back.

The solids used were quartz sand (minimum fluidizing velocity
u_{mf} = 1.35 cm/s) and spent cracking catalyst (u_{mf} = .20 cm/s).
The various measurement techniques which have been described
in detail elsewhere [4-6] are shown schematically in Fig. 2.
Basis of the methods used are miniaturized capacitance
probes (A,B,C). Within the fluidized bed such a probe registers
the variations of the solids concentration within the measuring
volume, as a function of time. Bubbles striking the probe as
they rise, cause electric pulses. With a single probe A one
is able to measure the local average bubble pulse duration
and the average number of bubbles encountered by the probe
per unit of time.

With two probes A and B arranged vertically above each other
it is possible to obtain via a measurement of the cross-
correlation function of the probe signals the local mean
bubble rise velocity. From the quantities $\bar{v}_b$, T_b and k the

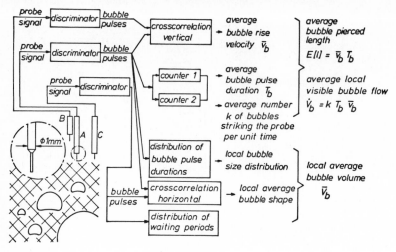

Figure 2.

local average bubble pierced length and the local visible
bubble flow may be computed.

Furthermore, with a relation based on geometrical probability
theory it is possible to obtain from a measurement of the
distribution of the bubble pulse durations the local bubble
size distribution. Using two probes arranged in a common
horizontal plane permits via the measurement of a cross-
correlation function to determine the local average bubble
shape which makes it then possible to compute further
characteristic properties of the local bubble assemblage,
as for example the local average bubble volume. Finally,
the measured distribution of the waiting periods between the
arrival of successive bubbles at the probe tip gives some
information about the spatial arrangement of rising bubbles.

For each fluidized bed system a large number of measurements
at different locations within the bed has yielded a detailed
picture of the bubble development with height above the
distributor as well as of the spatial distribution of bubbles
within the bed.

For both solids basically the same characteristic spatial
distribution of bubbles which has been described already in a

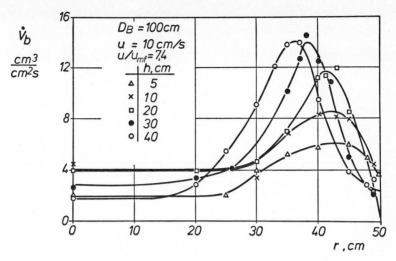

Figure 3.

previous paper [7] is found, i.e. a zone of increased bubble development close to the distributor near the wall which is displaced towards the vessel centre with increasing height above the distributor (Fig. 3).

Bubble growth, however, turns out to be quite different (Fig. 4). In the quartz sand bed a rapid increase in the average bubble size d_v with height h above the distributor is registered

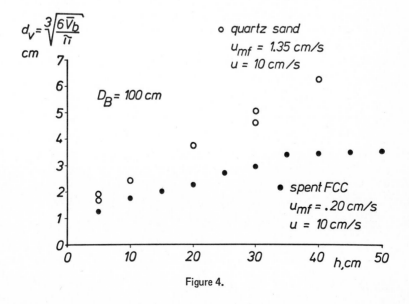

Figure 4.

whereas in the FCC bed the rate of bubble growth is found to
decrease with increasing height. The results indicate that
in the case of the FCC bed bubble growth is strongly influenced
by the mechanism of bubble splitting, which is in accordance
with previous findings of Matsen [8].

For heights greater than 30 or 40 cm the experiments suggest
the existence of a quasi-steady state of bubble development
which is characterized by a dynamic equilibrium between bubble
coalescence and splitting.

THEORY AND EXPERIMENTAL RESULTS

The statistical model presented in this paper is dealing
with the coalescence mechanism only. The additional influence
of bubble splitting will be treated in a separate publication.
The model is based on the fundamental work of Harrison and
Leung [9] and of Clift and Grace [10-11] who have shown that
coalescence of a pair of bubbles takes a well-defined period
during the greater part of which the two coalescing bubbles
can be observed as separate entities. Following their results
the velocity u_A of the following bubble relative to the
leading one (Fig. 5) is given by the Davies-Taylor equation [12]

$$u_A = \frac{\sqrt{2}}{3} \sqrt{gD} \qquad\qquad (1)$$

and the time T_c required for complete coalescence of two
equal-sized bubbles rising in vertical alignment is

$$T_c = \frac{D}{u_A}. \qquad\qquad (2)$$

The idea of the statistical model of bubble coalescence proposed here is that in a freely bubbling bed bubbles are getting
into contact with other bubbles at random times. From the
randomness of the beginning of individual coalescence processes it follows immediately that at a given height h above
the distributor in the course of time a random distribution

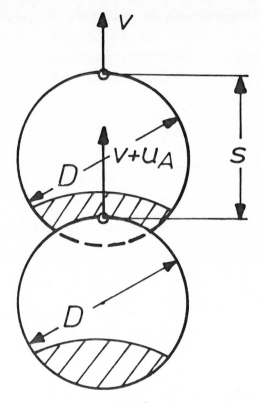

Figure 5.

of the stages of the individual coalescence processes will be
registered. In the case of the coalescence of bubbles of
equal sizes the instantaneous nose-to-nose separation s is
a measure for the remaining duration of the corresponding
coalescence process.

Thus there follows from the randomness of the beginning of
individual coalescence processes a random distribution of the
nose-to-nose separations s registered at the measuring level h.

This statistical conception of bubble coalescence has in a
first step been tested with a simple fluidized bed system,
i.e. the chain of bubbles injected into an incipiently
fluidized bed. Taking into consideration that bubble formation
at an orifice is not a completely deterministic process but
has instead a certain stochastic component it follows that in

the height h sufficiently remote from the orifice the process
of arrival of the bubbles in the measuring plane h is a
Poisson process and hence the distribution of the nose-to-nose
separations s between successive bubbles can be described by
a negative exponential distribution,

$$Q(s) = 1 - e^{-\frac{s}{E[s]}} \tag{3}$$

where $E[s]$ denotes the expectation value of s,

$$E[s] = \frac{\bar{v}_b}{n} . \tag{4}$$

With the coalescence time T_c given by equ. (2) it follows for
the relative change of the local bubble frequency n with
height h above the nozzle

$$\frac{\Delta n}{n} = - \frac{u_A}{v\bar{v}_b} \ n \ \Delta h \tag{5}$$

where v denotes the local rise velocity of the non-coalescing
bubbles and of the leading bubbles among the coalescing ones.
$\bar{v}_b$ is the rise velocity averaged over all bubbles of the local
bubble assemblage. With equ. (5) the statistical coalescence
model predicts the change of bubble development with height
on the basis of properties characterizing the local state of
fluidization.

Coalescence in a chain of rising bubbles injected into an
incipiently fluidized bed, has been investigated by Botterill,
George and Besford [2] by means of X-ray photography. They
measured local frequency, average volume and rise velocity of
the bubbles for different injection rates and different heights
above the injection nozzle.

From their data local values $u_A/(v\bar{v}_b)$ have been computed which
are plotted against the frequency n in Fig. 6. As is shown
by this plot it holds as an approximation

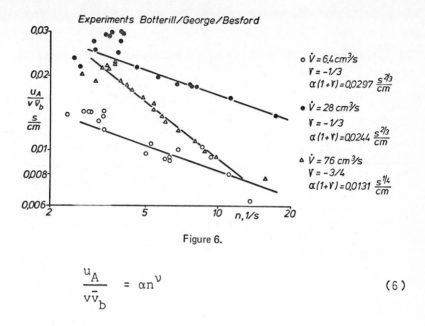

Figure 6.

$$\frac{u_A}{v\bar{v}_b} = \alpha n^{\nu} \tag{6}$$

With equ. (6) it follows from equ. (5)

$$\frac{1}{n^{1+\nu}} = \alpha\,(1+\nu)\,h + const. \tag{7}$$

The statistical coalescence model thus postulates a linear relationship between the quantity $(1/n^{1+\nu})$ and the height h above the injection nozzle. This is confirmed by the graphs in Fig. 7, where on the left-hand side measured bubble frequencies are plotted as $(1/n^{1+\nu})$ against height h. The slopes of the solid lines which have been computed from the plots in Fig. 6 are seen to be a good approximation to the measured decrease of bubble frequency with height.

The deviation from the linear relationship observed for heights greater than about 15 cm can be explained by the effect of bubble splitting observed by Botterill et al. toward the top of the bed.

On the right-hand side of Fig. 7 measured values of the bubble frequency n plotted against height h are compared to the

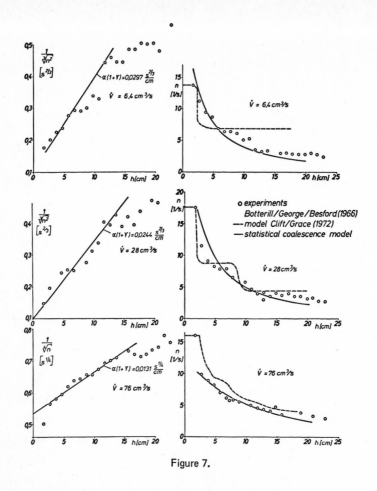

Figure 7.

predictions of the statistical coalescence model and of the model by Clift and Grace [13].

The deterministic model by Clift and Grace assumes a strictly periodic bubble formation at the orifice which results in a stepwise decrease of the computed bubble frequency with height for low injection rates, whereas the agreement between the predictions of the statistical model and the experimental values clearly demonstrates the existence of a stochastic component in the process of bubble formation at an orifice. Due to this stochastic component the statistical model being based on the assumption that bubbles are randomly arriving in the measuring plane h proves to be capable of describing the coalescence pattern in the immediate vicinity of the nozzle. As the good

agreement between the predictions of the model and the experimental results is demonstrating the usefulness of the statistical conception of bubble coalescence, the same conception
will now be applied to the case of the freely bubbling bed.

Converting the time dependent state of fluidization over the
measuring surface inside a freely bubbling fluidized bed to a
spatial array yields the picture of randomly distributed bubbles
shown on the left-hand side of Fig. 8. In a first step this
random distribution of bubbles of different sizes may be replaced by a regular array of equal-sized bubbles of mean
diameter where L denotes the average distance between the
centers of rising bubbles given by

$$L = \sqrt[3]{\frac{\overline{v}_b}{f}} \qquad (8)$$

The statistical model of coalescence suggested here is now
postulating that the change in bubble characteristics between
the heights h and h + Δh may be described by the coalescence
processes occurring in the bubble chains that result if the
distances L between bubbles of the vertical rows are replaced

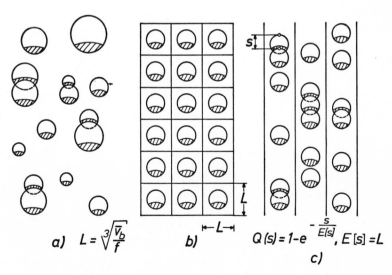

$$a) \quad L = \sqrt[3]{\frac{\overline{v}_b}{f}} \qquad b) \quad \leftarrow L \rightarrow \qquad Q(s) = 1 - e^{-\frac{s}{E[s]}}, \; E[s] = L$$
$$c)$$

Figure 8.

by randomly distributed distances. The distribution of the
nose-to-nose separations s then is given by a negative
exponential distribution with the expectation value of s
being identical to L.

Thus the problem of bubble coalescence in a freely bubbling bed
is reduced to the problem of coalescence in a bubble chain and
consequently the relative change of bubble frequency f with
height h above the distributor is given by

$$\frac{\Delta f}{f} = - \frac{u_A}{v} \quad \frac{\Delta h}{L} \tag{9}$$

where v again denotes the rise velocity of non-coalescing bubbles
and of the leading bubbles among the coalescing ones. An
evaluation of the measurements has yielded the result that v
as well as $\bar{v}_b$ are approximately proportional to the square root
of the mean bubble diameter D. Thus it follows with equations
(1) and (8) from equ. (9) as an approximation

$$\frac{\Delta f}{f} \approx - \text{const.} \quad \sqrt[3]{f} \; \Delta h \tag{10}$$

from which it follows after integration

$$\frac{1}{\sqrt[3]{f}} = \text{const}_1 \; h + \text{const}_2 \tag{11}$$

The statistical model of bubble coalescence thus postulates
a linear relationship between the quantity $(1/\sqrt[3]{f})$ and the
height h above the distributor. As is shown by corresponding
plots of measured bubble frequencies, examples of which are
given in Fig. 9 the relationship predicted by equ. (11) indeed
gives an excellent description of the measured decrease of
bubble frequency with height.

Representing the bubble frequencies for all gas velocities
investigated in a common plot (Fig. 10) yields the result
that as a simple approximation the dependence of bubble

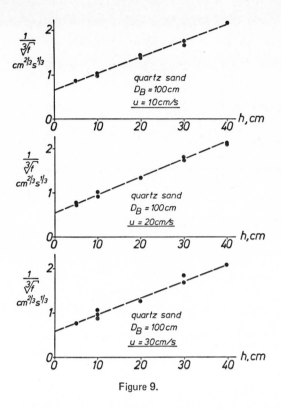

Figure 9.

frequency on height above the distributor can be described by
a common relation for all gas velocities,

$$\frac{1}{\sqrt[3]{f}} = 0.039 \, h + 0.57 \quad cm^{2/3} \, s^{1/3} \qquad (12)$$

where h is to be inserted in cm.

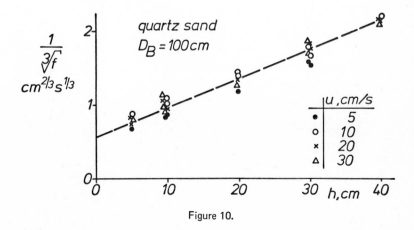

Figure 10.

Up to now only one aspect of bubble coalescence has been treated, namely the reduction of the number of bubbles with height as a consequence of coalescence processes. For a description of bubble growth due to coalescence the observed increase of visible bubble flow with height has to be taken into account. In Fig. 11 the visible bubble flow $\bar{V}_b$ averaged over the bed's cross-sectional area is plotted against height h above the distributor for two different gas velocities.

As may be seen from this graph the visible bubble flow is considerably increasing with height in the vicinity of the distributor plate.

Taking into account experiments with coalescing bubble pairs carried out by Clift [14], Grace [15] and Grace and Venta [16] this increase of the visible bubble flow may be interpreted as the result of a net inflow of gas into the bubbles during coalescence processes. Thus a model for bubble growth between the heights h and h + Δh can be formulated as follows:

  (increase of the total bubble gas volume between h and h + Δh)

    = (number of coalescence processes among the bubbles
        passing the measuring surface in h)

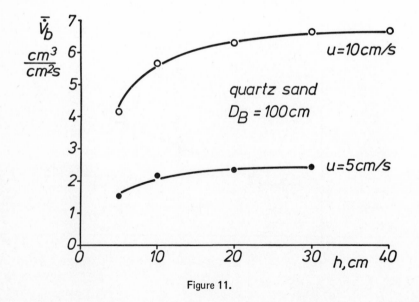

Figure 11.

x (net inflow of gas during an individual coalescence
process)

x $\left(\dfrac{\text{mean residence time of bubbles in (h, h + } \Delta h)}{\text{mean duration of an individual coalescence process}}\right)$

The plausible assumption that the net inflow of gas, V_c, during an individual coalescence process is proportional to the local average bubble volume $\bar{V}_b$, i.e.

$$V_c = \gamma \bar{V}_b \qquad (13)$$

then leads to a simple first order differential equation the solution of which is

$$\bar{V}_b = \bar{V}_{bo} \left(\frac{f_o}{f}\right)^{1+\frac{2}{3}\gamma} \qquad (14)$$

As shown by Fig. 12 where the local average bubble volume $\bar{V}_b$ is plotted against the inverse bubble frequency on a logarithmic grid, the same general proportionality

$$\bar{V}_b \sim \left(\frac{1}{f}\right)^{1.21}$$

is holding for all gas velocities investigated. The exponent 1.21 means that the net inflow of gas V_c during an individual coalescence process equals on an average over 30 % of the local mean bubble volume, i.e. the volume of the combined bubble is about 15 % larger than the sum of the two bubbles' volumes before coalescence.

With equ. (1) and a simple empirical correlation for the bubble volume $\bar{V}_{bo}$ corresponding to $f_o = 1 \text{ cm}^{-2} \text{ s}^{-1}$,

$$\bar{V}_{bo} = 2.5 + 0.68 \ (u-u_{mf}) \ \text{cm}^3 \qquad (15)$$

where $(u-u_{mf})$ is to be inserted in cm/s it follows from

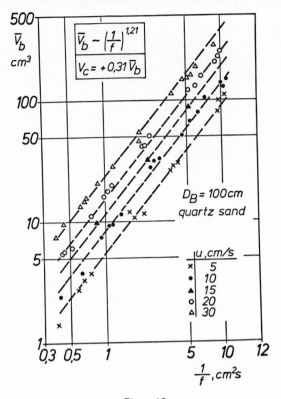

Figure 12.

$$\bar{V}_b = \bar{V}_{bo} \left(\frac{f_o}{f} \right)^{1.21} \tag{16}$$

for the volume equivalent bubble diameter d_v,

$$d_v = \sqrt[3]{\frac{6 \bar{V}_b}{\pi}} \quad , \tag{17}$$

$$d_v = 0.853 \left[1 + 0.272 \left(u - u_{mf} \right) \right]^{1/3} \left(1 + 0.0684 \, h \right)^{1.21} \text{ cm} \tag{18}$$

where $(u - u_{mf})$ and h are to be inserted in cm/s and cm, respectively.

This correlation contains as variables the height h above the distributor and the excess gas velocity $(u - u_{mf})$, the latter one

characterizing the extent of bubble development for a given
fluidized bed system.

In Fig. 13 measured bubble sizes plotted against height h above
the distributor are compared to bubble growth predicted by the
empirical correlation, equ. (18). The agreement between
experimental and computed bubble sizes is quite good.

In Fig. 14 where theoretical bubble sizes are plotted against
measured ones the empirical growth correlation is applied to
further fluid bed systems investigated in the course of the
present work. It turns out that the correlation developed here
is able to describe bubble growth over the whole range of
parameters investigated, i.e. for bed heights up to 1 m, for
gas velocities up to 30 cm/s and for bubble diameters up to 15 cm.

The bubble growth correlation presented here has been derived
from measurements in fluidized beds of quartz sand exclusively.
The fact that this same correlation is also describing bubble
growth for the fluidization of glass spheres indicates that
the influence of the properties of the fluidized solids has

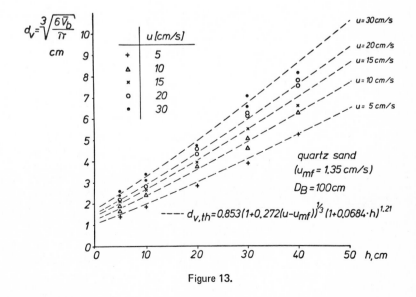

Figure 13.

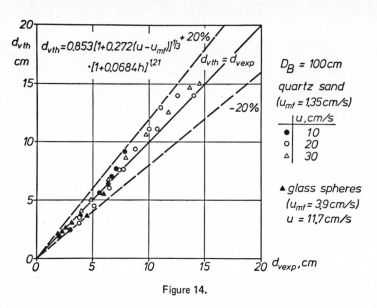

Figure 14.

sufficiently been considered by introducing the excess gas velocity $(u-u_{mf})$.

This conclusion is confirmed by an evaluation of measurements published by Rowe and Everett [3] who have determined bubble sizes by X-ray photography for different solid materials. In the comparison between theory and experiments shown in Fig. 15 only measurements for excess gas velocities greater than 5 cm/s

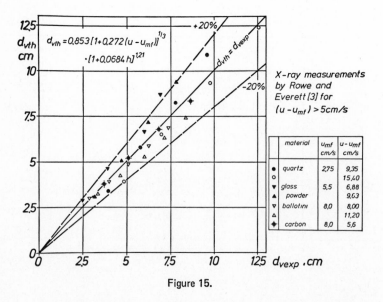

Figure 15.

have been used, because the growth correlation developed here
presupposes a sufficiently high number of coalescences occurring
in the bed and thus is certainly not applicable to some very
small excess gas velocities investigated by the authors.

The presentation of data in Fig. 15 yields the result that
for all fluidized bed systems with excess gas velocities
greater than 5 cm/s investigated by Rowe and Everett the
bubble growth correlation developed here gives a good description of the measured bubble development.

The author's own experimental results together with the
measurements of Rowe and Everett [3] suggest as preliminary
limits of application for the bubble growth correlation,
equ. (18):

bed diameter:	D_B > 20 cm
distributor:	porous plate
mean particle diameter:	$100 \leq d_p \leq 350$ µm
particle shape:	spherical, angular, sharp-edged
minimum fluidizing velocity:	$1.35 \leq u_{mf} \leq 8$ cm/s
excess gas velocity:	$5 \leq u-u_{mf} \leq 30$ cm/s

SUMMARY AND CONCLUSIONS

The bubble characteristics in fluidized beds of 45 and 100 cm
diameter have been investigated for quartz sand and spent FCC.

Although the spatial distribution of the bubbles within the
beds exhibits basically the same characteristic features for
both materials bubble growth, however, turns out to be quite
different. A rate of bubble growth decreasing with height
indicates a considerable influence of bubble splitting in the
case of the cracking catalyst. This result means that bubble
growth for both materials is governed to a quite different
extent by the coalescence mechanism and the splitting mechanism,

respectively, leading us to the conclusion that a description
of bubble growth in fluid beds should be based on the relevant
physical mechanisms. Empirical correlations for bubble growth
in which the influence of the solids properties is only described by the mean particle diameter or by the minimum fluidizing velocity can therefore necessarily only be valid over a
limited range of solids properties for which the mechanisms
governing bubble growth are essentially the same.

For the case of bubble growth dominated by the coalescence
mechanism a statistical coalescence model has been derived which
is based on the physical characteristics of the coalescence
process. A test of this model for the special case of coalescence in a chain of bubbles injected into an incipiently
fluidized bed results in a good agreement with the X-ray
measurements of Botterill et al. Further application of this
model to the case of the freely bubbling bed yields a simple
correlation for the decrease of bubble frequency with height
above the distributor.

The experiments are further demonstrating that a net inflow of
gas from the suspension phase into the bubble phase in the course
of coalescence processes represents a considerable contribution
to bubble growth. Due to this effect the volume of a combined
bubble is shown to be about 15 % greater than the sum of the
two bubbles' volumes before coalescence. The permeability of
the bubble boundaries during the coalescence process to
convective gas transport seems to be of major importance in
the modelling of catalytic reactors the more so since further
investigations [17] have shown that not only a net inflow but
also under certain flow conditions a considerable net outflow
of gas during coalescence processes can be observed.

Taking into account the relation between bubble frequency and
height derived from the statistical coalescence model and the
inflow of gas during coalescence an empirical correlation for
bubble growth in large diameter freely bubbling beds has been

derived which has been shown to be valid over a broad range of fluidizing variables.

LITERATURE CITED

[1]   J. Werther; AIChE Symp. Ser. 7o(1974) 141, 53.

[2]   J.S.M. Botterill, J.S. George and H. Besford; Chem. Engng. Progr. Symp. Ser. 62(1966) 62, 7.

[3]   P.N. Rowe and D.J. Everett; Trans. Instn. Chem. Engrs. 5o(1972), 55.

[4]   J. Werther and O. Molerus; Int. J. Multiphase Flow 1(1973), 1o3.

[5]   J. Werther; Trans. Instn. Chem. Engrs. 52(1974) 149.

[6]   J. Werther; Trans. Instn. Chem. Engrs. 52(1974) 16o.

[7]   J. Werther and O. Molerus; Int. J. Multiphase Flow 1(1973), 123.

[8]   J.M. Matsen; AIChE Symp. Ser. 69(1973) 128, 3o.

[9]   D. Harrison and L.S. Leung; Symp. Interaction between Fluids & Particles. Instn. Chem. Engrs. 1962, 127.

[10]   R. Clift and J.R. Grace; Chem. Engng. Progr. Symp. Ser. 66(197o) 1o5, 14.

[11]   R. Clift and J.R. Grace; AIChE Symp. Ser. 67(1971) 116, 23.

[12]   R.M. Davies and G.I. Taylor; Proc. Roy. Soc. A 2oo(195o), 375.

[13]   R. Clift and J.R. Grace; Trans. Instn. Chem. Engrs. 5o(1972), 364.

[14]   R. Clift; Ph.D. dissertation, McGill University, Montreal 197o.

[15]   J.R.Grace; AIChE Symp. Ser. 67(1971) 116, 159.

[16]   J.R. Grace and J. Venta; Can. J. Chem. Engng. 51(1973), 11o.

[17]   J. Werther; to be published.

PREDICTING THE EXPANSION
OF GAS FLUIDIZED BEDS

D. GELDART

NOMENCLATURE

a	Variable in equation 1	cm
A	Cross sectional area of bed	cm^2
b	Variable in equation 1	-
d_A	Arithmetic mean of adjacent sieves	μm
d_B	Frontal diameter of a bubble	cm
d_{Bh}	Frontal diameter of a bubble at any level h	cm
d_{Bmax}	Maximum stable bubble size (eqn. 11)	cm
$\bar{d}_p$	Mean particle size	μm
D	Bed diameter	cm
g	Acceleration due to gravity	981 cm/s^2
h	Distance above the gas distributor	cm
H_F	Average height of a fluidized bed	cm
H_{max}	Maximum height of a slugging bed	cm
H_{mf}	Height of bed at incipient fluidization	cm
N	Number of holes per unit area of distributor	
Q_B	Volumetric flow rate of gas appearing as bubbles	cm^3/s
R	Expansion ratio (= H_F/H_{mf})	
U	Superficial gas velocity	cm/s
U_B	Rise velocity of a single isolated bubble	cm/s
U_{BC}	Rise velocity of a constant sized bubble	cm/s
U_{Bh}	Rise velocity of a bubble at level h	cm/s

U_{mf}	Superficial gas velocity at incipient fluidization	cm/s
U_t	Terminal velocity of particles	cm/s
$\bar{V}_B$	Mean bubble hold-up in bed	cm^3
X	Weight fraction of particles on each sieve	-
Y	Actual visible bubble flow rate ÷ gas flow rate in bubble phase predicted by two phase theory	-
ρ_p	Particle density	gm/cm^3

1. Summary - In a 12" dia tube, we have measured bubble sizes (d_B), and visible bubble flow rates (Q_B) for sand-like powders of various mean sizes $(\bar{d}_p)$ and bed depths (H_{mf}). Based on these results we give equations which predict the expansion ratio (R) of freely bubbling beds of similar materials. R increases with $U-U_{mf}$, and at equal values of $U-U_{mf}$, decreases with increasing H_{mf} and d_p.

2. Relevant Theoretical and Experimental Background - The fluidization characteristics of powders fall into 4 broad groups identi-

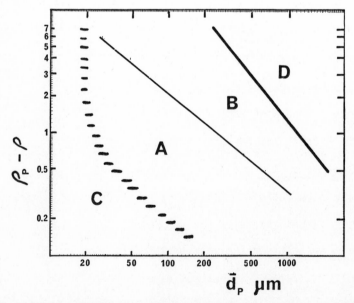

Figure 1. Classification of fluidization properties according to size and density of the powder.

fied by $\bar{d}_p$ and ρ_p (Fig. 1)[1]. Cracking catalysts are typical group A powders; bed expansion is large, bubble size is limited. Group B powders (e.g. sand) are of greater size and density and d_B may be predicted[2] at any level h by:

$$d_{Bh} = a + bh \quad \cdots \quad \cdots \quad \cdots \quad \cdots \quad \cdots \quad \cdots \quad (1)$$

where $b = .027 (U-U_{mf})^{0.94}$. For porous plate distributors $a = 0.915$ $(U-U_{mf})^{0.4}$ and for a distributor with N holes/cm^2: $a = \dfrac{1.43}{g^{0.2}}\left\{\dfrac{U-U_{mf}}{N}\right\}^{0.4}$. This equation is valid providing that $d_{Bh} < \frac{1}{2}$ bed diameter D. Units must be centimeters and seconds.

According to the simple two-phase theory: $Q_B/(U-U_{mf})A = Y = 1 \cdots \quad \cdots \quad (2)$ but this has been questioned[3,4,5]. Our measurements[6,7] on freely bubbling beds of sands indicate that Y decreases as $\bar{d}_p$ increases (Fig. 2A). Since we found d_B to be independent of $\bar{d}_p$ (equ.1), this can only mean that the number of bubbles/second also decreases as $\bar{d}_p$ increases, and this was also verified experimentally (Fig. 2B).

3. Bed Expansion in Freely Bubbling 'Sand-Like' Powders (Group B) (Fig. 3) - In such powders the average bed expansion is equal to the mean bubble hold-up $(\bar{V}_B)$ i.e:

$$\bar{V}_B = (H_F - H_{mf})A \text{ and since } R = H_F/H_{mf}, \quad \bar{V}_B = (R-1)H_{mf}A \cdots \quad \cdots (3)$$

The residence time of bubbles in a small element Δh is $\Delta h/U_{Bh}$, where U_{Bh} is the average bubble rise velocity at height h. The volume of bubbles, ΔV_B, in the element is $\dot{Q}_B \cdot \Delta h/U_{Bh}$ so over the whole bed,

$$\bar{V}_B = \int_0^{H_F} \dot{Q}_B \cdot \frac{dh}{U_{Bh}} \quad \cdots \quad \cdots \quad \cdots \quad \cdots \quad \cdots \quad (4)$$

There is doubt as to whether:

$$U_{Bh} = U_B = \sqrt{g \cdot \frac{d_{Bh}}{2}} \quad \cdots \quad \cdots (5) \quad \text{or} \quad U_{Bh} = U_B + U - U_{mf} \quad \cdots \quad \cdots (5A)$$

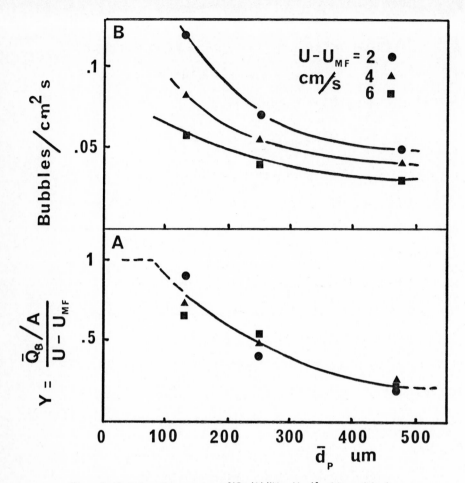

Figure 2. Variation of parameters $\gamma [(Q_B/A)/(U - U_{mf})]$ with particle size.

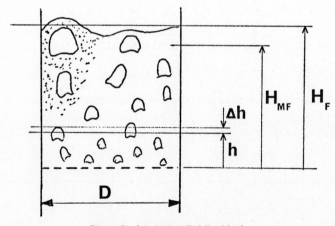

Figure 3. A bubbling fluidized bed.

For group B powders $U_B \gg U - U_{mf}$ so the simpler equ.(5) will be used. This may slightly overestimate R. Q_B increases slightly with h but for bed expansion we are interested in the average value, $\bar{Q}_B$. Combining equ.(1) to (5) and integrating:

$$R - 1 = \frac{2}{b}\sqrt{\frac{2}{g}} \cdot Y(U - U_{mf}) \left\{ \frac{(a + b\,H_F)^{\frac{1}{2}} - a^{\frac{1}{2}}}{H_{mf}} \right\} \quad \cdots \quad \cdots \quad \cdots \quad (6)$$

H_F is an unknown, but the equ.is readily solved by writing initially H_{mf} for H_F. R - 1 has been calculated for porous plates and Y = 1, and values are plotted for various bed heights in Fig. 4. This can be used together with Fig. 2A to calculate R - 1 for any $\bar{d}_p$. Alternatively equ 6.can be used in conjunction with measured values of R to calculate values of Y for group B powders other than sands. Although the literature contains much data on the expansion of small

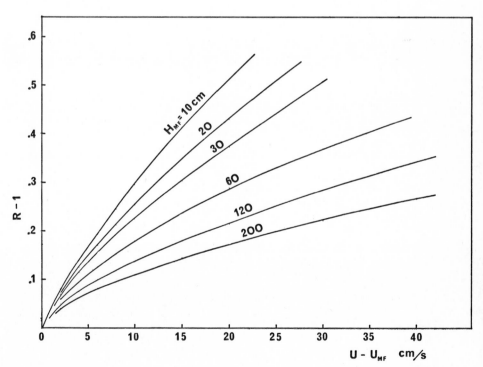

Figure 4. Plot of equation 6 for various initial bed heights assuming porous distributor and $\gamma = 1$.

diameter slugging beds[8], there is little information concerning freely bubbling beds. However theory and experiment are compared in Fig. 5 and agreement is generally good.

4. <u>Bed Expansion when Bubble Rise Velocity is Constant</u> – In the general case of constant bubble velocity, U_{Bc}, equations 3, 4 and 5A give:

$$1 - \frac{1}{R} = \frac{Y(U - U_{mf})}{U_{Bc}} \quad .. \quad .. \quad .. \quad .. \quad .. \quad (7)$$

There are two situations in which this can occur. The first is when the bed is slugging, and in this case $U_{Bc} = 0.35\sqrt{gD} + (U - U_{mf})$ .. ..(8) The substitution for U_{Bc} from (8) into (7), and $Y = 1$ leads to:

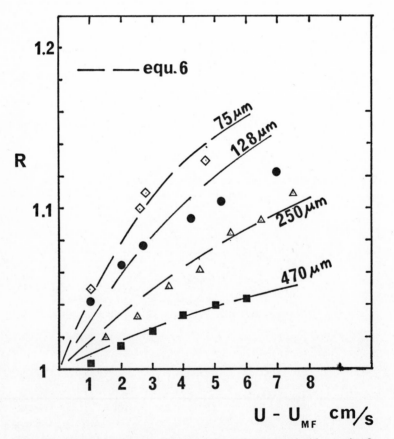

Figure 5. Experimental values of bed expansion ratio compared with equation 6.

$$R = \frac{U - U_{mf}}{0.35\sqrt{gD}} + 1 \quad \cdots \quad \cdots \quad \cdots \quad \cdots \quad \cdots \quad \cdots \quad (9)$$

It has been shown[8] that in this case $R = H_{max}/H_{mf}$.

The second situation occurs in a freely bubbling bed if the maximum stable bubble size d_{Bmax} is reached. This can happen in group A powders, and U_{Bc} is given by equ 5A. Moreover $Y = 1$ in fine powders (see Fig. 2A) so that:

$$R = \frac{U - U_{mf}}{\sqrt{\frac{g}{2} \cdot d_{Bmax}}} + 1 \quad \cdots \quad \cdots \quad \cdots \quad \cdots \quad \cdots \quad \cdots \quad (10)$$

Equ.(10) has been used to estimate d_{Bmax} from expansion data but at present there is no reliable method for predicting d_{Bmax}, though data in a recent paper[9] suggests that

$$\sqrt{\frac{g}{2} \cdot d_{Bmax}} = U_t \quad \cdots \quad \cdots \quad \cdots \quad (11)$$

where U_t is the terminal velocity of particles twice the mean size of the powder.

5. <u>Suggested Procedure for Estimating Bed Expansion Ratio</u> -

   (i)   Calculate the mean sieve size from $\bar{d}_p = 1/\Sigma \frac{X}{d_A}$ where X is the weight fraction of powder of sieve size d_A.

 (ii) Knowing the particle density, use Fig. 1 to find the powder group.

(iii) If group A, calculate an approximate d_{Bmax} from equ.(11). If this is larger than D/2 the bed may slug so use equ.(9) to calculate R. If less than D/2 use equ.(10).

 (iv) If the powder belongs to group B, use Fig. 2A to estimate Y and then equ.(6) to calculate R. Alternatively use Fig. 4 to estimate R-1 for $Y = 1$ and multiply the answer by the value of Y from Fig. 2A.

<u>Note</u>: Nomenclature and references appear in the supplemental sheets.

REFERENCES

1. Geldart, D., Powder Technol., 1973, $\underline{7}$, 285.

2. Geldart, D., Powder Technol., 1972, $\underline{6}$, 201.

3. Baumgarten, P.K. and Pigford, R.L., A.I.Ch.E.Jl., 1960, $\underline{6}$, 115.

4. Lockett, M.J., Davidson, J.F., Harrison, D., Chem. Engng. Sci., 1967, $\underline{22}$, 1059.

5. Grace, J.R. and Harrison, D., Chem. Engng. Sci., 1969, $\underline{24}$, 497.

6. Geldart, D., Ph.D. Thesis, Univ. of Bradford, 1971.

7. Baeyens, J. and Geldart, D., "Application of Fluidization", Conf. Toulouse, 1973.

8. Matsen, J.M., Hovmand, S. and Davidson, J.F., Chem. Engng. Sci., 1969, $\underline{24}$, 1743.

9. Matsen, J.M., A.I.Ch.E., Symp. Ser. 1974, $\underline{69}$, no. 128, 30.

MASS TRANSFER FROM A SLUG
TO A FLUIDIZED BED

J. R. F. GUEDES DE CARVALHO AND D. HARRISON

It is doubtful whether experimental studies that have been reported so far provide satisfactory data with which to test theoretical
models of mass transfer between bubbles and dense phase in fluidized
beds. In some early work (1,2,3), the measured rates of mass transfer
appear to have been dominated by end effects. Chiba and Kobayashi (4)
unfortunately interpret their interesting results with a model in which
bubble sizes are sometimes larger than the bed diameter. More recently
Drinkenburg and Rietema (5) measured bubble concentration as a function
of bed height, but the irregularity of bubble shape and gas adsorption
on the solids made their results difficult to interpret.

This paper describes an attempt to overcome these difficulties by
measuring the mass transfer from single slugs in a fluidized bed.

EXPERIMENTAL

Fig.1 shows the experimental system. The fluidized bed was contained in a 2m.perspex tube (I.D. 0.051m) fitted with a brass sinter
plate distributor. Glass ballotini particles of narrow size range
(d_{av} = 77μm) were fluidized by nitrogen; U_{mf} = 0.0088 m/s, ε_{mf} = 0.43.
At several heights up the tube wall, holes were drilled to allow for a
sampling capillary tube connected to a syringe. A slug of tracer gas
of known volume was injected into the bed just above the distributor,
using a solenoid valve and pressure vessel.

For each run a particle charge was used to give a bed level
about 0.05-0.08m below the nearest sampling point. A disc of "rigi-
mesh" covered with a thin layer of rubber foam was placed perpendicular
to the column axis some way above this sampling point. The position of
the disc was adjusted so that, after each slug injection, the bed level
would be just below it. The disc offered virtually no resistance to

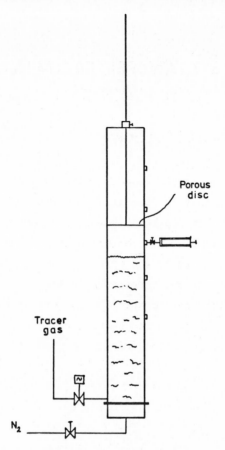

Figure 1. Diagram of apparatus.

the flow of the fluidizing gas, but it largely prevented macromixing of
the slug reaching the surface with the gas above it.

An experimental run consisted of injecting a tracer slug into
a fluidized bed at minimum fluidization and taking a sample
(10^{-5} m^3) when the slug reaches the bed surface. Experiments
have been carried out at several bed heights, analysing argon and
helium tracers by gas chromatography. Before each run nitrogen was
bubbled through the bed to remove all tracer gas in it.

ANALYSIS OF RESULTS

We assume that the experimental concentration profile obtained
represents the variation in tracer concentration for a slug of the
same volume as it rises through the bed. The theory of Hovmand and
Davidson (6) may then be used to interpret the results.

A material balance on a slug gives

$$(q + k_G S)(C_p - C_b) = V \frac{dC_b}{dt} = U_b V \frac{dC_b}{dy} \tag{1}$$

where $q = \frac{\pi D^2}{4} U_{mf}$ (2)

and $\quad k_G S = \frac{\varepsilon_{mf}}{(1-\varepsilon_{mf})} 4(\pi D_G)^{\frac{1}{2}}(\frac{g}{D})^{\frac{1}{4}} D^2 I$ (3)

if we use the simplified theory in which the diffusion and throughflow terms are additive; the latter was always less than 20% of the total in our experiments. C_p and C_b are the tracer concentrations respectively in the particulate phase and in the slug at height y; V is the slug volume and U_b its velocity of rise. D is bed diameter, D_G is the diffusion coefficient of tracer in fluidizing gas and I is an integral along the slug surface (6); g is acceleration of gravity.

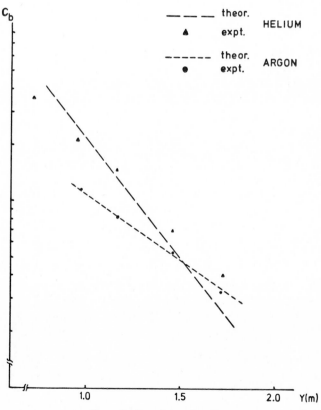

Figure 2. Experimental results.

We have $C_p = 0$, therefore (1) gives

$$\frac{d \ln C_b}{dy} = -\frac{q + k_G S}{U_b V}$$

Fig.2 shows a plot of $\ln C_b$ against y for our experiments, and the results are summarised in the following Table

Tracer	$D_G \times 10^4$ (m^2/s)	$V \times 10^6$ (m^3)	$\frac{d \ln C_b}{dy}$ (theor.) (m^{-1})	$\frac{d \ln C_b}{dy}$ (expt.) (m^{-1})
Argon	0.183	314	-1.61	-1.61
Helium	0.668	306	-2.92	-2.16

DISCUSSION

There is excellent agreement between theory and experiment for argon, but for helium the experimental slope falls considerably below that predicted. Nevertheless, it seems clear for beds of fine particles that gas diffusivity is an important parameter in determining the exchange rate, in line with the theory of Hovmand and Davidson (6). Our experimental programme is being extended to include further slug volumes and other tracer gases.

ACKNOWLEDGEMENT

One of the authors (JRFGC) is grateful for financial support from the Calouste Gulbenkian Foundation during the course of this work.

REFERENCES

1.   Szekely, J. Proc.Symp.on Interaction between Fluids and
         Particles, Inst.Chem.Engrs, 197, June 1962.

2.   Davies, L. and Richardson, J.F. Trans.Inst.Chem.Engrs
         44 (1966) T 293.

3.   Stephens, G.K., Sinclair, R.J. and Potter, O.E.
         Powder Technology 1 (1965) 157.

4.   Chiba, T. and Kobayashi, H.
         Chem.Eng.Sci. 25 (1970) 1375.

5.　　Drinkenburg, A.A.H. and Rietema, K.
　　　　　Chem.Eng.Sci. <u>28</u> (1973), 259.

6.　　Hovmand, S. and Davidson, J.F. in Fluidization, ed.
　　　　　J.F.Davidson and D.Harrison, Academic Press 1971.

INTERPHASE MASS TRANSFER
IN A FLUIDIZED BED

CLAUDE CHAVARIE AND JOHN R. GRACE

SCOPE

Of all the phenomena affecting the performance of bubbling fluidized beds as chemical reactors, interphase mass transfer is probably of primary importance in most cases. While models exist to predict the rate of transfer between a bubble and the dense phase, experimental data are scarce and none of the models is widely accepted. There is therefore a need to obtain reliable data which will allow the testing of mass transfer models and lead to realistic estimates for interphase transfer coefficients in fluidized bed chemical reactor models.

OBJECTIVES

(1) To measure interphase mass transfer rates for isolated gas bubbles rising in a fluidized bed together with relevant hydrodynamic features of the bubbles.

(2) To compare the mass transfer coefficients obtained with mass transfer models in order to offer guidance on the interphase rates to be used in applying fluidized bed chemical reactor models.

RESULTS

Single bubbles containing ozone were injected at regular time intervals (approximately every 3 seconds) into

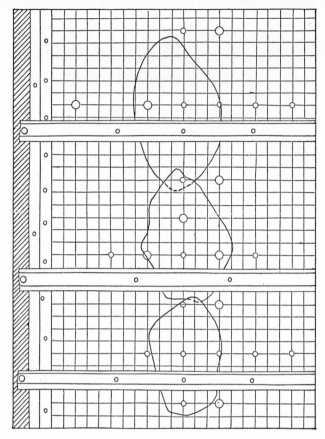

Figure 1. Traces of the outline of an isolated bubble rising through the two-dimensional bed.

an ozone-free "two-dimensional" fluidized bed of height
245 cm, width 56 cm and thickness 1 cm. The background
gas velocity was maintained at 5 to 30% in excess of that
required for incipient fluidization and the bubbles had to
be rather large (typically 20 cm in diameter) in order to
minimize bubble splitting. Bubble phase ozone concentrations were measured at different levels in the bed by means
of ultraviolet absorption. Bubble velocities, areas, and
perimeters were measured from ciné photographs of the
bubbles. The measurement technique was therefore complete-

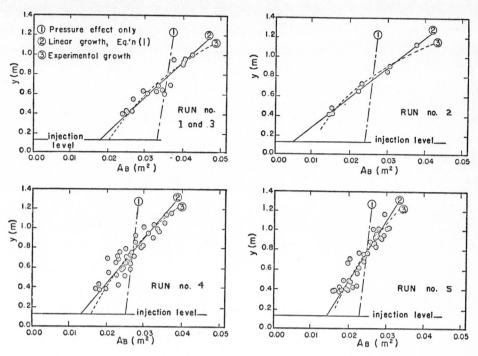

Figure 2. Growth of isolated bubbles as they rise through the bed and comparison with growth due to hydrostatic pressure effect only and linear growth.

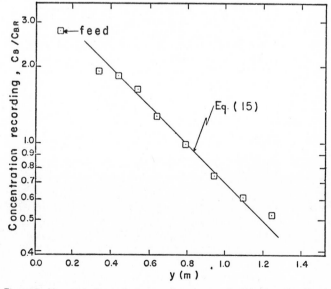

Figure 3. Mass transfer analysis assuming constant bubble size. Run No. 2.

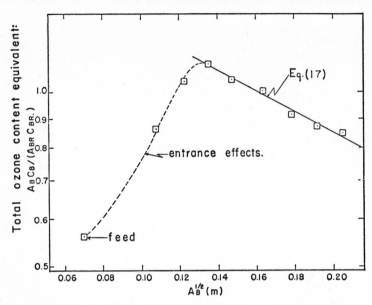

Figure 4. Mass transfer analysis with bubble growth accounted for. Run No. 2.

ly non-interfering. By combining the above measurements,
it was possible to calculate mass transfer coefficients
which were not subject to entrance effects.   Appropriate

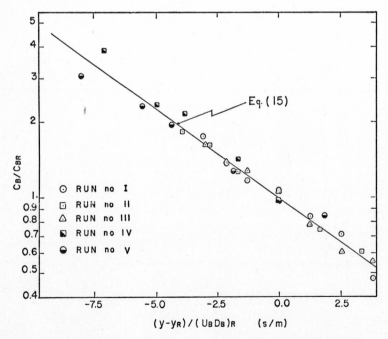

Figure 5. Mass transfer analysis for the data as a whole assuming constant bubble size.

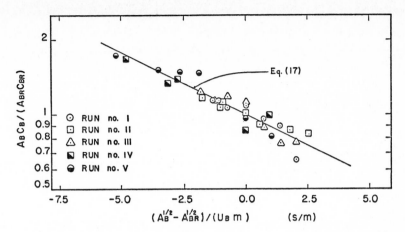

Figure 6. Mass transfer analysis for the data as a whole with bubble growth accounted for.

allowance was made for unsteady bubble growth which was
found to occur as bubbles rose freely through the bed.
(Typically bubbles doubled in volume in a distance of
about 80 cm.)   For the conditions employed in this work
(i.e. glass beads with U_{mf} = 5.3 cm/s, ϵ_{mf} = 0.46, bubbles
of area 200 to 400 cm^2 and velocity 40 to 60 cm/s), the
overall mass transfer coefficient was found to be 1.6 cm/s.
Porous alumina particles were added to the glass beads in
some of the experimental runs, but no effect of gas
adsorption on paticles could be detected in the observed
mass transfer rates for the conditions investigated.

SIGNIFICANCE

In most previous experimental studies of interphase mass transfer in fluidized beds, it has not been
possible to eliminate entrance effects.   Moreover, it has
generally been assumed that bubbles maintain constant size
during their ascent.   In the present study it was found
that failure to account for the observed unsteady growth

Table 1 Experimental mass transfer results

Run No.	U_B cm/s	D_{BR} cm	m cm	k_{GT} assuming constant D_B	k_{GT} allowing for bubble growth
I	55.0	21.8	2.8	4.2 cm/s	2.0 cm/s
II	48.4	18.8	3.3	4.0 "	1.1 "
III	55.0	21.8	2.8	4.2 "	1.9 "
IV	48.8	19.0	2.2	4.7 "	1.5 "
V	45.5	18.0	1.6	3.6 "	1.7 "
Overall				4.1 cm/s	1.6 cm/s

of bubbles would lead to an overestimate of the mass
transfer coefficient by a factor of 2.5, and this could
lead to serious errors in reactor modelling. Such growth
is almost impossible to detect in three-dimensional
columns unless X-rays are employed.

The amount of data obtained in the present study
is insufficient to allow discrimination between competing
mass transfer models from the point of view of proper
dependence on physical properties such as molecular diffusivity, bubble diameter, particle size, etc. The rates
of transfer measured fall between predictions of those
models where transfer is based purely on diffusion at the
cloud boundary and those which assume that the resistance
at the bubble interface is rate-controlling with diffusive
and convective (throughflow) effects assumed to be additive.

COMMENTS

The overall mass transfer rate obtained in the present investigation is consistent with a model which assumes convective transfer from bubbles as predicted by the hydrodynamic model of Murray and diffusive transfer to be obliterated by transpiration or by the convective component. However, other models could also be devised which would predict the experimental value of the mass transfer coefficient obtained in the present work.

Table 2   Predictions of mass transfer models

Reference	Basis of Model	Predicted k_{GT} cm/s
Higbie, Trans. AIChE 31, 365 (1935)	Penetration theory at cloud bndy; No resistance at bubble bndy	0.50
Partridge & Rowe, Trans. IChE, 44, 335 (1966)	Cloud acts like cylinder in liquid; No resistance at bubble bndy	0.20
Chiba & Kobayashi, Chem. Eng. Sci., 25, 1375 (1970)	Diffusion at cloud bndy; No resistance at bubble bndy	0.30
Kunii & Levenspiel, Fluidization Engng. Wiley, 1969	Penetration theory at cloud bndy; Thruflow & mol. diffn. at bubble bndy	0.51
Davidson & Harrison, Fluidized Particles, Cambridge, 1963	No resistance at cloud bndy; Thruflow & mol. diffn. at bubble bndy	4.1
Hovmand & Davidson, Trans. IChE, 46, 190 (1968)	No resistance at cloud bndy; Thruflow & mol. diffn. without interaction at bubble bndy	5.0
This work	No resistance at cloud bndy; Thruflow at bubble bndy as predicted by Murray; Mol. diffn. obliterated by thru-flow	1.7

Further experiments are required to allow the best model
to be chosen unambiguously.

The measured mass transfer coefficient for
single isolated bubbles is in good agreement with the
range of values required to optimize the best chemical
reactor model for parallel chemical reaction studies
carried out using ozone decomposition in the same bed
with almost identical particles. This close agreement
may be fortuitous, but the agreement implies that the
transfer rates for freely bubbling beds are at least of
the same order of magnitude as for isolated bubbles.

APPENDIX

Equations Referred to in the Figures

Eqn. (1): - Linear bubble growth

$$A_B = my + n$$

where A_B = bubble area

y = height above distributor

m, n = constants

Eqn. (15): - Concentration variation assuming constant bubble size

$$\ln \frac{C_B}{C_{BR}} = - \frac{4k_{GT}}{U_B D_{BR}} (y - y_R)$$

where C_B = tracer concentration in bubble

k_{GT} = mass transfer coefficient

U_B = bubble velocity

D_B = bubble diameter

Subscript R refers to reference level

Eqn. (17): - <u>Concentration variation with linear bubble growth</u>

$$\ln \frac{A_B C_B}{(A_B C_B)_R} = - \frac{4k_{GT} \sqrt{\pi}}{U_B m} \left[A_B^{\frac{1}{2}} - A_{BR}^{\frac{1}{2}} \right]$$

MASS TRANSFER FROM A GRID JET IN A LARGE GAS FLUIDIZED BED

L. A. BEHIE, M. A. BERGOUGNOU AND C. G. J. BAKER

NOMENCLATURE

A	Cross-sectional area of fluid bed
a_b	Transfer areas from bubbles to emulsion per unit volume.
a_j	Transfer area from grid jets to emulsion per unit volume.
C	Gas concentration in grid jet.
C_a	Value of C on grid jet axis.
C_b	Value of C in emulsion around grid jets.
C_o	Value of C on jet axis at $x = 0$.
d_b	Bubble diameter.
D_o	Grid nozzle diameter.
FCC	Fluidized cracking catalyst.
Fr	Froude number.
G	Gas mass flux through reactor.
g	Gravitational constant.
h	Depth of jet penetration.
H	Fluidized bed height.
k_b	Mass transfer coefficient, bubbles to emulsion ($kg/m^2 \cdot s$).
k_j	Mass transfer coefficient, grid jets to emulsion ($kg/m^2 \cdot s$).
L_e	Catalyst loading in emulsion (m^3/m^3)
$\frac{M}{Mo}$	Normalized axial momentum - see reference 3.
No	Nozzle number (x/D_o)
Q	Gas mass flow through reactor.
Q_b	Mass flow through reactor as bubbles.
Q_e	Mass flow through emulsion.

r Radius.

R Rate of reaction per unit weight of catalyst (kg/kg·S).

Re Reynolds number $(D_o u_o / \phi)$

t Mass dissipation time, ms.

t_o x_o / u_o

u_{mf} Minimum fluidizing velocity (1.53 cm/s)

u_o Grid jet velocity.

u_s Superficial gas velocity through bed.

V_e Volume of emulsion

x Distance from Grid.

x_o See equation 5.

Greek Symbols

α Dimensionless group $\dfrac{k_j a_j A h}{Q}$

β Dimensionless group $\dfrac{k_b a_b A h}{Q_b}$

ϕ Kinematic gas viscosity

γ Dimensionless group $\alpha + \beta\left(\dfrac{H}{h} - 1\right)$

δ Fraction of flow through emulsion Q_e / Q $(\delta \ll 1)$

ε_b Fraction of reactor volume above h occupied by bubbles $(\varepsilon_b = 0.25)$.

ε_{mf} Void fraction of emulsion at incipient fluidization $(\varepsilon_{mf} = 0.5)$.

η Fractional conversion.

μ Dimensionless rate constant, $\dfrac{\nu \, L_e \rho_s V_e}{Q}$

ν Rate constant (kg/S·kg catalyst)..

$\nu^{'}$ Effective rate constant, $\nu \, L_e \, \rho_s$ (kg/S·m^3 emulsion).

ρ_o Density of jet gas at nozzle mouth.

ρ_s Density of catalyst particle.

ΔC $(C - C_b)$

ΔC_a $(C_a - C_b)$

ΔC_o $(C_o - C_b)$

$\Delta T / \Delta T_o$ Normalized axial temperature - see reference 5.

ABSTRACT

Mass transfer from a vertical grid jet within a 61.0 cm diameter and 122.0 cm deep fluidized bed of cracking catalyst was studied. A five mole percent mixture of CO_2 in air issued from a test nozzle having a diameter varying from 6.4 mm to 19.1 mm and a nozzle velocity from 15.2 to 91.5 m/s. Axial concentration data hve been related to a Froude, a nozzle, and a Reynolds number in the following correlation:

$$\ln \, (\Delta C / \Delta C_o) \; = \; -1.92 \; Fr^{-0.504} \; No^{0.905} \; Re^{0.068}$$

In addition, a simple model gave mass transfer coeffic- ients for grid jets (k_j) in the range of 2500 to 7000 kg/m^2·s. When this simple model was combined with a

well known model for a bubbling fluid bed, high conversions were predicted in the grid region for fast
reactions.

SCOPE

For the engineer involved in the design of commercial fluidized
bed reactors, it has become increasingly evident that their
performance is affected by the arrangement for introducing the
reactant gases[1]. It is not uncommon to find grid characteristics reflected all the way up to the bed surface when the
solids do not have ideal fluidizing properties. A poor grid
not only affects conversion, but also leads to excessive bed
carryover, poor mixing, undersirable temperature gradients,
mechanical vibrations, and a host of similar problems.

Moreover, in the case of fast reactions, results from industrial
pilot plant experiments by Botton[2] have shown that the grid
region can account for a large fraction of the total conversion.
A pseudo-first order reaction was carried out in a 45.7 cm
diameter reactor with various sized catalysts. In a typical
experiment with a superficial column velocity of 18.3 cm/s,
the conversion was over 98 percent with a bed height of 274.4 cm
and 80 percent when half the solids were removed from the column.
Consequently, there is a great incentive to try and characterize
the grid region by measuring mass transfer rates within it.

OBJECTIVES

To date, the grid region of a gas fluidized bed has not been
modelled successfully. In fact, in most bed models the grid
region is ignored. The main objectives of this work were to
measure mass transfer rates from grid jets having the same
velocities and diameters as those found in industrial scale
reactors and to model the grid region such that it could be
used in conjunction with bubble models.

EXPERIMENTAL

The fluid bed column used in this study was of pilot plant scale being 61.0 cm in diameter and 7.93 m high. A complete description of the apparatus can be found elsewhere [5,6].

One of three brass test nozzles, each 38.1 cm long and having an inside diameter of either 6.1 mm, 12.7 mm or 19.1 mm, was placed on the axis of the column. Oil-free compressor air passed through one of three calibrated rotameters, a flexible plastic tube and then the test nozzle into the fluid bed of fresh cracking catalyst (FCC). The geometric mean diameter of the catalyst was 60 microns by weight. Auxiliary air from the compressor passed through a calibrated orifice plate in a 12.7 cm diameter line into the 61.0 cm diameter by 61.0 cm high acrylic windbox below the grid. The air then went through a 6.4 mm thick perforated grid around the test nozzle. The grid had 232 holes each of diamater 1.6 mm on an equilateral triangle pattern. The auxiliary gas served two purposes. First it kept the bed around the test nozzle well fluidized by contributing 1.27 cm/s to the column superficial velocity. Secondly, it acted as a diluant for the system so that steady-state operation could be achieved.

The probe tip used for sampling the jet gas consisted of a piece of 1.6 mm diameter by 11.4 cm long brass tubing which was fastened inside a brass tube holder. A 9.58 mm long fibrous steel packing filter was fitted inside flush with the end of the streamlined nose. The probe tip screwed into a 12.7 mm diameter by 94.0 cm long hollow stainless steel probe which was mounted in a probe support mechanism [6]. A 1.6 mm diameter by 4.57 m long plastic tube connected the probe tip to a seven port stainless steel continuous gas sampling valve mounted on a Varian Aerograph (series 150) gas chromotograph equipped with a thermal conductivity detector. The detailed experimental procedure can be found in reference (6).

RESULTS AND DISCUSSION

The mass transfer study described in this paper is a contin-
uation of studies on momentum[3,4] and heat[5] dissipation
from grid jets in a gas fluidized bed. In the former study,
axial and radial profiles of longitudinal momentum were
measured for grid jets issuing from single and milti-holed
grids in a 27.9 cm diameter column. In the latter study,
axial and radial temperature profiles for single jets issuing
into a prefluidized bed were measured in a 61.0 cm diameter bed.

Axial Mass Dissipation

In Figure 1, a comparison is made between the axial concentration
profiles for a 30.5 m/s grid jet issuing from the 19.1 mm nozzle

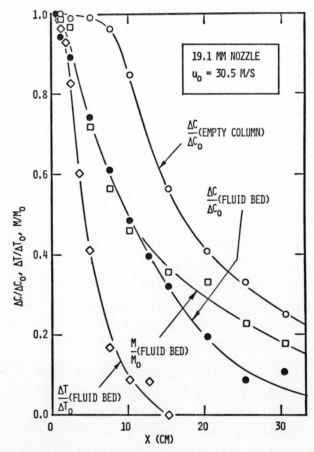

Figure 1. Comparison of axial profiles of concentration, temperature, and momentum for a low velocity grid jet issuing into a fluid bed of FCC.

into the fluid bed of FCC and into the empty 61.0 cm column.   In
both cases, the auxiliary gas velocity was maintained at 15.2
cm/s. As in all experiments after steady state operation was
achieved, the concentration of CO_2 measured in the fluid bed C_b
varied less than five percent.   Thus, the axial concentration C
at a point in the grid jet was expressed in terms of an elevated
concentration ΔC, where

$$\Delta C = C - C_b \dots\dots\dots\dots\dots\dots\dots\dots\dots\dots\dots (1)$$

For the jet conditions of Figure 1, the axial concentration
profile in the fluid bed lies well below that of the empty column.
However, this was not the case in general.   For smaller nozzles
or higher jet velocities, the axial concentration profiles moved
closer rogether as shown in Figure 2.   For the 12.7 mm nozzle
with an initial velocity u_o of 76.2 m/s, the axial concentration
profiles in the empty column and in the fluid bed show a characteristic cross-over. Although the CO_2 dissipates faster for $0 < x < 7.6$
cm, the concentration profile lay slightly below that in the fluid
bed when $x > 12.7$ cm.   Also shown in Figure 1 are axial profiles
of temperature[5] and longitudinal momentum[3,4] measured in a
fluid bed of FCC.   The temperature profile lies much below both
the concentration, and momentum profiles which were indistinguish-
able up to $x = 15.0$ cm .   For $x > 15.0$ cm, the momentum profile
was higher than that for the concentration.   For higher velocity
jets, this trend changed as illustrated in Figure 2.   For
$u_o > 61.0$ m/s, the only difference between the axial temperature
and momentum prfiles in the fluid bed was that the temperature
profile went to zero, whereas the momentum profile leveled off to
the value of the rising bubble swarm.   As can be seen in Figures
1 and 2, there is a significant reduction in the axial mass concentration of CO_2 in the grid region.

Correlation of Results

A statistical analysis was performed on the axial concentration
data for initial jet velocities u_o between 15.2 and 91.5 m/s,
nozzle diameters D_o between 6.35 mm and 19.1 mm and axial distances

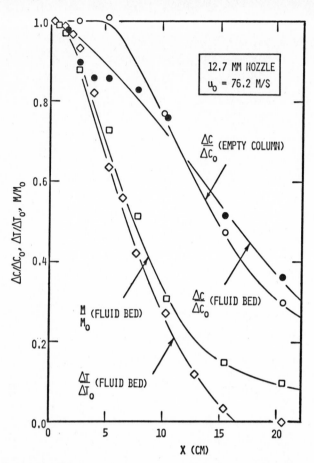

Figure 2. Comparison of axial profiles of concentration, temperature, and momentum for a high velocity grid jet issuing into a fluid bed of FCC.

downstream x between 0 and 30.5 cm. The following relationship resulted:

$$ln\ (\Delta C/\Delta C_o) = -6.38\ u_o^{-0.939}\ D_o^{-0.837}\ x^{1.41} \ldots\ldots\ldots (2)$$

where u_o is in m/s, D_o in mm and x in cm. The standard deviation between experimental values of $\Delta C/\Delta C_o$ and those calculated by equation (2) is 0.046 for 173 data points.

In Figure 3, the effect of initial jet velocity on the axial mass dissipation for a grid jet in the fluid bed is shown for the 12.7 mm nozzle. The curves in the plot are those given by equation 2

and illustrate the goodness of fit. Equation 2 can be expressed
in dimensionless form as follows;

$$ln \ (\Delta C/\Delta C'_o) \ = -1.92 \ \bar{Fr}^{0.504} \ No^{0.905} \ Re^{0.0680} \ \ldots\ldots\ldots\ldots \ (3)$$

where $Fr = (u_o{}^2/gx)$, $No = (x/D_o)$ and $Re = (D_o u_o/\phi)$ are the Froude,
nozzle and Reynolds numbers respectively. The kinematic velocity ϕ
was not varied in this study.   In addition, a change in the
auxiliary gas velocity between 7.6 and 22.9 cm/s had no effect
on the axial profiles.

Grid Region Model

Mass transfer from a grid jet was analyzed by a simple model
similar to that used in the heat transfer study[5]. The model

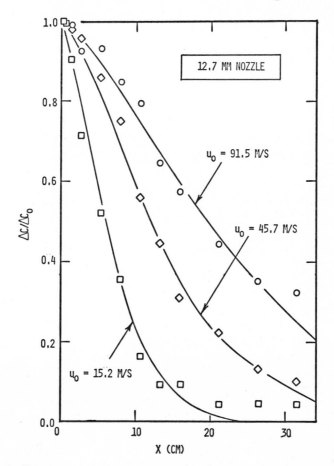

Figure 3. Effect of initial jet velocity u_0 on axial mass dissipation for a grid jet in FCC.

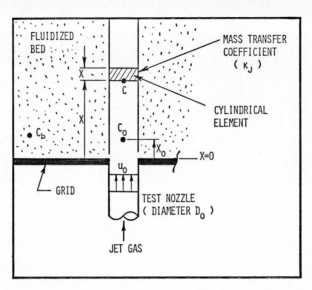

Figure 4. Simple mass transfer model for grid jet issuing into a fluidized bed.

which is illustrated in Figure 4, is based on the following
assumptions:

1.   The gas jet maintains its cylindrical form after its entry
      into the fluidized bed.

2.   The mean concentration of CO_2 in a jet cross-section is
      given by the concentration measured at the centre of the
      jet.

A mass transfer coefficient k_j between the gas jet and the
fluidized bed may be defined by means of a mass balance on the
cylindrical element as shown in Figure 4.   The element is a
distance x downstream from the nozzle mouth and has a diameter
and a thickness of D_o and Δx respectively.   At steady-state,
the mass flow into the element is equal to the mass flow out.
The following equation results:

$$ln\ (\Delta C/\Delta C_o)\ =\ -(4k_j/D_oG)\ (x-x_o),\ x \geqslant x_o \ \ldots\ldots\ldots\ldots (5)$$

The usefulness of the model comes from the fact that a straight
line results when the experimental data are plotted as $ln(\Delta C/\Delta C_o)$

against x This means that the mass transfer coefficient is independent of x for $x \geqslant x_o$ and can be evaluated from the slope
of the line.   Because the straight line does not go through the
point $(\Delta C/\Delta C_o = 1.0, x = 0)$, the model indicates that the jet
concentration remains unchanged from the nozzle mouth up to a
distance x_o. Figure 5 shows a plot of $ln(\Delta C/\Delta C_o)$ against x for the
19.1 mm nozzle and three jet velocities. Table 1 lists the values of
the mass transfer coefficients together with values of x_o for all
experiments.

It is of interest to develop a relationship which would give the
time required to lower the jet concentration to a certain percent-
age of its initial value.   Equation (5) can be manipulated to yield:

$$ln \ (\Delta C/\Delta C_o) = -(4 \ k_j/D_o\rho_o) \ (t-t_o), \ t > t_o \ \dots\dots\dots (6)$$

where
$$t_o = x_o/u_o \ \dots\dots\dots\dots\dots\dots (7)$$

and ρ_o is the density of the gas exiting in the nozzle.   If the mass
dissipation time is defined as the time required for the elevated

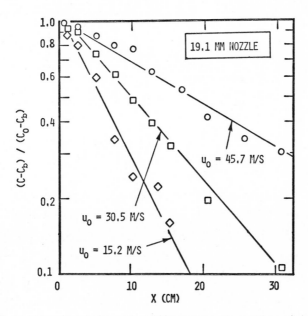

Figure 5. Effect of initial jet velocity on the grid jet mass transfer coefficient (k_j)—see Table 1.

Table 1   Mass transfer coefficients (k_j) for grid jets in a fluid bed of FCC

D_o (mm) → / u_o m/s ↓	6.4	12.7	19.1
15.2	$k_j = 2520$ $x_o = 7.6$	$k_j = 3610$ $x_o = 5.1$	$k_j = 4000$ $x_o = 10.2$
30.5	$k_j = 3720$ $x_o = 27.9$	$k_j = 3840$ $x_o = 12.7$	$k_j = 6460$ $x_o = 7.62$
45.7	$k_j = 5410$ $x_o = 43.2$	$k_j = 6010$ $x_o = 40.6$	$k_j = 3840$ $x_o = 5.1$
61.0	$k_j = 5790$ $x_o = 40.6$	$k_j = 4870$ $x_o = 30.5$	—
76.2	$k_j = 5670$ $x_o = 43.2$	$k_j = 5910$ $x_o = 38.1$	—
91.5	$k_j = 6950$ $x_o = 55.9$	$k_j = 6060$ $x_o = 40.6$	—

x_o is in mm, k_j in $k_g/m^2 \cdot hr$.

concentration ΔC_o, to reduce to 10% of its initial value ΔC_o, then the dissipation times are those given in Table 2.

Behie and Kehoe [7] have taken this development a step farther by generating volumetric mass transfer coefficients ($k_i \cdot a_j$) for the grid region and combined the grid region model with the simple bubbling bed model of Orcutt [7] to yield what is termed a GRID MODEL. It as assumed the reaction was first order and there were no diffusional limitations in the emulsion so that the rate was R=vy.   The final equation that resulted was -

$$\eta = \frac{\mu \, (1 + e^{-\gamma})}{\mu + 1 - e^{-\gamma}} \qquad \dots\dots\dots\dots\dots\dots\dots (8)$$

where η was fractional conversion, $\eta = \dfrac{\nu L_e \rho_s V_e}{Q}$ a dimensionless rate constant and $\gamma = \alpha + \beta (H/h-1)$ a dimensionless group. Values of jet penetration h and the diameters of bubbles d_b breaking away from the jets have been calculated by the empirical equations of Basov[9]. Values of volumetric mass transfer coefficients in the bubble regime $k_b \cdot a_b$ were calculated from the euqations of Kunii and Levenspiel[10]. Figure 6 shows the effect on conversion for both a fast and slow reaction in a fluidized bed. The effective rate constant $\nu' = \nu L_e \rho_s$ were taken from reported data. For the fast reaction, the value $\nu' = 9.63$ kg/m^3·S was taken from the data of Cooke et al[11] for a fluidized bed carbonizer. For the slow reaction, $\nu' = 9.63 \times 10^{-2}$ kg/m^3·s is within the range reported by Hovmand et al[12]. The curves are computed for a grid with 12.7 mm holes and a jet velocity of 45.7 m/s. The predictions of the Bubble Model and Grid Model (i.e. combined models) are shown for each reaction. For the fast reaction,

Table 2   Mass dissipation times for grid jets in a fluid bed of FCC

Nozzle (mm)	u_o (m/s)	C_o (mole %)	Concentration after mass dissipation time (mole %)	Mass dissipation time (ms)
19.1	45.7	3.08	0.69	12.3
12.7	61.0	4.87	0.86	6.98
12.7	15.2	4.60	0.60	9.06
6.4	45.7	4.60	0.58	3.86
6.4	91.5	4.94	0.67	2.88

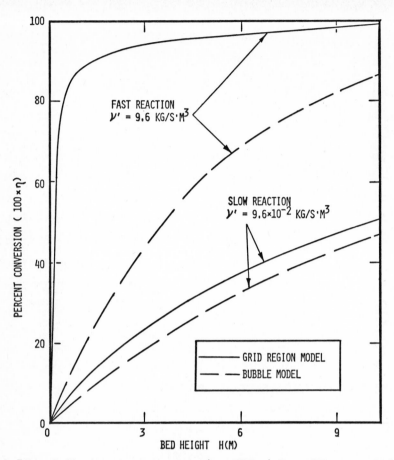

Figure 6. Effect of grid region on reactor conversion ($u_0 = 45.7$ m/s, $D_0 = 12.7$ mm, $u_s = 61.0$ cm/s).

the two predictions are significantly different. With a bed
0.6 m deep, the Grid Model predicts a conversion of 84%, while
the Bubble Model predicts only 11%. On the other hand, for the
slow reaction the difference between the two models is much less
significant, since mass transfer does not limit the rate severely.

Also, the model is consistent with general experimental observations in that it predicts an increased conversion by decreasing
grid hole size or increasing the number of holes.

Radial Concentration Profiles

Figure 7 shows a comparison betwen radial mass concentration distributions measured in the fluidized bed of FCC and in the empty

61 cm column for x = 5.1 cm.   The normalized radial concentration
for a given x is expressed as:

$$\Delta C / \Delta C_a = (C-C_b)/(C_a-C_b) \dots\dots\dots\dots\dots\dots\dots (8)$$

where C and C_a are respectively the concentrations of CO_2 measured
at the radius and on the jet axis at the same distance downstream
of the nozzle.   The axial concentration was 4.1 mole percent in
the fluid bed and 4.7 mole percent in the empty column.   As can be
seen, the profiles are reasonably symmetrical.   It is of interest
to note that the mass spreads out much farther laterally in the

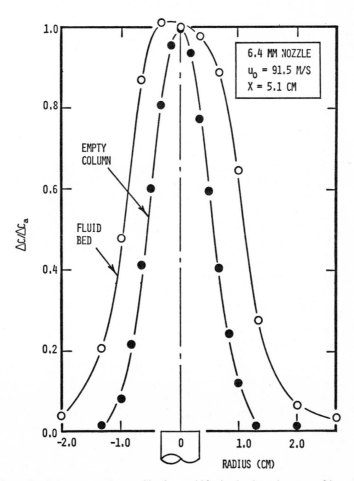

Figure 7. Comparison of radial concentration profiles for a grid jet issuing into the empty 61 cm column and into a
fluidized bed of FCC.

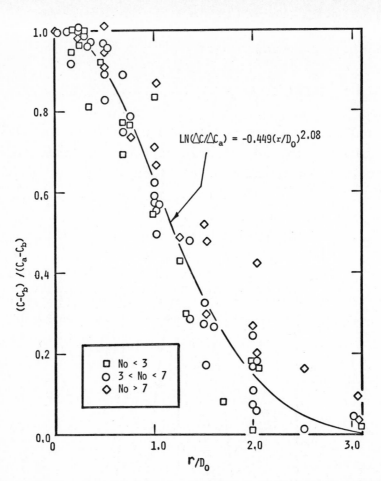

Figure 8. Radial concentration distributions for a grid jet issuing into a fluid bed of FCC.

fluid bed. At $r = 1.27$ cm, the profile in the empty column has dropped to zero, whereas that in the fluid bed is over 20% of the axial value.

Data for twelve radial profiles are combined in Figure 8. Details of the experimental conditions can be found elsewhere[6]. In general, three grid jet regions were observed. First for nozzle numbers $No < 3$, the radial profiles were very peaked. Secondly, for $3 < No < 7$ was a zone in which the grid jets disspiated most of their mass. The radial profiles in Figure 7 which are in this region can be described to a good approximation by the equation:

$$ln\ (\Delta C/\Delta C_a) = -0.449\ (r/D_o)^{2.08} \quad \dots\dots\dots\dots\dots (9)$$

Thirdly, for $No > 7$, the radial profiles are flatter.

CONCLUSIONS AND SIGNIFICANCE

For a vertical grid jet of air containing 5 mole percent of CO_2 as tracer and issuing into a fluidized bed of cracking catalyst, the axial concentration profiles were described by a statistically fitted curve of the form:

$$ln \ (\Delta C/\Delta C_o) = -1.92 \ Fr^{-0.504} \ No^{0.905} \ Re^{0.0680}$$

Within the jet mass dissipation zone, $3 < No < 7$, the radial concentration profiles were described by the equation:

$$ln \ (\Delta C/\Delta C_a) = -0.449 \ (r/D_o)^{2.08}$$

A simple grid mass dissipation model gave jet mass transfer coefficients between 2500 and 7000 $kg/m^2 \cdot hr$ and mass dissipation times between 2 and 13 ms.

It is apparent that for fast reactions, the grid arrangement can have a significant effect on conversion. The critical factor in determining the conversion in the grid region is the value α. If α is greater than 3, the conversion in the grid region is close to its maximum value and little improvement can be effected by changing either the size or number of holes. On the other hand, if α is much less than 3, improvements in conversion may be obtained by modifying the grid.

ACKNOWLEDGEMENTS

The authors would like to thank the National Research Council of Canada for a grant in aid of research to Dr. M.A. Bergougnou and for a post-graduate scholarship to Dr. L.A. Bhie.

REFERENCES

1.   Zenz, F.A., Tripartite Chemical Eng. Conf., Montreal (Sept. 1968).

2.   Botton, R., Chem. Eng. Progress, Symp. Series, 66 (101), 8 (1970).

3.   Behie, L.A., M.A. Bergougnou,C.G.J. Baker, and W. Bulani,
        Can. J. Chem. Eng., 48, 158 (1970).

4.   Behie, L.A., M.A. Bergougnou,C.G.J. Baker, and T.E. Base,
        Can. J. Chem. Eng., 49, 557 (1971).

5.   Behie, L.A., M.A. Bergougnou , and C.G.J. Baker,
        Can. J. Chem. Eng., (accepted for publication).

6.   Behie, L.A.,Ph.D. Thesis, The University of Western
        Ontario, Canada. (1972).

7.   Behie, L.A., and P. Kehoe, AICHe J., 19 (5), 1070 (1973).

8.   Orcutt, J.C., Ph.D. dissertation, University of Delaware,
        Newark. (1960).

9.   Basov, V.A., V.I. Markhevka, T.Kh. Melik-Akhanzarov and
        D.I. Orocho, Int. Chem., Eng., 9, 263 (1969)

10.   Kunii, D. and O. Levenspiel, Fluidization Engineering,
        Wiley (1969).

11.   Cooke, M.J., W. Harris, J. Highley and D.F. Williams,
        Tripartite Chem. Eng. Conf., Montreal (Sept. 1968).

12.   Hovmand, S., W. Freedmand and J.F. Davidson, Trans., Instn.
        Chem. Engs., 49, 149 (1971).

MASS TRANSFER FROM SINGLE RISING BUBBLES TO THE DENSE PHASE IN THREE DIMENSIONAL FLUIDIZED BEDS

K. RIETEMA AND J. HOEBINK

NOMENCLATURE

C_b bubble concentration

C_d dense phase concentration

d_e equivalent bubble diameter

d_o mean vertical bubble size

h height

K mass transfer coefficient

S_b specific surface

U_b bubble velocity

V_b bubble volume

$\emptyset$ angle of flattening of spherical cap bubble

SCOPE

Published work on mass transfer between bubbles and dense phase in fluidized beds is characterized by a great variety of models and correlations, which all have a very limited validity. Most experimental results are far from complete, since never all important influences have been considered in the same experiments. Most striking fact which follows from the literature in this field is the experimentally observed increase of mass transfer coefficient with increasing bubble diameter whereas most theories predict the opposite influence.

INTRODUCTION

The best-known method to determine mass transfer coefficients from single bubbles is to measure the bubble con-

centration as a function of height in the bed after injection of the bubble in a homogeneously fluidized bed.
The overall mass transfer coefficient can then be determined from:

$$- U_b \frac{d(V_b C_b)}{dh} = K.S.V_b (C_b - C_d)$$

with initial condition: h = 0, $C_b = C_{bo}$. This equation
may be considered as definition of the overall mass transfer coefficient h = 0 is an arbitrary reference height.
From the literature it follows which processes play a
role in the total mass transfer:
- diffusive and convective mass transfer via the cloud-bubble
boundary;
- diffusive transfer via the cloud - dense phase boundary;
- transfer via cloud shedding;
- transfer to solids in the cloud and solids raining
through a bubble in case they are porous or adsorbing.

All these factors are assumed here to be part of the overall mass transfer coefficient.
It follows also from the literature which factors have to
be considered in the experimental technique:
- experiments must include a method to select only bubbles
which are unsplitted
- since part of the injected gas escapes into the dense
phase, the bubble volume differs from the injected volume;
- the bubble volume changes with height, due to gas leakage
to the dense phase or gas extraction from the dense phase
in the lower region of the bed. This latter fact may show
an apparent increase of the mass transfer coëfficiënt;
- the bubble form differs widely from a spherical form,
and moreover changes its form continuously when rising up;
- any detection device must not disturb the bubble before
detection.
In the experiments to be described below all points as
mentioned above have been considered.

EXPERIMENTAL WORK

Experiments have been carried out in a 45 cm diameter and 95 cm high fluidized bed. Gas bubbles were injected in the bed center through a 0,8 cm diameter pipe, inserted horizontally in the bed at 3 cm above the metal distribution plate. Bubble detection took place with a small device placed vertically above the injection point that included two miniaturized capacitance probes and a capillary. With the capacitance probes local bubble velocity and vertical size were measured; the technique has been described by Werther and Molerus (1). With the capillary gas samples were sucked into a flame ionisation detector at reduced pressure which technique is taken from Drinkenburg and Rietema (2). Both probes and capillary could be inserted at various heights in the bed. Details of the equipment will be given in an appendix. Bubble concentration, velocity and vertical size have been measured at four heights in the bed. Entrance effects caused by bubble injection have been avoided by choosing a suitable referance height above the injection point. Unsplitted bubbles have been selected by comparing the local bubble velocity with the rising velocity calculated from the distance between injection and detection height, and the time between injection and detection moment. Since there is a strong spread in the measured vertical size of the bubbles, for each data point 50 unsplitted bubbles have been detected. Bubble size has been varied during experiments. Two powders have been used: glass beads and spent cracking catalyst, data of which are shown in table 1. The superficial gasvelocity was always = 0,25 cm/sec.

Table 1 Solids properties

	cracking catalyst	glass beads
mean particles size	66 μ	51 μ
particle density	1400 kg/m^3	2700 kg/m^3
incipient fluidization	0,20 cm/s	0,22 cm/s
bubble point	0,55 cm/s	0,45 cm/s

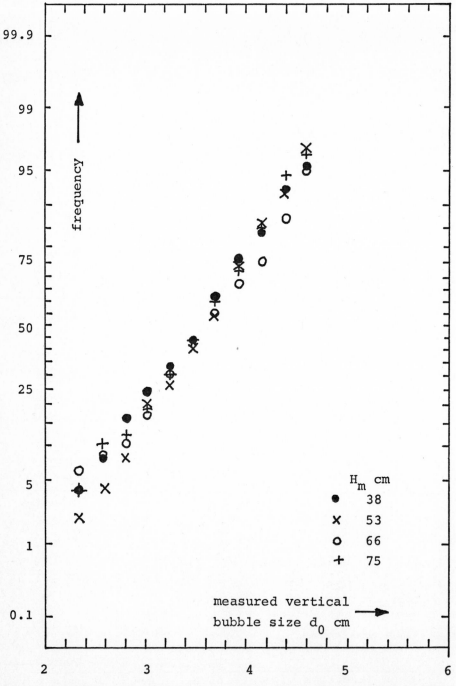

Figure 1. Distribution of vertical bubble sizes. H_m is distance between injection and detection point.

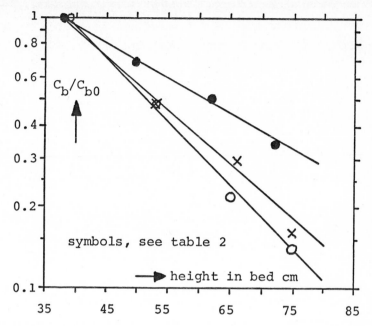

Figure 2. Concentration measurements with glass beads. Reference heights ● 38, X 38, ○ 39 cm.

RESULTS

Figure 1 shows the distribution of measured vertical bubble sizes, when a constant bubble volume is injected in
a bed of glass beads; plots are given for different heights
in the bed on cumulative normal probability paper. The
measured size appears to be normally distributed. In prin-
cipe spread may be caused by two effects. Firstly the bubble form changes continuously when rising up in the bed.
Secondly the bubble may carry out a zig-zag movement a-
round a vertical axis. As mean and spread are constant
over the bed height, it is concluded that bubbles rise
strictly vertically, that no preferential dense phase flow
occurs in the bed, and that no volume changes take place.

Table 2 Glass beads

d_e cm	d_o cm	U_b cm/s	S_b 1/cm	$\emptyset$ rad	K cm/s	
3,82	2,53	39,0	1,79	2,86	0,65	●
4,96	3,55	44,4	1,36	2,73	1,50	X
6,68	4,68	51,6	1,01	2,73	2,82	○

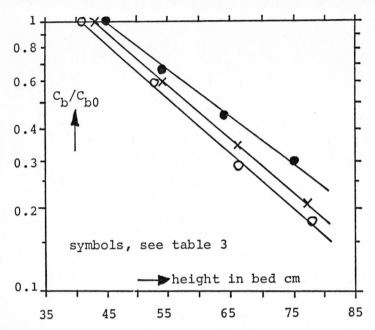

Figure 3. Concentration measurements with spent cracking catalyst. Reference heights ● 45, X 43, ○ 41 cm.

Results of concentration measurements are presented in
figure 2.
The mass transfer coefficient has been calculated from
the slope of the curves, assuming the bubble to have a
spherical cap form, while use is made of the relation
$U_b = 0,7 \sqrt{g\ d_e}$. Details are given in the appendix.
Table 2 summarises the results on glass beads. It is
seen that the mass transfer coefficient increases with
bubble diameter. Moreover the bubble form appears to correspond approximately to the top half of a sphere with
radius equal to the mean measured vertical bubble size.
Figure 3 and table 3 present similar results with spent
cracking catalyst. Although the concentration plot is
linear , as with glass beads, it has to be remarked here

Table 3 Spent cracking catalyst

d_e cm	d_o cm	U_b cm/s	S_b 1/cm	∅ rad	K cm/s	
4,69	2,80	43,2	1,58	3,34	1,07	●
5,90	3,50	48,4	1,31	3,57	1,70	X
7,64	4,99	55,2	0,90	2,90	2,82	○

that both vertical bubble size and bubble velocity have been observed to depend on height in the bed during the experiments with catalyst.

LITERATURE

1.. J. Werther, O. Molerus, Int.J.Multiphase flow 1, 103 and 123 (1973).
2. A.A.H. Drinkenburg, K. Rietema, Chem.Eng.Sc. 27, 1765 (1972) and 28, 259 (1973).

APPENDIX I

Calculation of the Mass Transfer Coefficient from Experimental Data

Assuming that a bubble has a spherical cap form (figure) the following quantities can be calculated:

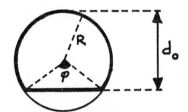

maximum vertical height of a bubble:

$$d_o = R(1 + \cos \phi/2) \tag{1}$$

bubble volume

$$V_b = \pi R^3 (\tfrac{2}{3} + \cos \phi/2 - \tfrac{1}{3} \cos^3 \phi/2) \tag{2}$$

bubble's specific volume

$$S_b = \frac{2}{R} \frac{1 + \cos \phi/2 + \tfrac{1}{2} \sin^2 \phi/2}{\tfrac{2}{3} + \cos \phi/2 - \tfrac{1}{3} \cos^3 \phi/2} \tag{3}$$

The maximum vertical height d_o of a bubble as well as its velocity U_b are determined experimentally. The bubble volume is calculated from the rising velocity according to $U_b = 0,71 \sqrt{g \, V_b^{1/3}}$.

R and ϕ are calculated from equation (1) and (2). From the slope of the concentration plot (see figure 2 and 3 of the paper), which equals $- \dfrac{KS_b}{U_b}$, the mass transfer coefficient K is determined.

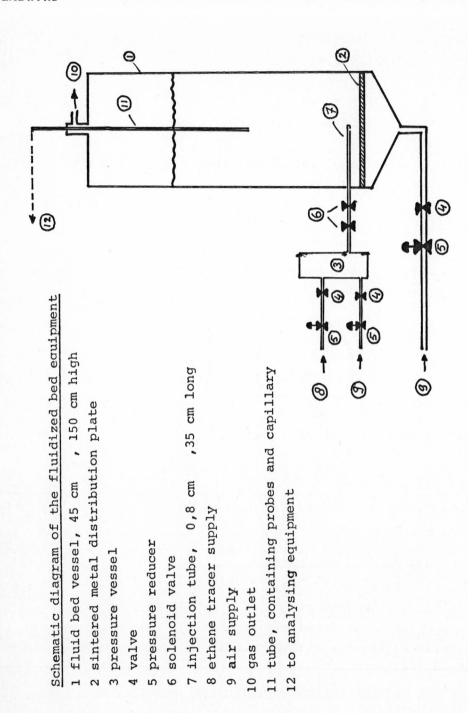

Schematic diagram of the fluidized bed equipment

1 fluid bed vessel, 45 cm , 150 cm high
2 sintered metal distribution plate
3 pressure vessel
4 valve
5 pressure reducer
6 solenoid valve
7 injection tube, 0,8 cm ,35 cm long
8 ethene tracer supply
9 air supply
10 gas outlet
11 tube, containing probes and capillary
12 to analysing equipment

APPENDIX IIb

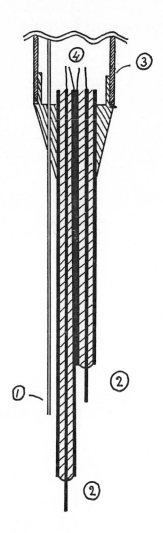

detection device

scale approximately   4:1

1. capillary

2 capacitance probe

3.tube

4.connections for coaxial cables

APPENDIX IIc

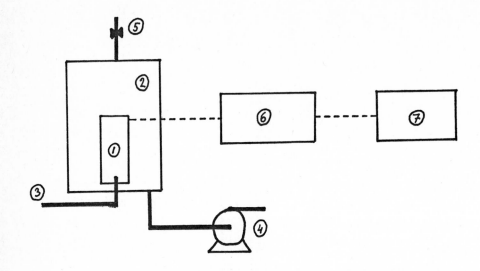

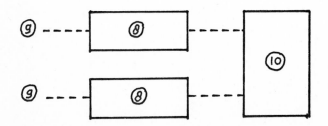

Block diagram of the analysing equipment
1. flame ionisation detector
2. vacuum house at 0,5 ata
3. capillary from detection device
4. pump
5. air leakage
6. flame convertor
7. recorder
8. capacitance meter
9. coaxial cable from capacitance probe
10. two channel storage oscilloscope

THE TRANSIENT PERIOD IN FLUIDIZED BEDS DURING TRANSFER FROM ONE TO ANOTHER FLUIDIZING GAS

K. RIETEMA AND J. HOEBINK

NOMENCLATURE

C	cohesion constant
c^o	gas concentration at atmospheric pressure
d_p	mean particle diameter
D_{eff}	effective diffusion coefficient
g	gravity acceleration
K	permeability
M	molecular weight
ΔP_D	pressure drop by diffusion
ΔP_F	pressure drop across fluidized bed
γ	constant
μ	gas viscosity
ρ_d	gas density
ρ	particle density
ω	molar fraction

INTRODUCTION

When in a free bubbling bed, fluidized with hydrogen, the gas flow is changed suddenly into an equal volumetric flow of nitrogen (without interrupting the flow itself), it will be observed that the fluidizing behaviour is disturbed for some time. In this transition period one finds that large particle conglomerates arise, and that strong chanelling occurs in the bed; the pressure drop across the bed decreases considerably and the free bubbling stops. After some time normal free bubbling behaviour and pressure drop are restored again.

This phenomenon is observed in a two dimensional bed filled with porous fresh cracking catalyst, but also in a bed
filled with solid glass beads of the same diameter; the
intensity of the phenomenon is the same in both cases.
If the flow was changed from nitrogen towards hydrogen, no
obstruction of fluidization was found.

THEORETICAL APPROACH

The phenomenon cannot be explained by the fact that one of
the gases is adsorbed more strongly than the other, since
in that case the intensity would be much more violent with
the catalyst, which has a much larger specific surface area
than the glass beads. In this paper two different explanations are presented, which both play an important role.
During the transfer period all gas A, present before the
transition, must be replaced by the new gas B. The first
explanation was suggested by Dr. C. van Heerden and starts
from the fact that immediately after the transition high
concentration gradients will arise in the lower part of the
bed. Because of solids movement and small disturbances the
concentration profile will not be flat, but horizontal
concentration gradients may occur locally, which cause a
diffusion process. If the molecular weight of gas B is
higher than that of gas A, this diffusion process corresponds with a net mass flow, which can be maintained only
when the concentration gradient is accompanied with a
pressure gradient. In regions where gas A is dominating,
the pressure will be decreased therefore, which may cause
compression of the dense phase and hence decreases the
mobility of this phase.
The second theory starts from the fact that gases of different viscosity show different expansions of the bed.
The bed porosity at the bubble point as well as the porosity of the dense phase in a free bubbling bed both de-
crease when the fluidization number $N_F = \rho_d^3 d_p^4 g^2 / (\mu^2 \gamma C)$ increases; this has been demonstrated by Oltrogge (1), and

by Rietema and Mutsers (2). A low dense phase porosity
will be coupled with a high apparent "viscosity" of the
dense phase as a whole. Now the phenomenon during the
transient period can be explained by analogy with the
well-known fingering effect, which occurs when a liquid
of high viscosity (comparable with a dense phase of low
porosity) is driven out of a medium with uniform permeability by a liquid with low viscosity (dense phase with
high porosity).

DIFFUSION EXPERIMENTS

The pressure drop ΔP_{ID}, due to diffusion, has been measured in a bed of 4 cm diameter, filled with 203 grams solid
glass beads with mean diameter of 38 μ. The whole vessel
was filled with gas A by fluidizing the solids. After
that the bed was vibrated to a height of 10,6 cm, corresponding with a bed porosity of 0,36. The flow of gas A
was stopped completely when a flow of gas B was installed
through a side tube just above the bed level. Because of
the equimolar diffusion process that starts now a pressure
drop across the bed arises when the molecular weights of
gas A and B are different. As is shown in the appendix
this pressure drop equals

$$\Delta P_{ID} = \frac{ID_{eff}}{K} (M_A - M_B) \ c^o \int_{,}^{o} \frac{\mu}{\rho} \ dw$$

The experimentally measured pressure drop reached a maximum in about 5 seconds which lasted several minutes; af-
terwards the pressure drop decreases to zero very slowly.
Results for several combinations of gases are presented
in table 1.

FLUIDIZATION EQUIPMENT

The transition period in a fluidized bed is assumed to be
characterized by the pressure drop across the bed ΔP_F as
a function of time after the moment of transfer.

Table 1

gas A	gas B	ΔP_{ID} theoretical mm w.c.	ΔP_{ID} experimental mm w.c.	max. change in ΔP_F mm w.c.	μ_A/μ_B	M_A/M_B
H_2	Ne	-35,5	-46,5	235	0,281	0,10
H_2	N_2	-20,2	-27,0	125	0,503	0,071
H_2	CO_2	-19,0	-24,3	110	0,607	0,045
C_3H_8	CO_2	0	- 0,5	50	0,559	1
H_2	C_3H_8	- 8,6	-10,0	30	1,086	0,045

These curves have been measured in a two dimensional bed
of width 26 cm and depth 1 cm. The mass of solid glass
beads (mean diameter 38µ) was 2439 grams, corresponding
with a quiescent height of the bed of 59 cm. Gases A and
B each were fed from gas cylinders and passed a pressure
reducer, a valve and a rotameter. By means of a three-
way valve one of them was fed to the bed. After adjusting
both flow rates to 3,0 cm/s, gas A flowed through the bed
for 10 minutes before transfer. Pressure drop changes and

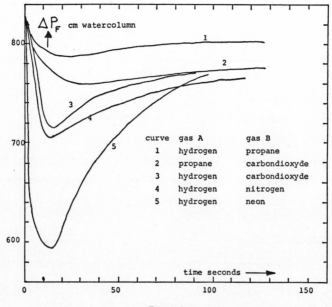

Figure 1

time were measured by videorecording a U-tube manometer
and a digital timecounter.
Results for several combinations of gases are given in
figure 1. Table 1 shows the maximum change in pressure
drop ΔP_F (experimental), and the viscosity and molecular
weight ratios.

CONCLUSIONS

When one compares the pairs of gases propane/carbondioxyde
and hydrogen/propane, it is seen that the transient phenomenon occurs when the molecular weights of gas A and B
are the same, as well as when the viscosities are the
same. So both explanations described in this paper play a
role. From the data it follows also that both effects may
enhance each other in certain situations, whereas in
other cases they counteract with each other.

LITERATURE

1. R.D.Oltrogge, thesis, University of Michigan, 1972
2. K.Rietema, S.M.P.Mutsers, Proc.Int.Symp."Fluidization
   and its applications", Toulouse, 1973, page 28.

APPENDIX I

Pressure Drop Across a Packed Bed, Due to Diffusion

Suppose a semi-infinite packed bed in the region x>0. At time t=0
the bed is filled with gas A at a concentration c^o; the region x<0
contains gas B only at the same concentration c^o. At times t>0 the
concentration of gas B in the region x<0 is kept constant at c^o.
An equimolar diffusion proces takes place after t=0. The concentration profile of gas B in the bed is given by:

$$\omega = \frac{c_B}{c^o} = 1 - \frac{2}{\sqrt{\pi}} \int_0^p e^{-p^2} \, dp, \quad p = \frac{x}{2\sqrt{\mathbb{D}_{eff} \cdot t}}$$

where $\mathbb{D}_{eff}$ is the effective diffusion coefficient of gas B in the
bed.
From the profile the molar flux N_B of gas B as well as the nett mass
flux Φ_t in the bed can be calculated:

$$N_B = - \mathbb{D}_{eff} \cdot \frac{\partial C_B}{\partial x} = c^o \, e^{-p^2} \sqrt{\frac{\mathbb{D}_{eff}}{\pi \cdot t}}$$

$$\mathbb{I}_t = (M_B - M_A) \, N_B = (M_B - M_A) \, c^o \, e^{-p^2} \sqrt{\frac{\mathbb{D}_{eff}}{\pi \, t}}$$

where M_B and M_A are the molecular weights of gas B and A. The pressure drop across the bed, necessary to maintain a mass flux $\mathbb{I}_t$ follows from d'Arcy's law:

$$\Delta P_{\mathbb{D}} = \int_o^\infty \frac{\mu}{K} \frac{\mathbb{I}_t}{\rho} \, dx$$

Here μ and ρ are the viscosity and density of the gas mixture respectively, and K is the permeability of the packed bed.
Since

$$\frac{d\omega}{dp} = - \frac{2}{\sqrt{\pi}} e^{-p^2} \qquad \text{it follows:}$$

$$\Delta P_{\mathbb{D}} = + \frac{\mathbb{D}_{eff}}{K} (M_B - M_A) \, c^o \int_1^o \frac{\mu}{\rho} \, d\omega \qquad (1)$$

For ideal gas mixtures the density may be calculated from:

$$\rho = c^o \left\{ \omega (M_B - M_A) + M_A \right\} \qquad (2)$$

The viscosity of a gas mixture is according to Bird, Stewart and Lightfoot*:

$$\mu = \mu_B \frac{\omega}{\phi_{BA} + (1-\phi_{BA})\omega} + \mu_A \frac{1-\omega}{1 + (\phi_{AB}-1)\omega} \qquad \cdot (3)$$

where ϕ_{BA} and ϕ_{AB} are constants.
With use of equations (2) and (3), direct integration of equation (1) yields the pressure drop across the packed bed due to diffusion:

$$\Delta P_{\mathbb{D}} = \frac{\mathbb{D}_{eff}}{2K} \left[\frac{\mu_B}{1-\phi_{BA}} \left\{ \ell n \, \phi_{BA} \frac{M_A}{M_B} - \frac{M_A(1-\phi_{BA}) + \phi_{BA}(M_A-M_B)}{M_A - \phi_{BA} M_B} \ell n \, \phi_{BA} \frac{M_B}{M_A} \right\} \right.$$

$$\left. + \frac{\mu_A}{\phi_{AB}-1} \left\{ \ell n \, \phi_{AB} \frac{M_B}{M_A} + \frac{M_B(\phi_{AB}-1) + \phi_{AB}(M_B-M_A)}{M_B - \phi_{AB} M_A} \ell n \, \phi_{AB} \frac{M_A}{M_B} \right\} \right]$$

*R.Bird, W.E.Stewart, E.N.Lightfoot, "Transport Phenomena", J. Wiley and Sons Inc., New York, 1960.

APPENDIX II

Data

gas	viscosity(1)	molecular weight
hydrogen	$0,88 \cdot 10^{-5} Ns/m^2$	2 kg/kmol
neon	$3,13 \cdot 10^{-5}$	20
nitrogen	$1,75 \cdot 10^{-5}$	28
carbondioxyde	$1,45 \cdot 10^{-5}$	44
propane	$0,82 \cdot 10^{-5}$	44

gas mixture A	B	molecular diffussion coefficient(1) m^2/s	viscosity constants(2) ϕ_{AB}	ϕ_{BA}
H_2	C_3H_8	$0,42 \cdot 10^{-5}$	3,6689	0,1535
C_3H_8	CO_2	0,09	0,7912	1,3665
H_2	CO_2	0,67	2,4969	0,1870
H_2	N_2	0,76	1,9213	0,2729
H_2	N_e	1,10	1,2725	0,4526

Packed bed permeability (3) $K = 1,17 \cdot 10^{-12} m^2$

Packed bed porosity $\varepsilon = 0,366$

Packed bed tortuosity (3) $\beta = 150/72$

effective diffusivity $\mathbb{D}_{eff} = \dfrac{\varepsilon}{\beta^2} \cdot \mathbb{D}_{molecular}$

(1) from International Critical Tables

(2) from R.Bird, W.E. Stewart, E.N. Lightfoot, "Transport Phenomena", J. Wiley and Sons Inc., New York, 1960

(3) from the Kozény relation

APPENDIX III

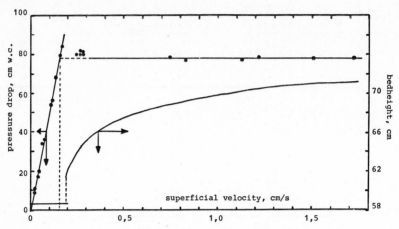

Fluidization characteristics of glass beads, fluidized with nitrogen

gas	min. fluidization velocity, cm/s
neon	--
hydrogen	0,46
nitrogen	0,16
carbondioxyde	0,18
propane	0,24

MEASUREMENT OF THROUGH-FLOW VELOCITIES IN A BUBBLE IN TWO DIMENSIONAL FLUIDIZED BEDS

X. T. NGUYEN, L. S. LEUNG AND R. H. WEILAND

NOMENCLATURE

q   throughflow velocity across a horizontal cross-section of

      bubble - cm/s

u_o minimum fluidization velocity - cm/s

u_B natural rate of rise of bubble (= solid downflow

      velocity) cm/s

u_{max} velocity measured of centre line - cm/s

θ angle defining position of q (Figure 2) - degrees

INTRODUCTION

This paper describes experimental measurements of the
throughflow velocity in a "simulated bubble" in a two dimensional
air-polystyrene fluidized bed.   The "simulated bubble" consists
of a stationary void enclosed by a gauze in a two dimensional bed
of falling particles.   The downward speed of the particles is
adjusted to that of the natural rate of rise of a bubble of the
same volume as the void.   There is some uncertainty about the
effect of the gauze boundary on throughflow velocities.   In our
previous measurements of pressure and voidage around the
simulated bubble using gauzes of different porosity and a non-

enclosed gauze cap, no significant effects of the gauze on these
measurements were observed   (Nguyen et al 1973).   In
particular, no difference in pressure was observed across the
gauze of the bubble.   This does suggest that the effect of the
gauze on throughflow velocity may be negligible.

EXPERIMENTAL
Apparatus

The main part of the apparatus (Figure 1) is a "falling"
two dimensional bed 122 cm. deep, 58 cm. wide and 0.635 cm.
thick, containing an artificial void.   The rate of downflow of
solid is controlled by adjusting the speed of a rotary valve
below the bed.   An independently metered air supply can be
passed into the bed near the bottom to maintain the falling
particles in the fluidized state.   The apparatus used here,
with the exception of continuous air injection at the bottom, is
similar to that reported by Drinkenburg and Rietema (1973).
Details of our apparatus have been presented previously
(Nguyen et al 1973, Nguyen 1974).

The simulated bubble consists of a cylindrical ring or a
kidney shape ring made from 325 mesh gauze clamped between the
walls of the falling bed.   Four different bubbles were employed:
two of cylindrical shape with radius of 3.81 cm. and 2.54 cm.;
and two of kidney shape with 3.81 cm. and 2.54 cm. radius of
curvature.   The kidney shape was adopted from a photograph of
a two-dimensional bubble presented by Rowe and Partridge (1966).

The velocity of gas inside the bubble was measured by a
Flow Corporation Measurement Model CTA 3 hot-wire anemometer

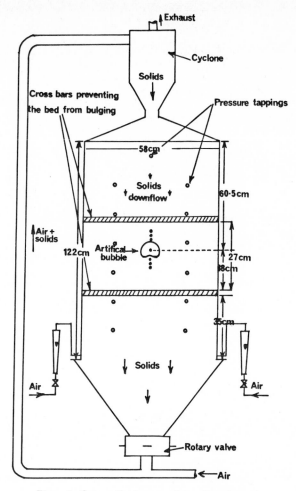

Figure 1. Schematic diagram of overall equipment.

operating in the constant temperature mode.   The hot wire used

was made of tungsten, 5μm diameter by 0.050 cm. long.

Calibration of the hot wire anemometer was carried out in a wind

tunnel following standard procedures (Owen and Parthurst, 1966).

The hot wire anemometer probe can be inserted at 13 positions in

the 3.81 cm. radius bubbles and 9 positions in the 2.54 cm.

radius bubbles (Figure 2);   it can be rotated in a plane

parallel to the plane of the wall of the bed to measure

direction as well as magnitude of gas flow.   The direction is

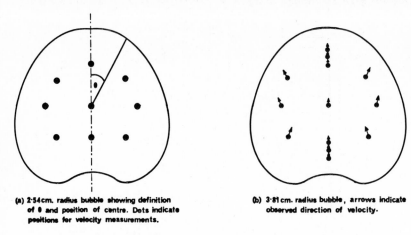

(a) 2·54cm. radius bubble showing definition
    of θ and position of centre. Dots indicate
    positions for velocity measurements.

(b) 3·81cm. radius bubble, arrows indicate
    observed direction of velocity.

Figure 2. Positions at which velocity was measured and direction of observed velocity.

readily measured at the mid-point between the two walls by

assuming no flow in the direction perpendicular to the wall.

The anemometer can also be traversed from one wall to the other

to measure wall effects.   The measured throughflow velocities

are reproducible to within 5%.

Polystyrene powder with specific gravity of 1.05, mean

particle diameter of 220 μm and a minimum fluidization velocity

in atmospheric air of 1.6 cm. s^{-1} was used as the solid.

Procedure

In an experimental run the solid downflow rate was adjusted

until it was equal to the natural rate of rise of the "simulated

bubble".   This velocity was set by gradually reducing the solid

downflow speed.   At high speeds, the void (or bubble), extended

well beyond the enclosed gauze ring but as the velocity was

reduced the bubble became smaller.   The natural rate of rise of

the simulated bubble is taken as the downward velocity at which

the bubble boundary has just withdrawn within the boundary

imposed by the gauze ring (Figure 3).   The solid downflow

velocity for the 2.54 cm. radius bubbles was about 32 cm. s^{-1} and that for the 3.81 cm. radius bubble was about 37 cm. s^{-1}.

While the solid downflow rate was adjusted the air flowrate into the bottom of the falling bed was also adjusted until the falling particles were in the incipiently fluidized state. This was checked by the pressure profile in the bed and by the presence of an occasional random bubble.

When the bed was adjusted the throughflow velocity was measured by the hot wire anemometer. The hot wire probe was placed mid-way between the walls and was rotated in the plane parallel to that of the walls of the fluidized bed to obtain the maximum current reading (corresponding to the maximum velocity) and the minimum current reading, the direction of the maximum velocity being obtained from the angle between the hot wire and the vertical plane. In separate experiments velocity profile was obtained by traversing normal to the wall. These velocity profiles have shown that wall effects do not extend to the central region between the two walls.

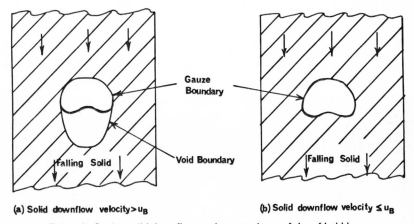

(a) Solid downflow velocity > u_B (b) Solid downflow velocity ≤ u_B

Figure 3. Setting solid downflow to the natural rate of rise of bubble, u_g.

Table 1 Comparison of measured throughflow velocity expressed in terms of minimum fluidization velocity with predicted values

Angle θ (defined in Figure 2)	Measured q/u_o				Davidson and Harrison	Murray; Rose and Partridge	Jackson; Leung et al
	3.81 cm. radius bubble		2.54 cm. radius bubble				
	cylindrical	kidney shape	cylindrical	kidney shape			
0					2	1.74	2.3
45	2.5	2.6			2	1.4	2.2
50	2.3	2.5			2	1.4	2.1
59	2.3	2.5	2.3	2.4	2	1.4	2.0
90	2.2	2.3	2.0	2.3	2	1.2	0.5
115	2.1	2.3	2.2	2.3	2		

(i) Flow direction

Typical results of measured direction of gas flow are summarised in Figure 2.   Along the bubble axis the direction of gas flow is vertically upwards as expected.   Elsewhere the flow directions are in qualitative agreement with predictions from the analysis of Pyle and Rose (1963).

(ii) Magnitude of throughflow velocity

For the 3.81 cm. radius kidney shape bubbles, the measured velocities at all points except the lowest position fall within the narrow range of 4.2 to 5.0 cm/s corresponding to 2.6 to 3.1 u_o.   The results obtained from the other three bubbles are similar.   It appears that within the range of this study, neither shape nor size of a bubble has any significant effect on the throughflow velocities in the bubbles.

The vertical throughflow velocity, q across any horizontal cross-section is obtained by taking the arithmetric average of the vertical component of the point velocities measured.   The results are summarised in Table 1 and compared with prediction from theoretical analyses.   The angle θ in Table 1 is defined in Figure 2 with the centre of the bubble.   Results are in reasonable agreement with prediction of $2u_o$ from Davidson's analysis.   Good agreement with Jackson's analysis was also obtained for θ less than 90°.

REFERENCES

Davidson, J.F. and Harrison, D., Fluidized Particles, Cambridge University Press (1963).

Drinkenburg, A.A.H. and Reitema, K., Chem. Eng. Sci., 28, 259, (1973).

Jackson, R., Trans. Inst. Chem. Eng., 41, 22, (1963).

Judd, M.R., Ph.D. Thesis, Cape Town University, (1965).

Leung, L.S., Nguyen, X.T. and Mak, F.K., Chem. Eng. Prog.,
Symposium Series, V.166, No. 104, 35, (1971).

Murray, J.D., J. Fluid Mech., 22, 57, (1965).

Nguyen, X.T., Leung, L.S. and Weiland, R.H., Proceedings of the
International Symposium in Fluidization and Its Applications,
Toulouse, October 1973, pp. 230-239, Societe de Chimie
Industrielle, (1974).

Nguyen, X.T., Ph.D. Thesis University of Queensland, (1974).

Owen, E. and Parthurst, R.C., "Measurement of Air Flow",
Pergammon Press, Oxford, 4th. Edition, (1966).

Partridge, B.A. and Rowe, P.M., Trans. Inst. Chem. Engrs, 44,
T349, (1966).

Pyle, D.L. and Rose, P.L., Chem. Eng. Sci., V.20., 25, (1963).

Rowe, P.N. and Partridge, B.A., Trans. Inst. Chem. Engrs., 43, 157,
(1965).

INFLUENCE OF THE INTERPARTICLE FLOW, INTRAPARTICLE DIFFUSION AND ADSORPTION OF THE GAS ON ITS DISTRIBUTION OF RESIDENCE TIMES IN FLUIDIZED BEDS

R. GOEDECKE AND K. SCHÜGERL

SUMMARY

The distribution of residence times and contact times of a two component tracer mixture (ar and CO_2) were measured in bench scale gas fluidized beds of 19 cm diameter with porous (γ-Al_2O_3) and/or non porous (quartz sand) particles for different bed height to diameter ratios, fluidization indexes u^+ as function of the temperature T (up to 300°C). The moments of the measured curves were used for the calculation of the first approximate of the model parameters to the fitting of the calculated curves to the measured ones by means of an axial dispersion model.

The optimization procedure of MARQUARDT and of GAUSS-SEIDEL are applied in the LAPLACE domain to evaluate the final model parameters. The shares of the partial mean residence times due to the interparticle porosity of the bed, intraparticle porosity of the particles and adsorption of the gas on the total mean residence times are separately evaluated. Also for Ar the adsorption effect cannot be neglected if porous particles are used. For CO_2 the share of the mean contact time due to adsorption on total mean residence time varies with increasing temperature (up to 300°) from about 0.9 to 0.5. At high temperatures the mean contact time also depends on the gas flow rate.

The temperature has only a slight effect on the over all Pe numbers of longitudinal dispersion. The partial Pe numbers due to the adsorption alone, Pe_{Ad}, decrease abruptly with increasing flow rate at the minimum fluidization region. The longitudinal dispersion coefficient due to adsorption alone, D_{Ad}, approaches zero for fluidi-

zation index $u^+ \rightarrow 1$, but for $u^+ > 1$ it increases considerably
with u^+ and slightly with T.

Pe_{Ad} and/or D_{Ad} are characteristic for the solids mixing in the
bed. The adsorption enthalpy which is calculated by means of the
temperature dependence of the mean contact time agrees fairly well
with the enthalpy, evaluated in a gas chromatograph-micro reactor.

INTRODUCTION

The distribution of residence times, DRT, of fluid elements pass
an isotherm reactor is usually used to characterize the transient
behaviour of the reactor. It is the purpose of this paper to investigate the influence of the properties of the particles as well as
of the fluid on the transient behaviour of fluidized beds at differe
temperatures.

EXPERIMENTALS

A fine steel apparatus of 19 cm diameter and 2500 cm over all height
consists of a gas heater, an evaporator, gas distributor and the
reactor (Figure 1) is used.

The tracer is given into the flow upstream to the gas distributor,
which consists of a conical packed bed and a fine steel porous plate
of 175 um mean pore diameter, by means of a magnetic valve at the
narrowest cross section of the tapered pipe. The reactor is equipped
by a vertical coaxial probe for gas sampling shich is disposed at
the narrowest cross section of the tapered bell. This bell and with
it the position of the sampling cross section can be shifted along
the axis of the reactor from the gas distributor plate to the upper
part of the reactor. At the lower edge of the bell a small capacity
probe is mounted by which the height of the bed is measured with
good accuracy during the runs.

The reactor is well insulated by glass wool and can be heated up
to 400°C. The composition of the gas is measured by a four channel
mass spectrometer (GD 150/4 of VARIAN MAT) at two different position
in the bed, directly above the gas distributor without solid particles (position 1) and directly above the top of the bed (position

A mixture of CO_2 and Ar (1:1) is used as tracer gas. The duration of
the opening time of the magnetic valve (i.e. the width of the input

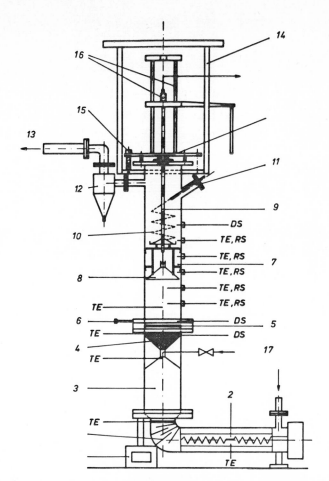

Figure 1. Reactor: 1. air inlet; 2. heater; 3. evaporator; 4. gas distributor; 5. gas distributor plate; 6. screw for discharge of the particles; 7. reactor; 8. bell for collecting gas sample from the entire cross section of the bad with possible longitudinal displacement and capacitor for measuring the bed height; 9. co-axial sampling tube; 10. electrical connection to the capacitor; 11. filling hole; 12. cyclon; 13. air outlet; 14. sampled gas; 15. lifting unit for for the reactor to exchange of the gas distributor plate; 16. mechanical device to displace the co-axial sampling tube 9 and the bell 8 along the reactor; 17. tracer inlet; TE, thermocouple; RS, radial sampling tube; DS, pressure gauge.

" δ-function") is 0.25 secs. The duration of the measurements are very different (from several secs up to several hours) depending on the tracer type (Ar or CO_2) and on the temperature. The analogue signal of the mass spectrometer is converted into digital values and stored by a data logger and evaluated by a computer. The frequency of read-in is varied according to the shape and duration of the curves. For each DRT about 1000 points are used to calculate the model parameters. Narrow fraction porous γ-Al$_2$O$_3$ and/or non porous

quartz sand particles of 285 μm mean diameter are used as solids.
The specific surface area of 244 m^2/g of the particles is mainly
due to the mesopores.
The adsorption behaviour (adsorption equilibrium constant and adsorption coefficients) of Ar and CO_2 on the γ-Al_2O_3 is measured in
an isothermal packed bed microreactor.

EVALUATION OF THE MODEL PARAMETERS

The axial dispersion model is applied to characterise the longitudinal mixing of the gas in the fluidized bed.
The axial dispersion model is described by the following dimensionless equation derived from the unsteady-state one dimensional materia
balance equation:

$$\frac{\partial c^*}{\partial \theta} + \frac{\partial c^*}{\partial x^*} = \frac{1}{Pe} \frac{\partial^2 c^*}{\partial x^{*2}} \tag{1}$$

with c^* dimensionless concentration C/C_0
    θ dimensionless time t/E
    x^* dimensionless longitudinal coordinate x/L
    $Pe = \frac{u_e H}{D_{ax}}$ Peclet number, which indicates the degree
        of longitudinal mixing in the reactor

The fitting of equation (1) to the measured curves are carried out
in the LAPLACE domain. The corresponding system transfer functions
depend on the boundary conditions of equ. (1). The transfer function
of equ. (1) is evaluated for different boundary conditions.
The best fit has been achieved by a model, which is closed for the
longitudinal dispersion at the entrance. If the DRT are measured at
the entrance and at the exit following transfer function prevails:

$$F(s^*, Pe) = \exp \frac{Pe}{2} (1 - q) \tag{2}$$

The model parameters were estimated by a least-squared error analysis of the transformed response data. For the determination of the
least-squares estimates the methods of GAUSS-SEIDEL [11] and MAR-
QUARDT [2] are applied.

Typical standard deviations of the fitted Peclet-number are for
Ar: $7.1.10^{-3}$.

RESULTS

The statistical mean residence times $\bar{t}$ are calculated from the first
moment of DRTs. The temperature has only a slight effect on the
mean residence time of Ar. In contrast to this behaviour the mean
residence times of the tracer CO_2 depend considerably on the temperature. With increasing temperature $\bar{t}$ decreases. The order of magni-
tudes of the mean residence times of Ar and CO_2 are different (e.g.
$\bar{t}$ = 3.8 to 11.5 secs for Ar and 28 to 96 secs for CO_2 at T = 105°C).
At room temperature the statistical mean residence times of CO_2
cannot be estimated because the tracer concentration in the bed di-
minishes extremely slowly (no end of the tailing can be found).

In fluidized beds of quartz sand no difference between the statistical mean residence times of Ar and CO_2 are found. The statistical
mean residence times $\bar{t} = M^{1.0}$ of the DRT and the fluid-dynamical
ones $\tau = \dfrac{H}{\bar{u}_e} = \dfrac{H}{u_e^o}$ are equal too.
A comparison of this behaviour of the statistical mean residence times
$\bar{t}$ with the ones in fluidized beds with γ-Al_2O_3 particles shows significant differences. In Figure 2 the statistical mean linear velocities $\bar{u}_e = \dfrac{H}{\bar{t}}$ and the fluid-dynamical effective mean velocities
$u_e = \dfrac{u_o}{\varepsilon}$ are compared for quartz sand and γ-Al_2O_3-particles, Ar
and CO_2 and for different temperatures.
Ar as well as CO_2 gives on quartz sand the relation I with the slope
of 45° (i.e. $\bar{u}_e = u_e$). Ar on γ-Al_2O_3-particles yields the relation
II, if the fluid-dynamical effective mean velocity is calculated
alone by means of the interparticle porosity ε , without considering
the porosity of the particles ε_{Po}. Ar on γ-Al_2O_3-particles provides the relation III, if the effective mean velocity is calculated
by the sum of the interparticle porosity ε and the particle porosity ε_{Po}:

$$u_e = \frac{u_o}{\varepsilon} \qquad (u_e)_{Fl + Po} = \frac{u_o}{\varepsilon + \varepsilon_{Po}} \qquad (3)$$

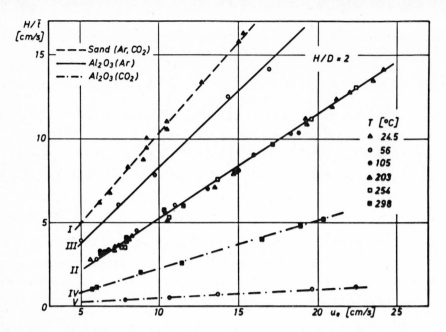

Figure 2. Statistical mean linear velocity $\bar{u}_e = H/\bar{t}$ as function of the hydrodynamical effective mean flow velocity.

The slope of the relation of $\bar{u}_e = f\,(u_e)_{Fl + Po}$ is nearer to the 45° than the one of $\bar{u}_e = f(u_e)$, but it still significantly deviates from it. Similar plotting for CO_2 on γ-Al_2O_3 depends on the temperature. The relation IV is valid for $298^\circ C$ and V for $56^\circ C$.
One can recognize from Figure 2 that the statistical mean velocity $\bar{u}_e$ of Ar and/or CO_2 is slightly and/or much lower than the flow velocity $(u_e)_{Fl + Po}$, which considers the interparticle porosity ε as well as the particle porosity ε_{Po}.

By the separation of the total gas volume in the bed V_{tot} into the partial volumes: interparticle (V_{Po}) and adsorbed (V_A) volumes, one can define the partial mean residence times

$$\tilde{\tau}_{Fl} = \frac{V_{Fl}}{u_o S}, \quad \tilde{\tau}_{Po} = \frac{V_{Po}}{u_o S} \quad \text{and} \quad \tilde{\tau}_{Ad} = \frac{V_A}{u_o S} \quad \text{and the total}$$

mean residence time $\tau = \dfrac{V_{tot}}{u_o S}$.

The partial residence times can be calculated by

$$\frac{M^{1.0}}{\tau} \doteq \frac{\bar{t}}{\tau} = 1 \tag{4}$$

where $\hat{\tau} = \hat{\tau}_{Fl} + \hat{\tau}_{Po} + \hat{\tau}_{Ad}$:

$\hat{\tau}_{Fl} = \dfrac{H\varepsilon}{u_o}$ fluiddynamical effective mean residence time by considering only the interparticle porosity of the bed ε. $\hat{\tau}_{Fl}$ is estimated by non porous particles.

$\hat{\tau}_{Po} = \dfrac{H\varepsilon_{Po}}{u_o}$ effective mean residence time by considering only the intraparticle porosity ε_{Po}. The sum $(\varepsilon_{Fl} + \varepsilon_{Po})$ is estimated by porous particles with non adsorbing gases. ε_{Po} can be evaluated by the difference $(\hat{\tau}_{Fl} + \hat{\tau}_{Po})$ and $\hat{\tau}_{Fl}$.

$\hat{\tau}_{Ad} = \dfrac{V_A}{S\,u_o}$ mean contact time, i.e. effective mean residence time by considering only the adsorbed volume of A of the gas and the volumetric flow rate $u_o S$. $\hat{\tau}_{Ad}$ is estimated by the relation $M^{1.0} - (\hat{\tau}_{Fl} + \hat{\tau}_{Po}) = \hat{\tau}_{Ad}$.

In Figure 3 these partial residence times for Ar are plotted as function of the fluid-dynamical effective mean flow rate u_e. One can recognize that the share of the partial residence times on the total residence time $\hat{\tau}$ changes only slightly by u^+ and does not change by T. It amounts: 0.55 for $\hat{\tau}_{Fl}$, o.3 for $\hat{\tau}_{Po}$ and 0.15 for $\hat{\tau}_{Ad}$, i.e. $\hat{\tau}_{Ad}$ cannot be neglected.

Because of the strong temperature dependence of the mean contact time $\hat{\tau}_{Ad}$ for CO_2 the share of the partial residence times on the over all residence time varies in a broad range. With increasing temperature the share of $\hat{\tau}_{Ad}$ decreases in the investigated cases from about 0.90 to about 0.50 at low fluidization indexes. $\hat{\tau}_{Ad}$ also decreases with increasing flow rate, especially at higher temperatures. At low temperatures this effect and also the influence of H/D on $\hat{\tau}_{Ad}$ is only slight. At higher temperatures $\hat{\tau}_{Ad}$ increases with growing H/D-ratio. The independence of $\hat{\tau}_{Ad}$ on u^+ and H/D

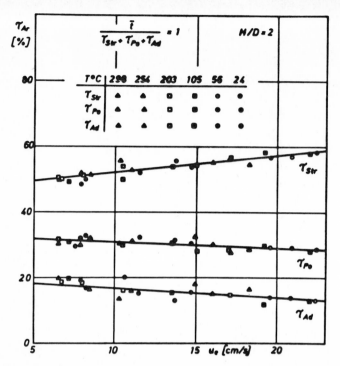

Figure 3. Shares of the partial mean residence times τ_{F1}, τ_{P0} and τ_{Ad} on the total mean residence time τ for Ar as function of the effective mean flow velocity u_e.

at lower temperatures and its dependence on them at higher temperatures indicates that in the latter case during the passage of the
tracer through the bed the adsorption cannot come to complete equilibrium.

The Pe-numbers are evaluated by fitting of the calculated curves
of the longitudinal dispersion model by applying the boundary conditions of equ. (2) and non linear optimization procedures. One can
recognize from figure 4 that the temperature generally does not have
a great influence on Pe. No influence of the temperature on D_{ax}
can be found either (Fig. 5).

In contrary to Ar, where the Pe numbers significantly depend on u^+,
the Pe numbers vary for CO_2 only slightly with u^+ (Figure 6), if
$u^+ \geq 1.6$.

For Ar as well as for CO_2 the Pe numbers are nearly independent on T.

DISCUSSION

The fluid-dynamical DRTs of CO_2 and Ar are also identical according
to the measurements made by sand. Therefore the mean contact time

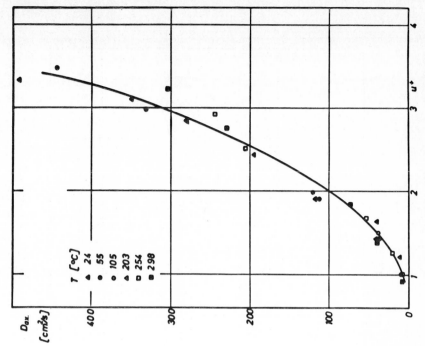

Figure 5. Longitudinal dispersion coefficients D_{ax} as function of the fluidization index u^+ for Ar; $H/D = 2$, γ-$A1_2O_3$.

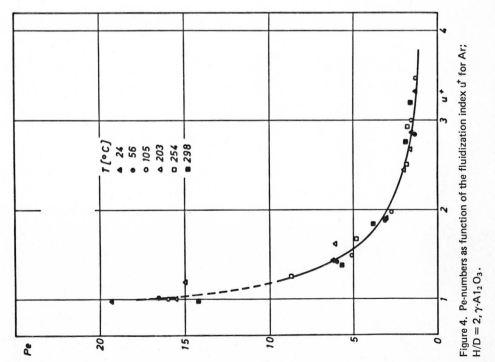

Figure 4. Pe-numbers as function of the fluidization index u^+ for Ar; $H/D = 2$, γ-$A1_2O_3$.

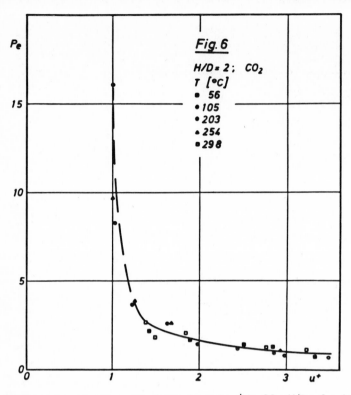

Figure 6. Pe-number as function of the fluidization index u^+ for CO_2; $H/D = 2$, γ-$A1_2O_3$.

of CO_2 is given by the difference of the first moments of the DRTs for CO_2 and Ar:

$$\bar{t}_K = \Delta \bar{M}^{1.0} = M_{CO_2}^{1.0} - M_{Ar}^{1.0}, \tag{5}$$

if the mean contact time of Ar is neglected.

The ratio of $H/\bar{t}_K$ gives a mean exchange rate, which is characteristic for the sorption process in the bed. From the dependence of these ratios on u_e the adsorption constant K_1 can be calculated, which is defined by:

$$K_1 = \frac{\Delta u_e}{H/\Delta \bar{t}_K}. \tag{6}$$

Plotting $\log K_1$ as function of the inverse temperature T the adsorption enthalpy is evaluated to be:

$$\Delta H_{ad\ 1} = 2.6\ \frac{kcal}{mol}\ .$$

The same adsorption constants and adsorption enthalpy are obtained
if the adsorption constant K_2 is defined as:

$$\frac{M^{1.0}}{\tau_{Po} + \tau_{Fl}} = 1 + \frac{K_2}{\varepsilon_{Po} + \varepsilon_{Fl}} \tag{7}$$

The adsorption constant K_A is estimated in a microreactor by evaluation of the first moment of the response function of an approximated δ -input function. From that adsorption enthalpy $\Delta H_{ad\ A}$ =
3.55 kcal/mol is obtained, which is in fairly good agreement with
$\Delta H_{ad\ 1}$. The difference is possibly due to the deviation from the
adsorption equilibrium in the bed at higher temperatures and flow
rates.

According to DAYAN and LEVENSPIEL [3] the second moments of the DRT
can be separated into partial moments.

$$(\frac{1}{Pe})_{over\ all} = (\frac{1}{Pe})_{Fl} + (\frac{1}{Pe})_{Po} + (\frac{1}{Pe})_{Ad} \tag{8}$$

For CO_2 the partial PECLET numbers Pe_{Ad} are plotted as function of
u^+ for different temperatures and for H/D = 3 on Figure 7. One can
recognize that the influence of the adsorption process on the dispersion is much smaller than its influence on the mean contact time.
Pe_{Ad} is larger for $u^+ < 1$ and it changes abruptly to low values at
$u^+ \sim 1$. This indicates that the low values of Pe_{Ad} due to the particle movement. Therefore the effective longitudinal dispersion coefficient D_{Ad} due to the adsorption must be characteristic for the
particle mixing. This dispersion coefficient is defined as:

$$D_{Ad} = \frac{H^2}{Pe_{Ad}\ M^{1.0}}\ (cm^2/s) \tag{9}$$

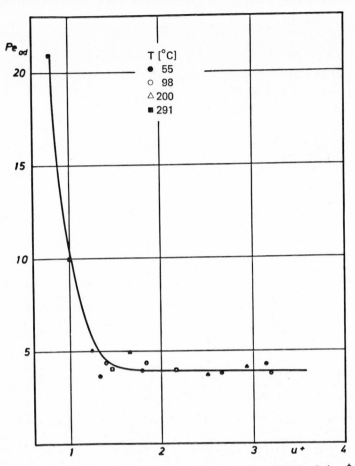

Figure 7. Partial Peclet-number due to adsorption, Pe_{Ad}, as function of the fluidization index u^+ for CO_2; H/D = 3, γ-Al$_2$O$_3$.

In Figure 8 D_{Ad} is plotted as function of u^+ for different temperatures. For $u^+ \to 1$ $D_{Ad} \to 0$ and in fixed bed D_{Ad} is negligible
small. For $u^+ > 1$ D_{Ad} becomes greater with increasing u^+ obviously due to the particle mixing. The temperature has only slight influence on D_{Ad} as expected.

D_{Ad} varies between 0 for $u^+ < 1$ and about 50 to 100 cm^2/s for $u^+ > 1$
(up to u^+ = 3.5). It is difficult to prove whether D_{Ad} is equal to
the longitudinal dispersion coefficients of solids D_{sol}, since no
comparable D_{sol} data are available yet [4]. The experimental conditions of DE GROOT [5] are the closest to the conditions of the present paper. Interpolating his data for bed diameter of 20 cm D_{sol} =
30 to 100 cm^2/s can be found for particle diameters: 70 - 300 μm

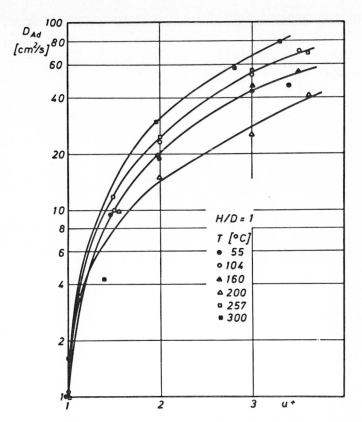

Figure 8. Longitudinal dispersion coefficient due to the adsorption, D_{Ad}, as function of the fluidization index u^+ for CO_2; $H/D = 1$, γ-$A1_2O_3$.

of silica for $H = 100$ cm and $u^+ = 6.7$. These D_{sol} - data are in the same range as the D_{Ad} values of the present paper. Further investigations are necessary, but one can presume that between D_{sol} and D_{Ad} a close correlation or maybe an identity prevails. If it turns out that they are identical, the measurement of D_{Ad} would offer for porous particles an easy method to estimate D_{sol}.

ACKNOWLEDGEMENT

The authors gratefully acknowledge the financial support of the Deutsche Forschungsgemeinschaft.

LITERATURE CITED

1. HIMMELBLAU, D.M.:   Process Analysis by Statistical Methods, John Wiley (1970)

2. MARQUARDT, D.W.:   Chem.Eng.Progr. 55 (6) (1959) 65

3. DAYAN, J.; LEVENSPIEL, O.: Chem.Eng.Sci. 23 (1968) 1327

4. POTTER, O.:   "Mixing" in "Fluidization", Ed. J.F. Davidson
                         D. Harrison, Acad. Press London 1971, p. 342

5. DE GROOT, J.H. in   "Proceedings of the International Symposium o
                         Fluidization" Eindhoven, Ed. A.A.H. Drinken-
                         burg, Netherlands Univ. Press, p. 353

PART III

Liquid Phase Fluidization

THE VOIDAGE FUNCTION IN
A LIQUID FLUIDIZED BED

B. SCARLETT AND M. J. BLOGG

INTRODUCTION

The expansion law for a bed of particulately fluidised particles was first given by Richardson and Zaki (1) as:-

$$\frac{V}{U} = \Sigma^n$$

V = superficial velocity

U = terminal velocity of a single particle

Σ = porosity of the expanded bed

The exponent n is a function of the Reynolds number and various values have been reported as the result of experimental measurements (2).

The most frequently reported values range from 4.7 to 2.3⁵ over the full range of Reynolds number.

An alternative method of relating the superficial velocity and the porosity of the bed is by means of the voidage function. The drag on a single particle is effectively increased by the presence of other particles, the increase being a function of the porosity, Σ. Thus, the effective drag on a single particle is given by the equation:-

$$F = f(\Sigma) \, C_D \, \frac{\rho V^2}{2} \cdot \frac{\pi d^2}{4}$$

EXPERIMENTAL RESULTS

The expansion of a bed of spherical glass particles fluidised by water was measured in a tube, 5 metres high, which was equipped

with pressure tappings and with sampling points at intervals of 8 cms.
Observations of both pressure profiles and particle segregation were
thus made for a number of sieve fractions. A typical result for a
nominal fraction of particles, 72-150 B.S. sieve, is shown in Figures 1 & 2.
Figure 1 shows the particle size distribution of samples taken at
various heights above the distributor and illustrates the strong
particle segregation which may occur. Figure 2 shows the corresponding pressure profile, illustrating the manner in which the pressure
gradient gradually decreases as the particle size decreases and the
bed porosity increases.

The results of these experiments lead to the following physical
picture of the structure of a particulate bed:-

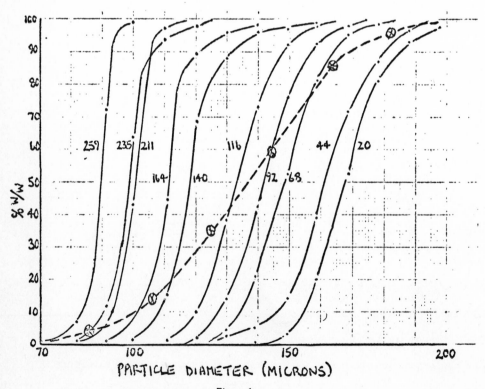

Figure 1.

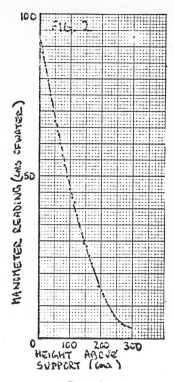

Figure 2.

a)   Particle Segregation   The particles are segregated by size.

There is, of course, a tendency for the particles to mix due to their

random motion but the segregation is also continually occurring. This

tendency is easy to accept if viewed from the concept of the voidage

function. Two particles of differing size cannot both be effectively

in equilibrium within a region of the same voidage, at any Reynolds

number, since the effective drag is dependent on the diameter of the

particle to a power which lies between 1 and 2.

There is, however, a possibility that the particles cannot form

a particulately fluidised bed. This behaviour occurs when the super-

ficial velocity is sufficient to cause fluidisation of a packed bed

of particles but is insufficient to maintain a stable voidage greater

than that of the packed bed. When the superficial velocity and particle size are in this relationship, the fluidised bed consists of a

circulating region or channeling region in which the particles are

continually circulating and mixing. Our observations show that a

bed may exhibit both types of behaviour. The smaller particles form-

ing a segregated and particulate region at the top of the bed and

the larger particles circulating in a well mixed region in the lower

part of the bed. This channeling behaviour is not dependent on the

distributor design. As the superficial velocity is gradually

increased, the relative proportion of the two regions decreases.

b)   Pressure Profile   The pressure profile reflects closely the

particle behaviour. The local pressure drop is approximately equal

to the bouyant mass of the particles and is thus a direct measurement

of that voidage. In the regions of bed where the particles are well

mixed the pressure gradient is linear, in the segregated regions the

gradient follows the local particle size and voidage. This effect is

illustrated by the pressure profile in Figure 2.

From our measurements, the pressure drop is not exactly equal

to the bouyant mass of particles but is slightly greater. At low

porosities this effect is most pronounced. The pressure drop may

exceed the bouyant mass of particles in our experiments by as much

as 4% of the total weight. We ascribe this additional pressure drop

to the loss of energy by interparticle collisions, this effect being

greatest at the lower porosities.

c)   Voidage Function   From the plots of particle size and pressure

drop against height, the voidage function for closely sized particle

fractions can be determined. This data is subject to the assumption

that the local pressure gradient is a measure of the local porosity.

On the other hand, the data has two important refinements:-

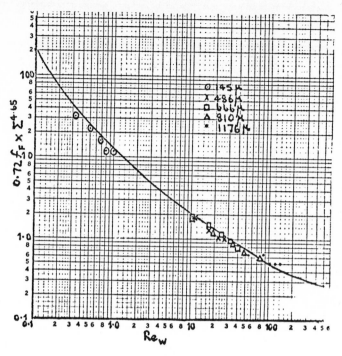

Figure 3.

1)   The particles are closely sized locally and thus there is no necessity to assume a mean particle size when correlating the results.

2)   The measurements are made only in the particulately fluidised region, the circulating region is excluded. Figure 3 shows the data for a range of particle beds correlated according to the voidage function relationship proposed by Lewis (3):-

$$f(\Sigma) = 0.72 \cdot \Sigma^{-4.65}$$

The results cover the Reynolds number range 0.3 to 150, being defined on the basis of superficial velocity and particle diameter. This function reduces the results to the drag coefficient curve for a single sphere with good accuracy over the whole range of Reynolds number. A voidage function of this form which is independent of Reynolds number is completely compatible with the Richardson and

Zaki law whose exponent reduces from 4.7 to 2.35 while the resistance changes proportionally from V^2 to V.

CONCLUSIONS

This paper presents experimental evidence that in the region of a fluidised bed which is truly particulate the following conditions apply:-

1)   The pressure drop is equal to the bouyant mass of particles

2)   The voidage function is independent of Reynolds number and has an exponent of -4.7

REFERENCES

1.   Richardson J.F. and Zaki W.M., Trans. Inst. Chem. Engrs. Vol 32
      p. 35 (1954)

2.   Wen C.Y. and Yu Y.H., C.E.P. Symp. Series No. 62, p.100 (1966)

3.   Lewis E.W. and Bowerman W.,Chemical Engineering Progress Vol. 48,
      p. 603 (1952)

MASS TRANSFER OF PHENOL FROM DILUTE AQUEOUS SOLUTION IN A FLUIDIZED ADSORBER

R. GANHO, H. GIBERT AND H. ANGELINO

NOMENCLATURE

a_v mass transfer area between fluid and particles per unit volume of bed $(L^2 L^{-3})$

a', b' coefficients in isotherm adsorption equation (eq. 2)

C concentration of solute in the fluid phase (ML^{-3})

C_o concentration of solute in the fluid phase entering the column (ML^{-3})

C_L concentration of solute in the fluid phase issuing the column (ML^{-3})

d_s mean particle diameter (L)

F volumetric flow rate $(L^3 T^{-1})$

K_f overall fluid phase mass transfer coefficient (LT^{-1})

k_f external diffusion mass transfer coefficient (LT^{-1})

K_p overall solid phase mass transfer coefficent (LT^{-1})

k_p internal combined solid phase mass transfer coefficent (LT^{-1})

m_c total amount of the dry adsorbant (M)

1 level inside the column (L)

L total height of the bed (L)

327

q   concentration of solute in the solid phase weight of solute/unit weight of dry adsorbant (MM^{-1})

S   cross sectional area of column (L^{2})

t   time (T)

Greek Letters

ρ_b bulk density $\dfrac{\text{weight dry activated carbon}}{\text{unit bulk volume of bed}}$ (ML^{-3})

$\left.\begin{array}{l} \alpha_1 \\ \\ \alpha_2 \end{array}\right\}$ slopes of the adsorption isotherm $(L^3\ M^{-1})$

Superscripts

*   equilibrium

1. INTRODUCTION

The sequences envolved in an adsorption operation can be grouped into the following steps [1-3] :

- fluid phase external diffusion,

- reaction at the solid surface which is generally very rapid,

- solid phase internal diffusion which concerns all transfer inside the external geometrical surface.

In the present study the adsorption of phenol on a liquid fluidized bed of activated carbon has been investigated in order to try to identify the controlling mechanism.

2. EQUIPMENT AND PROCEDURE

Figure 1 is a general diagram of the equipment. The aqueous phenol solution has been obtained by mixing a concentrated phenol solution with soften water. Phenol concentrations have been determinated by means of a U.V. spectrophotometer. Activated carbon employed was Filtrasorb 400. The different characteristics of the particle used have been experimentally determined and are presented

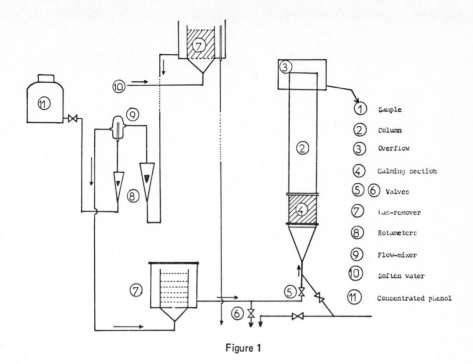

Figure 1

in table 1. The adsorption isotherms at 14°C and 18°C, the two
extremal temperatures   used in the present study, have been exta-
blished elsewhere : it exists no significant differences between
the corresponding results and so far only one curve has been
selected.[4].

In this study, the inlet phenol concentration has been
fixed equal to 94 – 98 mg/l and the influence of the liquid flow
rate has been investigated with three different sizes of particles.

Table 1

Mean particle diameter cm	Density in air g/cm^3	U_{om} cm/s	Particle specific area cm^2/g	Particle specific area cm^2/cm^3
0.129	0.68	0.40	68.6	44
0.093	0.68	0.31	96.3	58
0.071	0.68	0.21	123.0	73.5

Flow rates have been varied in the range 193 1/h (0.78 cm/s) to
578 1/h (2.34 cm/s). The evolution with time of the outlet solution
concentrations have been checked by taking samples and analysing
them.

3. THEORETICAL ANALYSIS

The definition of the various mass transfer coefficients
and the various relations are presented in the appendix. To write
and solve the equation the following asumption are made :

1) C is constant in a cross section area,

2) the perfect mixing of solid occurs and so "q" does not
change with position but is a function of time,

3) for a short internal of time a quasi-stationnary regime
is developped and C is only a function of "1",

4) the adsorption isotherm corresponds to a LANGMUIR
isotherm written as [4]

$$\frac{c^*}{q} = 5.13 \cdot 10^{-3} \, c^* + 4.06 \cdot 10^{-2} \quad (1) \quad c < c < 100 \text{ mg/l}$$

$$\text{or} \quad \frac{c^*}{q} = b'c^* + a' \quad (2)$$

Material Balance Relation

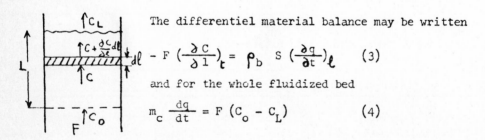

The differentiel material balance may be written

$$-F \left(\frac{\partial c}{\partial l}\right)_t d\ell = \rho_b \, S \left(\frac{\partial q}{\partial t}\right)_\ell \quad (3)$$

and for the whole fluidized bed

$$m_c \frac{dq}{dt} = F \, (C_o - C_L) \quad (4)$$

Fluid Phase Mass Transfer Coefficient

Using fluid phase mass transfer coefficient the following
equation may be written

$$\frac{K_f \, a_v}{\rho_b} \, (c - c^*) = -\frac{F}{\rho_b S} \frac{dc}{dl} \quad (5)$$

- "q" is independant from "l"
- C^* is the concentration at equilibrium with q, i.e. C^* value
remain the same at the inlet and at the outlet

$$K_f = \frac{F}{S\ L\ a_v}\qquad Ln\ \frac{C_o - C^*}{C_L - C^*}\qquad (6)$$

Solid Phase Internal Mass Transfer Coefficient

Using the different asumptions the following equation can
be written

$$K_p\ a_v\ (q^* - q) = -\frac{F}{\rho_b\ S}\ \frac{dC}{dl}\qquad (7)$$

and by means of equation (2)

$$K_p\ a_v = \frac{F}{m_c}\left[\frac{b'}{1 - b'q}(C_o-C_L) + \frac{a'}{(1 - b'q)^2}\ ln\ \frac{C_o - \frac{a'q}{1 - b'q}}{C_L - \frac{a'q}{1 - b'q}}\right]\qquad (8)$$

4. RESULTS AND DISCUSSION

For each experiment the values of C_o and C_L have been determinated every 5 min. and the average phenol concentration calculated
by means of eq. 4, then knowing the value of "q" and using the adsorption isotherm the corresponding equilibrium value of C^* has been
deduced and finally the values of K_f and K_p calculated.

Coefficient Kf

For each experiment the variations of K_f as a function of "q"
have been plotted (Cf fig. 2) : K_f is decreasing when "q" increases.
This fact establishes that external diffusion is not the only mechanism because one knows that for fixed experimental condition k_f is
constant. The values of k_f have been estimated using the curve K_f vs "q"
and extrapolating at q = 0. Under that conditions the gradient of concentration inside the activated carbon particle is zero and the internal
diffusion does not exist. The above results have been checked qualitatively using eq. 1a : if internal diffusion is acting while the
adsorption is developping, then $1/K_f$ is varying in the same manner as

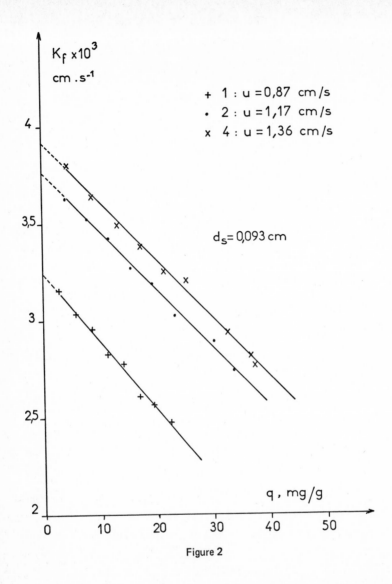

Figure 2

$1/\alpha_1 \rho_b k_p$. A tentative study of the variations of α_1 has been made calculating the slope α_1 at point (c^*,q) and taking into account the values of q and also their variations during one experiment : α_1 is decreasing while q is increasing.

In this study it was not intended to deduce an equation for the transfer but to confirm that external diffusion exists. In order to do that a comparison has been made between results obtained here and DAMRONGLERD's[5] for pure diffusion (Cf fig. 3) : the results of the present study fit well a straight line parallel to DAMRONGLERD's line, which has been obtained for the dissolution of benzoïc acid, but the results here are 30 % lower.

Coefficient K_p

It has been pointed out that, except at the beginning of the experiments that K_p remains almost constant. This result is valid for each particle size. This fact means that in equation 2a, α_2 must remain approximatively constant. A tentative verification has been made calculating α_2 at point (C, q^*), and taking for C the mean logarithmic value between C_o and C_L at the beginning and at the end of one experiment : is was established that α_2 is very low and remain constant.

From all the results the following values are proposed

$$d_s = 0.129 \text{ cm} \qquad\qquad K_p \, a_v = 8.2 \cdot 10^{-5} \text{ s}^{-1}$$

$$d_s = 0.093 \text{ cm} \qquad\qquad K_p \, a_v = 10.5 \cdot 10^{-5} \text{ s}^{-1}$$

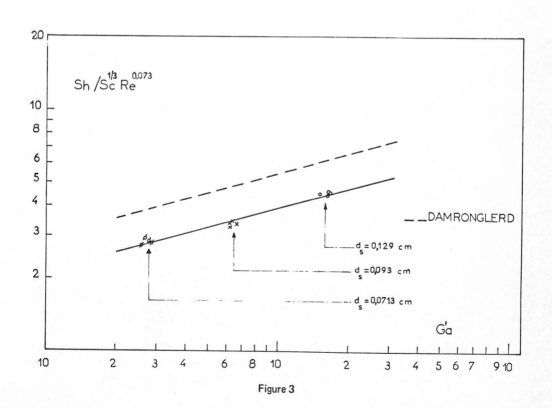

Figure 3

MILLER et CLUMP[6] studying, in agitated vessel, the adsorption of aqueous solution of phenol on activated carbon particle of 0.155 mm found $16.7 \cdot 10^{-5} \text{ s}^{-1}$: the order of magnitude is the same.

5. CONCLUSIONS

The kinetics runs concerning the adsorption of aqueous solution of phenol on three sizes of fluidized activated carbon particles allow to say :

- fluid phase external diffusion is the controlling mechanism at the beginning,

- then both fluid phase external diffusion and internal particle diffusion exist

- the use of eq. 2a is easier.

LITERATURE

1 – HIESTER N.K., VERMEULEN, Chem. Eng. Prog., 1952, 48, 505

2 – VERMEULEN T. in "Advances in Chemical Engineering", 1958, Academic Press, vol. II, p. 147.

3 – HIESTER N.K., VERMEULEN T., KLEIN G., in Chemical Engineer's Handbook, 7th Edition, section 16, Mac Grow Hill, 1963.

4 – GANHO R., Thèse de Docteur-Ingénieur n° 454, Univ. P. Sabatier, Toulouse, 1974.

5 – DAMRONGLERD S., Thèse de Docteur-Ingénieur n° 5, Univ. P. Sabatier, Toulouse, 1973.

6 – MILLER C., CLUMP C., Am. Inst. Chem. Eng. Jal, 1960, 16, 169.

APPENDIX

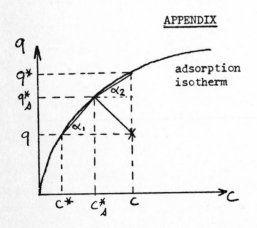

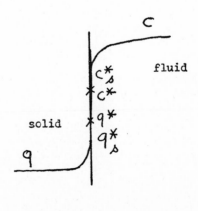

The following transfer equation can be written :

$$\left(\frac{dq}{dt}\right) = \frac{K_f a_v}{\rho_b}(c - c^*) = \frac{k_f\, a_v}{\rho_b}(c - c_s^*)$$

or

$$\left(\frac{dq}{dt}\right) = K_p\, a_v\, (q^* - q) = k_p\, a_v\, (q_s^* \cdots q)$$

Let us use $\quad \alpha_1 = \dfrac{q_s^* - q}{c_s^* - c} \quad$ and $\quad \alpha_2 = \dfrac{q^* - q_s^*}{c - c_s^*}$

then

$$\frac{1}{K_f} = \frac{1}{k_f} + \frac{1}{\alpha_1 \rho_b k_p} \tag{1a}$$

$$\frac{1}{K_p} = \frac{1}{k_p} + \frac{\alpha_2 \rho_b}{k_f} \tag{2a}$$

INFLUENCE OF TEMPERATURE ON MASS TRANSFER IN LIQUID FLUIDIZATION

V. VANADURONGWAN AND J. P. COUDERC

NOMENCLATURE

A   Transfer area

C_1 Entrance concentration

C_2 Exit concentration

C_S Solubility of benzoïc acid in water

d   Diameter of the dissolving spheres

D   Benzoïc acid diffusivity in water

g   Acceleration of gravity

k   mass transfer coefficient

m   mass flux

u   velocity

ρ density

μ viscosity

$$Ga = \frac{d^3 \rho_L^2 \, g}{\mu^2} \qquad \text{Galileo number}$$

$$Mv = \frac{\rho_S}{\rho_L} \qquad \text{density number}$$

$$Re = \frac{d \, u \, \rho_L}{\mu} \qquad \text{Reynolds number}$$

$$Sc = \frac{\mu}{\rho_L \, D} \qquad \text{Schmidt number}$$

$$Sh = \frac{k \, d}{D} \qquad \text{Sherwood number}$$

Subscripts

  L : liquid   S : solid

INTRODUCTION

Recent studies[1,2,3] have concluded that mass transfer between solid and liquid in a fluidized bed can be represented by correlations between five dimensionless groups, in the form

$$Sh = g(Ga, Re, Mv, Sc) \tag{1}$$

At the moment, only the influence of the hydrodynamical factors Ga and Re has been experimentally investigated[1,2]. To our knowledge, the influence of the physical properties of the system, expressed in the form of a Schmidt number has not yet been the subject of any study. We present here the results of an experimental study, relative to the dissolution of benzoic acid particles in water at various temperatures.

1. EQUIPMENT AND MODE OF OPERATION

The equipment and mode of operation are quite similar to those used by other workers[1,2,3].

1.1 Equipment

Figure 1 is a general diagram of the fluidization unit. It must simply be noted that the water circulating in the system is in closed loop. The temperature is controlled to within $\pm$ 0,1°C.

1.2 Mode of Operation

The mass transfer coefficient k is determined from the relation (3)

$$k = \frac{m}{A \, \Delta c_{mL}} \tag{2}$$

The mass flux m is determined by weighing the particles before and after the experiment. The transfer surface area A is determined for

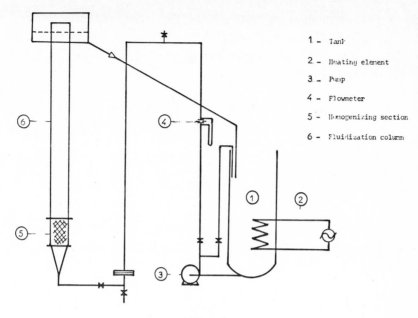

1 - Tank
2 - Heating element
3 - Pump
4 - Flowmeter
5 - Homogenizing section
6 - Fluidization column

Figure 1

the arithmetic mean of the diameters of benzoïc acid spheres at the beginning and at the end of the experiment. The logarithmic mean of concentration difference, Δc_{m_L} is :

$$\Delta c_{mL} = \frac{(c_s - c_1) - (c_s - c_2)}{\ln\left(\frac{c_s - c_1}{c_s - c_2}\right)} \qquad (3)$$

As water is recirculated, all the concentrations vary with time. Neverthless, since the mass of benzoïc acid dissolved m is small, and the total volume of circulating liquid is large, the variations are very small and we use time average values.

2. PHYSICAL PROPERTIES OF THE SYSTEM

Values of the density and viscosity of water and of the solubility of benzoïc acid in water have been taken from the literature[3,4,5]. Diffusivity of benzoïc acid in water has been experimentaly determined using a modified porous diaphragm cell method. A preliminary work in this field has already been published[6,7]. Modifications in the operating conditions allowed us to determine mean

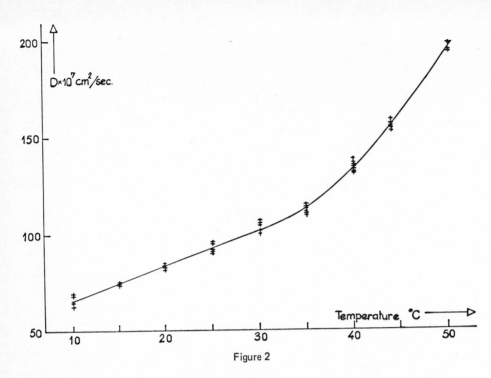

Figure 2

values of diffusivity between infinite dilution and saturation. This work will be published elsewhere[8] and we simply indicate the results obtained on figure 2.

3. EXPERIMENTAL ANALYSIS

Experiments have been organized in order to vary the Sherwood number as a function of temperature or Schmidt number at constant Reynolds, Galileo and density numbers.

Ten series of experiments have been done, varying the temperature in the range 16 to 50°C for each serie. Two examples of the results obtained are given on figure 3.

The influence of Sc on Sh can be represented by straight lines, which are practically parallel. The values of the slopes of these lines, determined by a least square method are presented in table 1. The average value is 0.436.

Knowing this new experimental result, all the data from DAMRONGLERD et al.[1,2,3] and from this work have been reworked. Two correlations are proposed

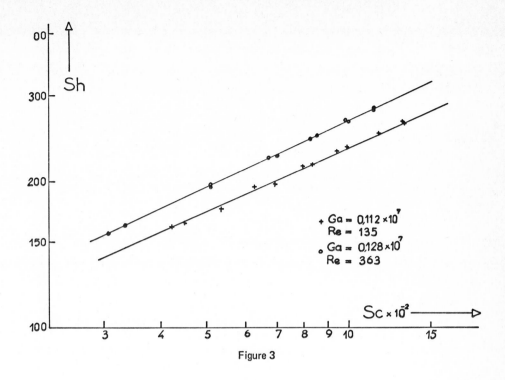

Figure 3

$$Sh = 0,147 \ Re^{0,011} \ Ga^{0,309} \ Sc^{0,436}$$

for $ReGa^{-\frac{1}{2}} < 0,405$

$$Sh = 0,078 \ Re^{-0,076} \ Ga^{0,390} \ Sc^{0,436}$$

for $ReGa^{-\frac{1}{2}} > 0,405.$

Table 1

Re	$Ga.10^7$	Slope		Re	$Ga.10^7$	Slope
688	0,117	0,4155		363	0,128	0,4406
626	0,121	0,4287		284	0,123	0,4481
556	0,120	0,4283		222	0,131	0,4412
503	0,126	0,4417		182	0,130	0,4374
429	0,125	0,4464		135	0,112	0,4308

The range of experimental conditions covered is :

$4,6 < d < 8,2$ mm

$180 < Re < 1300$

$8,3 . 10^5 < Ga < 44 . 10^5$

$300 < Sc < 1400$

COMMENTS

1 – The influence of temperature on mass transfer is important : at 50°C the rate is about two times greater than at 16°C. The study of this parameter is then of real concern.

2 – Analysis of the influence of temperature is long and tedious because, it is necessary to determine, with precision, the physical properties themselves and in particular the diffusivity. The quality of the results depends largely on the precision of these data.

3 – In the range covered, the exponant of the Schmidt number appears to be constant and equal to 0.436. This value lies in the range 1/3 – 1/2 corresponding for a fixed unique sphere to the two limiting theoretical cases of the boundary layer theory and of the penetration theory.

4 – The density number Mv does not appear in the proposed correlations. Its influence has not yet been investigated and it is included in the numerical coefficients.

5 – Like in previous work, it must be noticed that the Reynolds number has a very slight influence.

LITERATURE

1 – DAMRONGLERD S., COUDERC J.P. and ANGELINO H., Proceedings of the International Symposium "La fluidisation et ses Applications", Toulouse, 1973, p. 330.

2 – DAMRONGLERD S., COUDERC J.P. et ANGELINO H., Trans. Instn. Chem. Engrs (to be published)

3 - S. DAMRONGLERD, Docteur-Ingénieur thesis, Université Paul-Sabatier, Toulouse, 1973

4 - KIRK-OTHMER, Encyclopedia of Chemical Technology, 3, 422, 1964.

5 - International Critical Tables, Mc Graw Hill, 1933.

6 - J. LOZAR, Docteur de Spécialité Thesis, Université Paul-Sabatier, Toulouse, 1974.

7 - LOZAR J., LAGUERIE C. and COUDERC J.P., Can. J. Chem. Engrs. (to be published)

8 - V. VANADURONGWAN, Thesis in preparation.

FLUIDIZED BED CRYSTALLIZATION

T. BLICKLE, O. BORLAI, E. VASANITS, AND E. SIMON

NOMENCLATURE

A   constant of Clausius-Clapeyron's equation, $^{\circ}K$

B   constant of Clausius-Clapeyron's equation, kg/m^3

c   concentration of the solution, kg/m^3

$\bar{d}$ average diameter of the crystals, μ

G   capacity, kg/m^3h

T_1 temperature of the hot solution, $^{\circ}K$

T_2 temperature of the cold solution, $^{\circ}K$

T_3 common temperature of the solutions, $^{\circ}K$

v   linear velocity of liquid, mm/s

V_1 feeding rate of the hot solution, m^3/h

V_2 circulation rate of the cold solution, m^3/h

w   mass of crystals separating in the crystallizer, kg/h

w_1 mass of crystal separating in the cooler, kg/h

$\sigma/\bar{d}$ relative scattering

    The liquid fluidization was applied to investigation of different crystallization methods. The crystallization by cooling and the isotherm salting out crystallization were studied.

    With the mathematical model derived optimum mean temperature belonging to given initial temperatures

345

of the hot and cold solutions could be determined
for the mother lye departing from the crystallizer;
at this temperature the largest possible part of
crystals is separated in the crystallizer.

For the optimum common solution temperature determined theoretically, the ratio of crystals separated
in the crystallizer and in the whole cyclic process
has been determined.

The correctness of the values determined theoretically was checked by experiments and the effect
of cooling parameters on the grain size distribution
of crystalline product and on the performance of
crystallizer have been investigated.

For the investigation of cooling crystallization
by the mixing of cold and hot solutions an experimental apparatus has been constructed, which makes
possible the back-cooling of the mother-lye and thus
the circulation of the cold solution in the system.
Fig. 1. shows the connection diagram of elements
/1-4/ of the cyclic, fluidized-bed cooler-crystalli-
zer.

In the fluidized-bed crystallizer /1/, shown
schematically in Fig.1. the hot, saturated solution
of T_1 temperature and V_1 feeding rate is directly
mixed at fluidization flow rate with its cold solution of T_2 temperature, circulating at a rate of V_2.
From here, the mother-lye of T_3 common temperature

and a volumetric rate of V_1+V_2 is introduced after
the separation of the crystals produced into the
cooler /2/. A rotary film apparatus was used as
cooler, where the mother-lye was cooled back to the
inlet temperature T_2 of the cold solution. Crystals
separating during cooling were separated in settler
/3/ and the cold solution was pumped back with the
pump /4/ into the fluidized-bed crystallizer. The
crystals formed were withdrawn from the upper part
of conical design of the fluidized-bed crystallizer
/1/.

The salting out crystallization was investigated also in laboratory scale. The solution and the
reagent for salting out were fed at the lower third
part of reactor, the products were withdrawn at the
bottom, the mother lye was removed by suction and
was fed back to the crystallizer at lower part of
fluidized bed. Constant temperature within the system was maintained by a heat exchanger connected
into circulation system.

In crystallization by direct cooling the aim
was that the predominant part of the crystals should
separate in the fluidized-bed crystallizer /1/ /see
crystal quantity w in Fig.1./ In this case, intensive cooling can be applied in cooler /2/, where
only a small w_1 quantity of crystal nuclei of very
small size will be formed. This ensures on the one

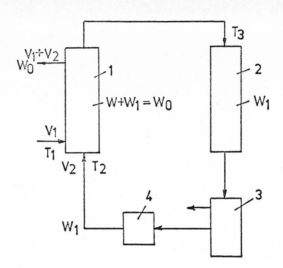

Figure 1. Scheme of the crystallization cycle: 1. fluidized-bed crystallizator; 2. cooler; 3. settler; 4. slurry pump.

hand the safe operation, free of clogging of the cyclic process, while on the other hand the small crystal nuclei getting back into the fluidized-bed crystallizer, further as seeding crystals crystallization.

Therefore our object in the solving of the theoretical model was to investigate the change of the ratio of the crystal quantity w formed in the crystallizer and of the total separated crystal quantity $w + w_1$, $\beta_o = \dfrac{w}{w + w_1}$, as a function of the cooling parameters.

On the basis of the solubility data in water of the model substance sulfaguanidine, the concentration of the saturated solution can be given by Clausius-Clapeyron's equation:

$$c = B\ e^{-\frac{A}{T}}\qquad\qquad /1/.$$

On the basis of equation /1/, the material
balance can be written for the crystal quantity separated in the fluidized-bed crystallizer:

$$V_1 \, B \, e^{-\frac{A}{T_1}} + V_2 \, B \, e^{-\frac{A}{T_2}} - /V_1 + V_2/ \, B \, e^{-\frac{A}{T_3}} = w \quad /2/$$

Material balance for the total crystal quantity
separated in the cyclic crystallizer:

$$V_1 \, B \, e^{-\frac{A}{T_1}} - V_1 \, B \, e^{-\frac{A}{T_2}} = w + w_1 \qquad\qquad /3/$$

One of the objects of our optimization task was
to express the ratio of crystal separation. β_o from
the ratio of material balances /2/ and /3/:

$$\beta_o = 1 - \frac{e^{-\frac{1}{T_3 \text{:opt}}} - e^{-\frac{1}{T_2'}}}{\frac{V_1}{V_1 + V_2} /e^{-1/T_1'} - e^{-1/T_2'}/} \qquad /4/$$

where: $\quad T' = \frac{T}{A}$.

The ratio of crystal separation, β_o approaches
the closer 1, the smaller is the difference between
the common solution temperature T_3 $_{opt}$ and the temperature T_2 of the cold solution. In this case the aim
is attained that almost the total quantity of crystals has separated in the fluidized bed crystallizer.

From equation /4/ β_o, the crystal separation
ratio belonging to the given solution temperatures

and feed rates has been determined for the model
substance sulfaguanidine, for which the values of
the constants of Clausius-Clapeyron's equation are:
$A = 5.66 \cdot 10^{3} {}^{\circ}K$ and $B = 2.33 \cdot 10^{8}$ kg/m^3.

Table I contains the results of the experiments,
carried out at constant V_2 circulation rate of the
cold solution. It can be seen that on increasing
the feed rate V_1 of the hot solution the temperature
T_3 of the mother-lye departing from the fluidized-
-bed crystallizer increases, and consequently the
value of β_0 descreases, that is to say less and
less crystals separate in the fluidized-bed crystallizer. However, this meant already the upper limit
of the feeding rate of the hot solution because in
this case supersaturated solution departed from the
fluidized-bed crystallizer, which might represent a
danger of clogging in the cooler of the cyclic
process. The ratio of crystal separation β_0, was
a relatively high value /see Table I/ and changed
only within a narrow interval, proving the intensive
cooling effect which can be a priori attained in
crystallization by direct cooling.

On the other hand, on increasing the circulation
rate V_2 of the cold solution at constant feeding
rate V_1 of the hot solution T_3, the temperature of
the mother-lye departing from the fluidized-bed
crystallizer decreased and consequently β_0, the

Table 1 Variation of parameters of direct cooling crystallization with feed rate of warm solution

V_1 / l/h /	V_2 / l/h /	$\dfrac{V_1}{V_1 + V_2}$	T_1 /K°/	T_2 /K°/	T_3 /K°/	β °	$\bar{d}$ / μ /	G / kg/m³h/
1,0	30	0,032	358,0	295,0	297,0	0,8475	32,43	20,3
4,0	30	0,117	359,0	299,0	304,5	0,8897	46,21	74,8
6,0	30	0,167	363,0	298,3	307,5	0,8351	51,02	113,2
12,0	30	0,286	363,5	296,5	312,0	0,8230	57,83	242,0
24,0	30	0,444	364,0	302,0	323,1	0,7991	58,60	487,0

ratio of crystal separation increased as can be seen
from Table II.

Data in Table I show that in crystallization by
direct cooling a product of small mean particle size
can be obtained. By increasing the feed rate of the
hot solution the average grain size of the sulfaguanidine crystals can be increased until a near
constant value of $d = 60 \mu$ is attained.

According to data in Table II. a product of
smaller mean particle size has been prepared by
increasing the feeding rate of the cold solution.
This can be explained by the higher cooling rate
and by the increased flow rate /i.e. comminuting
effect/ of the solution streaming upwards in the
fluidized-bed crystallizer.

The laboratory scale crystallizer of a volume
of about 10 litre has a high capacity: 250 kg/m^3h
and this can be increased near proportionally by increasing the feed rate of the hot solution of given
temperature.

The salting out was investigated for two models,
namely an inorganic salt and an organic pharmaceutical product. Depending on properties single crys-
talls were formed in the first case, while in the
second case agglomerates were gained.

Salting out of sodium chloride solution with
ethyl alcohol of the same volume was investigated

Table 2

V_1 /l/h/	V_2 /l/h/	$\dfrac{V_1}{V_1 + V_2}$	T_1 /K°/	T_2 /K°/	T_3 /K°/	ϕ °	$\bar{d}$ /μ/	G /kg/m³h/
12	8	0,600	364,0	303,2	334,5	0,6339	76,53	200
12	20	0,375	365,0	290,4	309,0	0,8057	58,92	212
12	30	0,286	363,5	296,5	312,0	0,8230	57,83	242
12	40	0,231	363,5	293,5	306,6	0,8630	56,89	202
12	50	0,194	362,0	300,4	308,5	0,8555	56,93	199

at 25°C in an apparatus of 3,4 litre. Ratio of circulation was 10-50 times greater than the feed rate,
while linear velocity /v/ of liquid relating to the
empty cross-section of apparatus varied from 6 8 to
13,4 mm/s. Effect of capacity and circulation rate
on the particle size distribution was investigated.
Data of experiments are summarized in Table 3.

With increasing capacity varying from 26,5-66,2
kg/m^{3}h the mean particle size decreased while inho-
mogenity of distribution increased. When the apparatus was operated with constant capacity the particle
size was increased with increased circulation rate
within the given range.

Table 3   Effect of capacity of apparatus and circulation on particle size
distribution of product

G /kg/h m^3/	v /mm/s/	$\bar{d}$ /μ /	/ $^-$ d
	6,8	219	0,31
26,5	10,2	228	0,31
	13,4	240	0,28
	6,8	217	0,31
39,7	10,2	227	0,30
	13,4	234	0,28
	6,8	207	0,36
53,0	10,2	203	0,37
	13,4	219	0,32
	6,8	201	0,36
66,2	10,2	210	0,35
	13,4	219	0,32

Effect of crystallization temperature on mean size of product is illustrated in the case of salting out by water of a pharmaceutical product dissolved in formic acid. The crystallization was carried out in an apparatus of 1,6 litre volume in temperature range 30-40°C.

The product has formed as agglomerate when crystallization was carried out with different ammounts of salting out reagent. Effect of temperature on the size and purity of the formed agglomerates was investigated. When crystallizer was operated at recirculation rate of 7 mm/s and capacity of 62 kg/m^3h, at 30-40°C the mean size of agglomerates was 240 μu and 270 μu, respectively. Increasing effect of temperature on the particle size is reasoned by change of supersaturation. When purity of product was compared with that of products consisting of single crystals gained by other processes under similar conditions the results were appropriate. In addition at the crystallization in fluidized bed the gained product can be filtered easily and has a pleasant form contrary to the product of earlier procedures, which consisted of small particles of 3-4 mm and being susceptible to adhesion.

MASS TRANSFER IN LIQUID FLUIDIZED BED AT LOW REYNOLDS NUMBERS

C. LAGUERIE AND H. ANGELINO

NOMENCLATURE

A   Surface area of the crystals in the bed

C   Solute concentration in the solution (kg of hydrate per kg of water)

C_1 Solute concentration in the solution at the entry of the column

C_2 Solute concentration in the solution at the exit of the column

c^* Solute concentration at saturation

$(c^* - c)_{lm}$ log mean concentration in the bed at time t

$$(c^* - c)_{lm} = \frac{(c^* - c_1) - (c^* - c_2)}{\ln \dfrac{c^* - c_1}{c^* - c_2}}$$

D   Diffusivity of the solute in the solution

f_s Surface shape factor

g   Acceleration of the gravity

k_D Mass transfer coefficient for the dissolution (kg of hydrate per hr . m^2X (kg of hydrate per kg of water)

K   Dimensional constant (equation 2)

L   Characteristic size of the crystals

M   Mass of the crystals in the bed

m   Exponent of U (equation 2)

n   Exponent of L (equation 2)

t   Time

U   Superficial velocity of the solution

Greek Letters

ε Voidage of the bed

μ Viscosity of the solution

ρ_1 Density of the solution

ρ_s Density of the crystals

Dimensionless groups

Ga   Galileo number $(Ga = \dfrac{L^3 \rho_1 (\rho_s - \rho_1) g}{\mu^2})$

Ga'   modified Galileo number $(Gà = \dfrac{L^3 \rho_1^2 g}{\mu^2})$

Re   Reynolds number of the crystals $(Re = \dfrac{L U \rho_1}{\mu})$

Sc   Schmidt number $(Sc = \dfrac{\mu}{\rho_1 D})$

Sh   Sherwood number $(Sh = \dfrac{k_D L (1+C)}{\rho_1 D})$

INTRODUCTION

It is generally accepted[1-3] that crystal growth results from two steps :

- a diffusional step, comparable to the dissolution, in which molecules of the solute move from the bulk solution to the surface of the crystal,

- a surface reactionnal step during which the molecules of the solute arrange themselves into the crystal lattice.

In the present study, the characteristics of the dissolution of monohydrate citric acid crystals in a fluidized bed at 25°C have been determinated and then compared with some preliminary results obtained for the crystallization of this acid under similar conditions. This standard procedure will permit to conclude whether the diffusional step is the most important step for the crystal growth of this acid under the taken conditions.

Because the citric acid is very soluble in water (207.2 grams of the hydrate per 100 grams of water at 25°C[4]) and the concentrated solutions are very viscous (20.6 cp at saturation at 25°C[4]) the

Reynolds and Galileo numbers will be very low (Re < 0.9 and Ga < 190) and the Schmidt numbers very high (Sc > 35,400). So, this study is relative to a particular case of mass transfer in a fluidized bed and comparisons with experimental results obtained by other authors in different ranges of Reynolds and Schmidt numbers will be very difficult. However a tentative comparison with the experimental work of DAMRONGLERD[5,6] concerning dissolution of benzoïc acid spheres in water has been done.

APPARATUS

The apparatus is shown schematically in Figure 1. It especially consists of a perspex column of 9.4 cm in diameter with a conic section at its lower part and a symetrical overflow at its upper part. The distributor is a 1.5 mm thickness perforated plate, the holes of which are inclined at 45° with respect to an horizontal plane in order to limit jet effects. This distributor can move around a diametral horizontal axis so that the magma can be removed through the column basis. The solution is circulated by a small stainless steel centrifugal pump. The flow rate can be fitted by means of a bypass and it is measured by a flow meter. A stired tank of 35

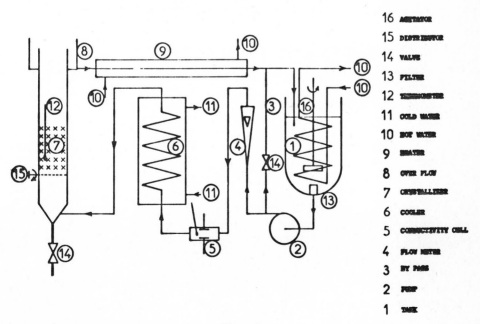

16	AGITATOR
15	DISTRIBUTOR
14	VALVE
13	FILTER
12	THERMOMETER
11	COLD WATER
10	HOT WATER
9	HEATER
8	OVER FLOW
7	CRYSTALLIZER
6	COOLER
5	CONDUCTIVITY CELL
4	FLOW METER
3	BY PASS
2	PUMP
1	TANK

Figure 1

liters is maintained at 30°C by hot water and permits the column inlet
concentration to stay constant for an experiment. Supersaturation is achieved
by cooling the solution in a coiled exchanger. Temperature of the bed is indicated on a thermometer graduated to 0.1°C and the cooling water flow rate is
regulated in order to maintain this temperature at about 25°C.

PROCEDURE

A solution of desired concentration was prepared by adding either
acid crystals or water in the stirred tank. Crystals to be dissolved or grown
were sieved, weighed and numbered from a sample of three hundred of them.
Then, they were introduced into the top of the column. The inlet concentration
was continuously measured by means of a conductimeter while the outlet
concentration was determined every six minutes by titration with soda solution.
At the end of the run, the crystals were filtered, washed in ethyl ether and
air dried. After drying, the crystals were weighed and numbered by taking a
sample of them.

ANALYSIS OF THE EXPERIMENTAL RESULTS

Most authors determine a mean value of the mass transfer coefficient k_D for each experiment by integrating the equation

$$\frac{dM}{dt} = k_D \; A \; (c - c^*)_{lm} \tag{1}$$

over the experiment using mean values for the mass transfer area A
and for the difference of concentration $(c - c^*)_{lm}$. This equation is
based upon two hypothesis :

- perfect mixing of the solid particles in the bed
- piston flow of the solution

However this method is correct only when the size of the crystals
remain nearly constant during the experiment. This condition is not
always verified. So, it is better to take into account the variation
of the crystal size and its influence on the mass transfer coefficient
and on the mass transfer area. The authors have already developped a
simple original method[7] which is summarized here. In order to inte-
grate relation (1), all the different variables M, k_D and A have to be
expressed as a function of L the characteristic size of a crystal

defined as the diameter of the sphere having the same volume as the crystal. Let us assume that the mass transfer coefficient k_D obeys the relation

$$k_D = K \ U^m \ L^n \tag{2}$$

On the other hand, the mass transfer area A is expressed by the following relation

$$A = N \ f_s \ L^2 \tag{3}$$

Then eq (1) can be integrated over the experiment. The dimensional coefficient K and the exponents "m" and "n" are determined by a numerical method of identification. This identifications for the dissolution experiments has been carried out by using a computer "MITRA 15". The derived relation is

$$k_D = 4.885 \ U^{0.217} \ L^{0.009} \tag{4}$$

The standard deviation is 5.3 % and the maximum deviation 9.9 %. This agrees with the experimental errors which have been estimated at 17 %. Some experimental values an included in Table 1.

These results may be presented in a dimensionless form similar to that proposed by DAMRONGLERD, COUDERC and ANGELINO[5,6]

$$Sh = 0.144 \ Re^{0.217} \ Ga^{0.264} \ Sc^{1/3} \tag{5}$$

for $0.15 < Re < 0.86$ $25 < Ga < 190$

$35,500 < Re < 37,900$ $\varepsilon < 0.85$

It must be noticed that as usual the value of the Sc exponent has been fixed equal to 1/3. This value was selected in concordance with theoretical studies concerning creeping flow around a sphere[8-11] but VANADURONGWAN and COUDERC[12] have established elsewhere in different conditions that 0.44 fit better their experimental results.

Variation of $Sh/Ga^{0.264} \ Sc^{1/3}$ versus Re are presented on figure 2. The values of the dimensionless numbers Sh, Ga and Re are average values for each expriment. Results concerning some crystallization experiments have also been plotted on this figure. The agreement between the

Table **1**

	u (cm/s)	L_1 (cm)	L_2 (cm)	Température (°C)	$C - C^*_{lm}$	t (mn)	k_D
Dissolution	0.434	0.2573	0.2247	24.9	0.0544	50	4.016
	0.154	0.1567	0.1497	24.9	0.0186	40	3.152
	0.244	0.1559	0.1540	25.0	0.061	27	3.463
	0.244	0.1989	0.1852	25.3	0.0425	30	3.627
	0.368	0.1979	0.1860	25.6	0.0340	30	3.914
	0.174	0.1974	0.1924	24.6	0.0169	30	3.310
	0.370	0.2476	0.2445	24.8	0.0088	30	3.918
	0.370	0.1569	0.1361	25.0	0.0600	30	3.468
Cryst.	0.301	0.1619	0.1798	25.3	0.0416	40	3.613
	0.382	0.1462	0.1729	25.0	0.0630	40	3.542
	0.392	0.2166	0.2484	25.4	0.0669	40	3.978
	0.300	0.2177	0.2381	25.5	0.0472	40	3.623

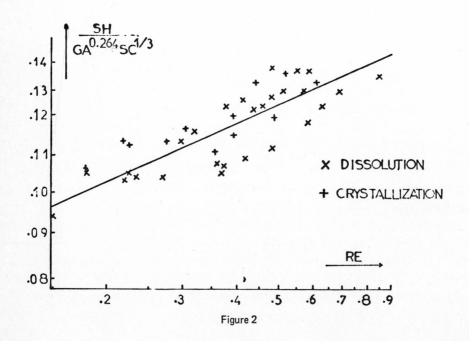

Figure 2

results obtained for the dissolution and the crystallization is fairly good.

DISCUSSION OF THE RESULTS

The agreement between the results obtained for the dissolution and the crystallization under similar conditions proves that in fluidized bed the growth of monohydrate citric acid is controlled by the diffusional step. This greatly differs from the conclusion obtained by KOCOVA and NYVLT[13] for the crystallization of monohydrate acid in an agitated vessel at 32°C. These authors have established that the resistance to surface reaction is of the same order as the resistance to diffusion. However it is very difficult to establish an exact comparison because the hydrodynamic conditions are different in a fluidized bed and in an agitated vessel.

As far as dissolution in liquid fluidized bed is concerned, only few experimental results can be found in bibliography and exclusively in a different range of experimental conditions, so a comparison is quite difficult. Nevertheless a tentative comparison with DAMRONGLERD's results[5,6] was made. In the case of dense bed ($\varepsilon < 0.85$) and using a perforated plate distributor, this author has established for the dissolution of benzoic acid the following equation

$$Sh = 0.474 \quad Re^{0.073} \quad Ga^{0.265} \quad Sc^{1/3} \qquad (6)$$

for $70 < Re < 2,300$ and $Sc < 2000$

Comparison of equations (5) and (6) points out that the influence of the Galileo number on mass transfer is the same, but that the Re exponent is different. Assuming that equation (6) is still valid to evaluate $Sh/Ga^{0.265} \ Sc^{1/3}$ under the particular experimental conditions developped here, leads to values being 5 times greater than those experimentally determined.

This conclusion seems to support the paper recently published by NELSON and GALLOWAY[11]. These authors have explained the "anomalously" small values of the mass transfer coefficient at low Re numbers in considering the differences in the flow regimes accross the bed according as Re number is higher or lower than 5. Furthermore, Le CLAIR and HAMIELEC[8] have observed that the ratio $Sh/Sc^{1/3}$ strongly decreases when Sc number increases for Re number less than 10. In fact, it seems that there is a need for new experiments in the intermediate range of Re and Sc values.

CONCLUSION

The comparative study in fluidized bed between the dissolution and the crystallization of monohydrate citric acid at 25°C under similar conditions permits to point out that the crystal growth of this acid is controlled by a diffusional mechanism.

The results for mass transfer are not easily comparable to other published results because the operating conditions developped in this study are quite particular. However the Galileo influence upon the mass transfer is the same as that determined by DAMRONGLERD[5,6].

LITERATURE CITED

1 – MULLIN J.W., "Crystallization", 2nd ed., Butterworth (London)1971.

2 – NYVLT J., "Industrial Crystallization from Solutions", Butterworth (London), 1971.

3 – STRICKLAND-CONSTABLE R.F., "Kinetics and Mechanism of Crystallization", Academic Press (London), 1968.

4 – LAGUERIE C., AUBRY M. and COUDERC J.P., publication to Jl. Chem. Eng. Data.

5 – DAMRONGLERD S., Docteur-Ingénieur Thesis, Université Paul-Sabatier, Toulouse, 1973.

6 – DAMRONGLERD S., COUDERC J.P. and ANGELINO H., will be published in Trans. Instn. Chem. Engrs, April 1975.

7 – LAGUERIE C. and ANGELINO H., publication proposed to Chem. Eng. Jl.

8 – PFEFFER R. and HAPPEL J., A.I.Ch.E. Jl., 1964, 10, 605

9 – LE CLAIR B.P. and HAMIELEC A.E., I.Ch.E., Symp. Series, Tripartite Chem. Engng Conf. (Montréal), 1968, 197.

10 – BIRD R.B., STEWART W.E. and LIGHTFOOT E.N., "Transport phenomena, Wiley (New York), 1960.

11 – NELSON P.A. and GALLOWAY T.R., Chem. Eng. Sci., 1975, 30, 1.

12 – VANADURONGWAN V. and COUDERC J.P., Communication presented here.

13 – KOCOVA H. and NYVLT J., Collection Czech. Chem. Commun., 1972, 37, 3669.

THE PERCOLATING POROUS ELECTRODE (P.P.E.) OF CONDUCTING FLUIDIZED GRAINS

F. COEURET AND P. LE GOFF

The Percolating Porous Electrode (P.P.E.) is a bed of conducting grains through which flows the electrolyte to be treated
(Figure 1). Among its potential applications to industry, the two
most important are :
- the metal extraction from dilute solutions obtained by leaching
  of ores.
- the elimination or recuperation of heavy metals from effluents.

      The electrolytic treatment of dilute solutions needs low
current densities and consequently large electrode surfaces to
permit large cell intensity or productivity. Owing to its granular
structure, the PPE presents a very high specific area, and the flow
of electrolyte through the bed increases the permissible current
densities by increasing the mass transfer coefficient between the
liquid and the electrode itself.

      The <u>fixed bed</u> (1) and the <u>fluidised bed</u> (2) of conducting
grains are two examples of PPE :
- the first one, a limiting case of the second, should operate periodically
- the second should allow continuous operations.

EFFICIENCIES IN A PPE

      Figure 1 shows a cathodic PPE and the concentration profile
in the electrolyte. The total current I_t is here the sum of the current I_p , corresponding to the reaction on the particles, and of the
current I_f, due to the reaction on the current feeder. It is possible to define 3 efficiencies R_t, R_p and R_f which correspond respectively to the currents I_t, I_p and I_f. Among the practical problems,
one is the minimization of the ratio R_f/R_p.

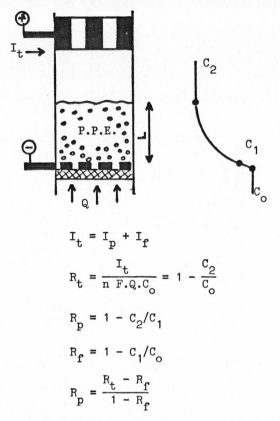

$$I_t = I_p + I_f$$

$$R_t = \frac{I_t}{n\,F.Q.C_o} = 1 - \frac{C_2}{C_o}$$

$$R_p = 1 - C_2/C_1$$

$$R_f = 1 - C_1/C_o$$

$$R_p = \frac{R_t - R_f}{1 - R_f}$$

Figure 1. Example of PPE and corresponding concentration profile.

COMPARISON OF SEVERAL ELECTRODES

Figure 2 compares values obtained for the limiting current in the cell of Figure 1 for several types of granular beds in contact with the current feeder, as a function of the electrolyte velocity in the empty column. This electrolyte is a mixture of $Fe(CN)_6K_4$ and $Fe(CN)_6K_3$ in NaOH as a conducting support.

The conducting grains are graphite or bronze microspheres (diameter 0,082 cm) electrolytically covered with nickel and gold respectively ; the non-conducting grains are glass microspheres (diameters 0,08 cm and 0,06 cm).

The fluidised PPE is less efficient than the fixed PPE ; it appears only interesting for porosities very near that of the fixed bed porosity, that is for conditions where the contribution of I_f to the total current I_t is very small.

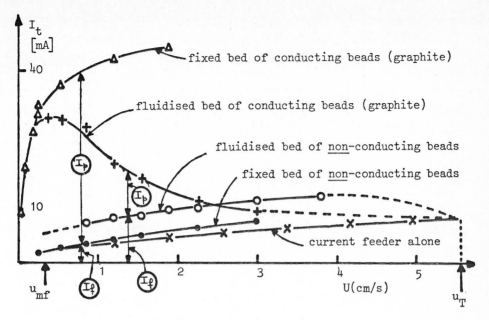

Figure 2. Comparison of several electrodes.

Through the comparison of these different electrodes, it can be seen that many phenomena can exist simultaneously in a fluidised PPE :

- the geometrical obstruction by the grains reduces the section free for the liquid flow and thereby increases the liquid velocity
- the agitation of the grains induces micromixing of the liquid
- the bed conductivity by formation of unsteady bridges of grains between the current feeder and the different parts of the bed
- the bed conductivity by turbulent diffusion of charged grains. Initialy charged on the feeder, a grain would lose progressively its charge during collisions with other grains.

LOCAL ELECTRODE POTENTIAL IN THE PPE

Distributions of local electrode potential along the height of the fluidised PPE show that the current feeder becomes the most active part of the bed as soon as a porosity of about 0,6 is reatched. Figure 2 is an illustration of the influence of bed expansion on the potential distribution for bronze gold-plated particles.

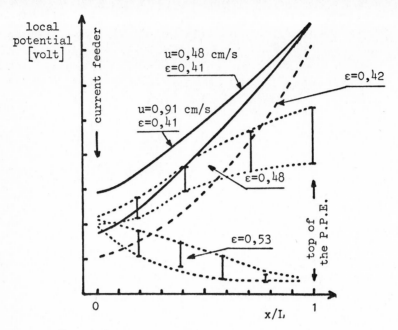

Figure 3. Evolution of electrode potential with bed expansion.

From a practical point of view, the conclusion is that the "volumic" fluidised PPE only exists at low porosities corresponding to dense beds.

REFERENCES

(1) D.N. BENNION and J. NEWMAN
     J. of Appl. Electrochem., 2, 113-122, (1972)
(2) P. LE GOFF, F. VERGNES, F. COEURET and J. BORDET
     Ind. Eng. Chem., 61, 8-17, (1969)

LIQUID FLUIDIZATION: MASS TRANSFER BETWEEN THE LIQUID AND AN IMMERSED SURFACE

F. COEURET, P. LE GOFF, A. STORCK, G. VALENTIN AND F. VERGNES

It is well known (1) that the fluidisation of _inert_ particles around or near a surface immersed in a liquid increases the mass transfer rate to that surface.

At a given value of the liquid flow rate in the column, the presence of particles decreases the cross-section and thus increases the interstitial velocity of the fluid. Furthermore the fluidised particles act as turbulence promoters keeping the surface free for reaction.

MEAN VALUES $\bar{k}$ OF THE MASS TRANSFER COEFFICIENT

Figure 1 shows the experimental variations of $\bar{k}$ at a cylindrical surface (diameter 1,5 cm ; length 6 cm) axialy immersed in a liquid fluidised bed of glass beads (diameter 0,2 cm). The method uses the cathodic reduction of potassium ferricyanide. The diagram gives also results for the same surface respectively immersed in the liquid alone, in the bed maintained fixed and in a bed hydraulicaly transported.

The conclusions drawn from a similar study are :
- at low velocities, corresponding to fluidised beds, the presence of particles increases, by a factor of 10 to 15, the mass transfer coefficient, as compared to the pure liquid.
- at high velocities, corresponding to hydraulic transport, the presence of particles with a maximum volumic concentration of 20 %, does not increase the value of $\bar{k}$.
- values of $\bar{k}$ of the same order of magnitude may be obtained either with a fluidised bed or with a hydraulic transport, but in this latter case, the flow-rate is about 50 times greater. As a conse-

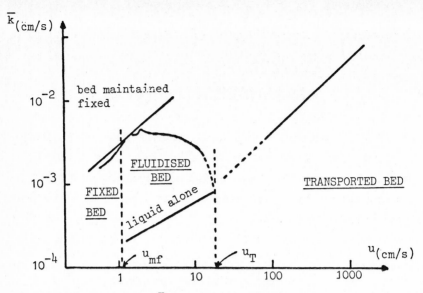

Figure 1. Variation of $\bar{k}$ with the liquid velocity in different systems.

quence the power consumption is much smaller in the fluidised bed than in the fast flowing pure liquid.

FLUCTUATIONS OF THE MASS TRANSFER COEFFICIENT

Fluctuations of the local mass transfer coefficient at a microelectrode have been analysed. Their mean frequency varies with

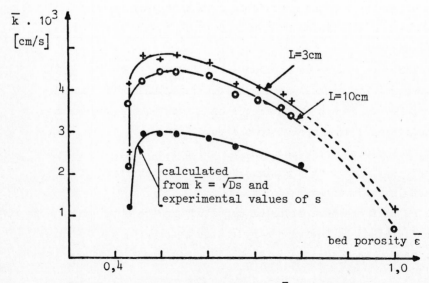

Figure 2. Experimental and calculated values of $\bar{k}$ in a fluidized bed.

the bed expansion similarly to the mean value $\bar{k}$. This analogy may be used to interpretate the mass transfer mechanism with a "random surface renewal model" at high values of s and a "stagnant boundary-layer model" at low values of s. Figure 2 compares experimental values of $\bar{k}$ for two electrode lengths (3 cm and 10 cm) and glass beads of diameter 0,1 cm with the curve calculated from the renewal model ($\bar{k} = \sqrt{D.s}$). The maximum values of $\bar{k}$ appear at the same bed porosity $\varepsilon (= 0,5)$. Furthermore the absolute values of experimental and theoretical mass transfer coefficients are not very different.

These results agree with a previous study (2) where we have analysed the fluctuations of the local fluidised bed porosity by means of conductivity probes.

REFERENCES

(1) P. LE GOFF, F. VERGNES, F. COEURET et D. MAESTRI
     Communication at the Spring Meeting of the Electrochemical
     Society, Chicago, 13-18 May 1973.
(2) J. BORDET, F. COEURET, P. LE GOFF and F. VERGNES
     Powder Technology, 6, 253-261, (1972).

BASIC RELATIONS FOR THE LIQUID PHASE SPOUT-FLUID BED AT THE MINIMUM SPOUT FLUID FLOWRATE

H. LITTMAN, D. V. VUKOVIC, F. K. ZDANSKI, AND Z. B. GRBAVCIC

NOMENCLATURE

A	=	area
d_N	=	spout inlet tube diameter $\cong d_S$
d_p	=	particle diameter
d_S	=	spout diameter
D	=	column diameter
$\delta(h\frac{z}{H})$	=	$\frac{3}{2}(h\frac{z}{H})^2 - (h\frac{z}{H})^3 + \frac{1}{4}(h\frac{z}{H})^4$
H	=	bed height
H_M	=	maximum spoutable height
h	=	H/H_M
$\hat{h}$	=	$1 - (1 - h)^3$
$P_a(z)$	=	annular pressure at z
ΔP_{mF}	=	pressure drop in bed fluidized
ΔP_{mSF}	=	pressure drop at V_{mSF} (annulus)
ΔP_{mS}	=	pressure drop at V_{mS} (annulus)
U	=	superficial fluid velocity
$U_a(z)$	=	fluid velocity in annulus at z (superficial)
$U_S(z)$	=	average fluid velocity in spout at z (superficial)

$U_r(z)$ = leakage velocity from spout to annulus (superficial)

V = volumetric flowrate

$[V_a(0)]_{mSF}$ = inlet annular flowrate at V_{mSF}

$[V_N(0)]_{mSF}$ = inlet spout flowrate at V_{mSF}

z = column height variable

ν = kinematic viscosity

Subscripts

a   = annulus

c   = column

s   = spout

mF = minimum fluidizing

mSF = minimum spout fluidizing

mS = minimum spouting

SUMMARY

Equations have been developed to predict the minimum spout-fluid flowrate, annular flowrate and pressure drop for liquid phase spout-fluid beds. The equations predict the results for spouted beds and pressure drop in spout-fluid beds.

INTRODUCTION

Spout-fluid beds have characteristics which are a blend of fluidization and spouting and therefore provide an important link between these two phenomena. Of equal importance, they offer the designer of processing units an important option in handling his fluid-particle processing problems.

Recently data were published [1] which show that the minimum spout-fluid flowrate in a liquid phase bed follows the simple linear relationship.

$$V_{mSF} = V_{mF} - \left(\frac{V_{mF}}{V_{mS}} - 1 \right) \left[V_N(0) \right]_{mSF} \tag{1}$$

Since correlations for V_{mF} and V_{mS} exist and because they are relatively easy quantities to measure, it is possible to predict or measure V_{mSF} in a relatively simple manner. This linear relation is in agreement with the data of Ray and Sarkar [2] for gas phase systems but disagrees with Nagarkatti and Chatterjee [3].

The annular pressure drop at V_{mSF} was obtained by extending the theory of Mamuro and Hattori [4] for spouted beds. A major result of that analysis, confirmed experimentally, is that ΔP_{mSF} is a linear function of $[V_a(0)]_{mSF}$. Unfortunately the equation for ΔP_{mSF} given in reference 1 is incorrect.

In this paper, a theory is developed for spout-fluid beds giving the spout and annular velocities, and the annular pressure drop at the minimum spout-fluid flowrate. The theory is experimentally confirmed using pressure drop measurements.

THEORY

Across any horizontal cross-section of the spout, the fluid has an average velocity, U_S. By continuity

$$- \frac{dU_S}{dz} = \frac{4}{d_S} U_r \tag{2}$$

where U_r represents the velocity of the fluid leaking from the spout to the annulus. Based on the work of Mamuro and Hattori [4] this velocity is represented as

$$U_r = c_1 (z - H_M)^2 \tag{3}$$

from which it is apparent that $U_r = 0$ only for beds at the maximum spoutable height.

The data in Figure 1 show that when $H \geq H_M$, $V_{mSF} = V_{mF}$ for any combination of spout and annular flow rates and that when $H < H_M$, $V_{mS} < V_{mF}$ so that $V_{mSF} < V_{mF}$. It is logical to assume and it is experimentally verified that the fluid velocity at the top of the spout is equal to the minimum fluidizing velocity.

Combining equations 2 and 3 and integrating over the length of the spout with $U_S(0) = [V_N(0)]_{mSF}/A_S$ and $U_S(H) = U_{mF}$, c_1 is evaluated as

$$c_1 = \frac{3d_S}{4\ H_M^3 \mathring{h}} \ (U_S(0) - U_{mF}) \tag{4}$$

where $h = H/H_M$ and $\mathring{h} = [1 - (1 - h)^3]$. Thus

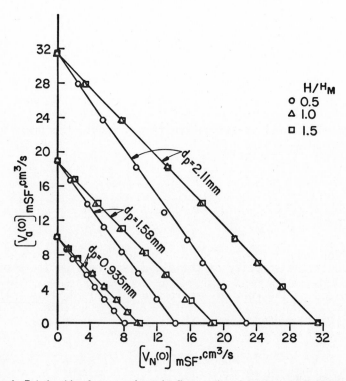

Figure 1. Relationship of spout and annular flows at the minimum spout-fluid flowrate.

$$U_r(z) = \frac{3d_S h}{4H_M \dot{h}} (U_S(0) - U_{mF})(1 - h\frac{z}{H})^2 \tag{5}$$

An important fact to note is that U_r decreases as the bed height is lowered.

The fluid velocity in the spout is obtained using equations 2 and 5.

$$\frac{U_S(0) - U_S(z)}{U_S(0) - U_{mF}} = \frac{1 - (1 - h\frac{z}{H})^3}{h} \tag{6}$$

Across any horizontal cross-section of the annulus, the fluid has an average velocity, U_a. By continuity

$$\frac{dU_a}{dz} = \frac{\pi d_S}{A_a} U_r \tag{7}$$

with $U_a(0) = [V_a(0)]_{mSF}/A_a$.

Integration of equation 7 gives the annular velocity

$$\frac{U_a(z) - U_a(0)}{U_a(H) - U_a(0)} = \frac{1 - (1 - h\frac{z}{H})^3}{\dot{h}} \tag{8}$$

where $U_a(H) - U_a(0) = A_S/A_a(U_0 - U_{mF})$ from a material balance on the bed.

It has been found experimentally that the height of the bed does not significantly alter the fluid velocity in the bed at any bed level, z. Thus

$$\frac{U_a(H) - U_a(0)}{U_{mF} - U_a(0)} = \frac{1}{\dot{h}} \tag{9}$$

and equation 8 is simplified to

$$\frac{U_a(z) - U_a(0)}{U_{mF} - U_a(0)} \cong 1 - (1 - h\frac{z}{H})^3 \tag{10}$$

The pressure distribution in the annulus is obtained by assuming that
$- dP_a/dz = K U_a$. Integration of this equation using equation 10 for U_a gives

$$P_a(0) - P_a(z) = K U_{mF} H \frac{\phi(h\frac{z}{H})}{h} + K U_a(0) H \left[\frac{z}{H} - \frac{\phi(h\frac{z}{H})}{h}\right] \tag{11}$$

Since $KU_{mF}H$ represents the annular pressure drop if the annulus is fluidized
throughout, the annular pressure distribution is

$$\frac{P_a(0) - P_a}{\Delta P_{mF}} = \frac{\phi(h\frac{z}{H})}{h} + \frac{[U_a(0)]_{mSF}}{U_{mF}}\left[\frac{z}{H} - \frac{\phi(h\frac{z}{H})}{h}\right] \tag{12}$$

When $z = H$, the pressure drop ratio for the annulus

$$\frac{\Delta P_{mSF}}{\Delta P_{mF}} = \frac{\phi(h)}{h} + \frac{U_a(0)}{U_{mF}}\left[1 - \frac{\phi(h)}{h}\right] \tag{13}$$

is obtained.

Equation 13 shows that the ratio, $\Delta P_{mSF}/\Delta P_{mF}$, varies linearly with the
annular flowrate from its value at minimum spouting to unity in a fluidized
bed. When $H = H_M$

$$\frac{\Delta P_{mSF}}{\Delta P_{mF}} = 0.75 + 0.25 \frac{[U_a(0)]_{mSF}}{U_{mF}} \tag{14}$$

For beds deeper than H_M, the bed consists essentially of a lower spout
fluid bed and an upper fluidized bed zone so that

$$\frac{\Delta P_{mSF}}{\Delta P_{mF}} = \left[1 - \frac{0.25}{h} \left(1 - \frac{[U_a(0)]_{mSF}}{U_{mF}} \right) \right] \tag{15}$$

Verification of the Spouted Bed Equations and the Prediction of V_{mSF}

The equations required for spouted beds are obtained by setting $U_a(0) = 0$ in the above equations. In this way, the annular velocity, the leakage velocity of the spout to the annulus and the annular pressure drop ratio, $\Delta P_{mS}/\Delta P_{mF}$, are obtained from equations 10, 5, 13, 14 and 15.

$$\frac{U_a(z)}{U_{mF}} = 1 - \left[1 - h\,\frac{z}{H} \right]^3 \tag{16}$$

$$\frac{U_r}{U_{mF}} = \frac{3\,d_s h}{4\,H\,\dot{h}}\;\frac{U_S(0) - U_{mF}}{U_{mF}}\left(1 - h\,\frac{z}{H} \right)^2 \tag{17}$$

$$\frac{\Delta P_{mS}}{\Delta P_{mF}} = \frac{\mathit{f}(h)}{h} \qquad\qquad \left(H \le H_M \right) \tag{18}$$

$$= 0.75 \qquad\qquad \left(H = H_M \right)$$

$$= 1 - \frac{0.25}{h} \qquad\qquad \left(H \ge H_M \right)$$

Equations 16 and 18 have recently been verified for both liquid and gas phase spouted beds by Grbavčić et al [5]. They differ from previously published equations (4,6) because the leakage from the spout to the annulus is a function of h. Grbavčić et al also derive an equation for the ratio of U_{mS}/U_{mF} (essentially from equation 10 with $U_a(0) = 0$ and an overall material balance on the bed)

$$\frac{U_{mS}}{U_{mf}} = 1 - \frac{A_a}{A_c} (1 - h)^3 \tag{19}$$

which correlates data in the literature [5] quite well.

Combining equations 1 and 19 gives V_{mSF} as:

$$V_{mSF} = V_{mF} - \left[\frac{\frac{A_a}{A_c}(1 - h)^3}{1 - \frac{A_a}{A_c}(1-h)^3} \right] \left(V_N(0) \right)_{mSF} \tag{20}$$

The value of V_{mSF} is rendered completely predictable using only a literature correlation for U_{mF} [7,8].

The Maximum Spoutable Height

In liquid phase spouted beds at V_{mS}, the symbol, H_M, denotes the smallest bed height for which U_{mS} is essentially equal to U_{mF}. As found previously [1], the maximum spoutable height in a spout-fluid bed is approximately equal to H_M and in dependent of the ratio of the spout to annular flow rates. Grbavčić et al [5] give the following correlation for H_M for liquid phase spouting of spherical glass particles in cylindrical beds

$$\frac{H_M}{D} = 0.347 \left(\frac{D}{d_p} \right)^{0.31} \left(\frac{D}{d_N} \right)^{0.41} \left(\frac{\nu}{\nu_{H_2O}} \right)^{-0.125} \tag{21}$$

Experimental Apparatus and Bed Properties

The experiments were performed in a rectangular Plexiglas column 117 mm wide, 9.2mm. thick and 500mm. high with a vertically mounted centrally located, circular nozzle inlet, 9.3mm. long filled with 1mm. glass beads before entering the bed. A screen separated the particles in the calming section from those in the bed and prevented the bed particles from falling into the nozzle.

Table 1 Bed and particle data

Rectangular Column: 117 x 9.2 mm.

Spout Inlet Tube: 9.3 mm. dia.

d_p = 0.935 mm, ρ_p = 2680 kg/m³, H_M = 150 mm, U_{mF} = 0.930 cm/s

H mm	U_{mS} cm/s	$\dfrac{\Delta P_{mS}}{\Delta P_{mF}}$
20	0.430	0.782*
27	0.540	0.818**
40	0.642	
60	0.749	
75	0.800	
100	0.860	
120	0.884	
140	0.917	
150	0.927	
225	0.930	

d_p = 1.58 mm, ρ_p = 2680 kg/m³, H_M = 140 mm, U_{mF} = 1.75 cm/s

H mm	U_{mS} cm/s	$\dfrac{\Delta P_{mS}}{\Delta P_{mF}}$
20	0.770	0.763*
40	1.15	0.786**
60	1.33	
70	1.39	
80	1.47	
100	1.58	
120	1.65	
140	1.74	
160	1.75	
210	1.75	

d_p = 2.11 mm, ρ_p = 2480 kg/m³, H_M = 134 mm, U_{mF} = 2.84 cm/s

H	U_{mS}	$\dfrac{\Delta P_{mS}}{\Delta P_{mF}}$
20	1.16	0.725*
40	1.72	0.769**
60	1.99	
67	2.07	
80	2.32	
100	2.46	
120	2.63	
134	2.81	
140	2.84	
201	2.84	

*   Pressure tap at z = 0, r/D/2 = 0.5

** Pressure tap in spout inlet tube 20 mm below bed inlet

Table 2 Comparison of annular pressure drop with theory for spout-fluid bed

d_p mm	$\dfrac{H}{H_M}$	$[V_a(0)]_{mSF}$ cm³/s	$[V_N(0)]_{mSF}$ cm³/s	$\dfrac{U_a(0)}{U_{mF}}$	$-\dfrac{(\Delta P_a)_{mSF}}{\Delta P_{mF}}$* expt.	theory (eqn. 13)
0.934	0.5	0	8.33	0	0.585	0.531
		1.39	6.94	0.138	0.623	0.596
		2.78	5.83	0.277	0.678	0.661
		4.17	4.72	0.416	0.740	0.726
		5.56	3.61	0.555	0.814	0.791
		7.50	1.94	0.750	0.893	0.883
		8.61	1.11	0.861	0.942	0.935
		10.0	0	1	1	1
1.58	0.5	0	14.2	0	0.548	0.531
		2.78	11.9	0.147	0.610	0.600
		5.56	9.72	0.294	0.685	0.669
		8.33	7.78	0.441	0.755	0.738
		11.1	5.56	0.588	0.823	0.807
		13.9	3.61	0.735	0.886	0.876
		16.7	1.67	0.882	0.948	0.945
		18.9	0	1	1	1
2.11	0.5	0	22.8	0	0.510	0.531
		4.17	20.0	0.133	0.560	0.594
		6.94	17.5	0.221	0.595	0.635
		9.72	15.6	0.309	0.660	0.676
		13.9	12.8	0.442	0.721	0.738
		18.1	9.44	0.575	0.780	0.801
		23.6	5.56	0.752	0.875	0.884
		27.8	2.50	0.885	0.940	0.946
		31.4	0	1	1	1
0.935	1	0	9.72	0	0.782	0.750
		1.39	8.61	0.138	0.795	0.785
		2.78	7.22	0.277	0.827	0.819
		4.17	5.83	0.416	0.862	0.854
		5.56	4.44	0.555	0.900	0.889
		7.50	2.50	0.750	0.950	0.938
		8.61	1.39	0.861	0.972	0.965
		10.0	0	1	1	1

* Pressure tap at z = 0, r/D/2 = 0.5

Table 2 (continued) Comparison of annular pressure drop with theory for spout-fluid bed

d_p mm	$\dfrac{H}{H_M}$	$[V_a(0)]_{mSF}$ cm^3/s	$[V_N(0)]_{mSF}$ cm^3/s	$\dfrac{U_a(0)}{U_{mF}}$	$\dfrac{(\Delta P_a)_{mSF}}{\Delta P_{mF}}$ * expt.	theory(eqn. 14)
1.58	1	0	18.6	0	0.763	0.750
		2.78	15.3	0.147	0.778	0.787
		5.56	12.8	0.294	0.819	0.824
		8.33	10.3	0.441	0.865	0.860
		11.1	7.50	0.588	0.900	0.897
		13.9	4.72	0.735	0.945	0.934
		16.7	2.22	0.882	0.962	0.971
		18.9	0	1	1	1
2.11	1	0	31.4	0	0.725	0.750
		4.17	26.9	0.133	0.760	0.783
		6.94	24.2	0.221	0.783	0.805
		9.72	21.4	0.309	0.802	0.827
		13.9	17.2	0.442	0.834	0.861
		18.1	13.1	0.545	0.872	0.886
		23.6	7.78	0.752	0.923	0.938
		27.8	3.33	0.885	0.956	0.989
		31.4	0	1	1	1
0.935	1.5	0	10.0	0	0.868	0.833
		1.39	8.61	0.133	0.886	0.856
		2.78	7.22	0.277	0.900	0.880
		4.17	5.83	0.416	0.921	0.903
		5.56	4.44	0.553	0.954	0.926
		7.50	2.50	0.750	0.975	0.958
		8.61	1.39	0.861	0.992	0.977
		10.0	0	1	1	1
1.58	1.5	0	18.9	0	0.852	0.833
		2.78	16.1	0.147	0.856	0.858
		5.56	13.1	0.294	0.885	0.882
		8.33	10.6	0.441	0.924	0.907
		11.1	7.78	0.588	0.946	0.931
		13.9	5.00	0.735	0.970	0.956
		16.7	2.22	0.883	0.989	0.980
		18.9	0	1	1	1
2.11	1.5	0	31.4	0	0.815	0.833
		4.17	27.2	0.133	0.839	0.856
		6.94	24.2	0.221	0.858	0.870
		9.72	21.7	0.309	0.872	0.885
		13.9	17.5	0.442	0.900	0.907
		18.1	13.3	0.575	0.924	0.929
		23.6	7.78	0.752	0.943	0.959
		27.8	3.61	0.885	0.970	0.981
		31.4	0	1	1	1

* Pressure tap at z = 0, r/D/2 = 0.5

The nozzle and annular flows were metered separately using rotameters and the pressures measured using peizometer tubes. The pressure tap in the annulus was set just below the screen midway between the nozzle wall and the outer edge of the bed.

Particle size and density, bed height, minimum spouting and fluidizing velocities and maximum spout height data for the beds investigated are given in Table 1.

Experimental Test of Theory and Recommended Relations

Pressure drop measurements were obtained for three different particle sizes and bed heights in order to test the validity of equations 13 through 15. The agreement between the experimental data and the theory is excellent (Figure 2).

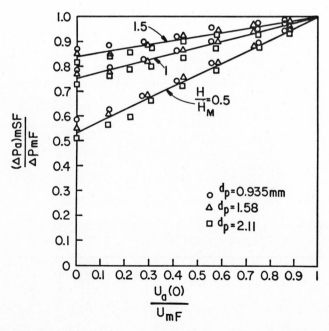

Figure 2. Experimental test of theory for predicting the annular pressure drop at V_{mSF}.

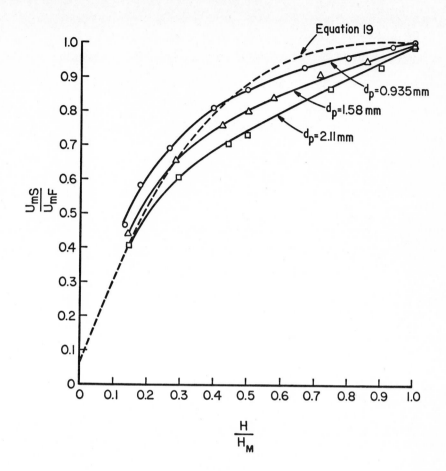

Figure 3. Comparison of theory and experiment for U_{mS}.

Equations 1, 10, 13, 14 and 15 are recommended for predicting the minimum spout-fluid flowrate, annular velocity and annular pressure drop. The correlation for H_M given in equation 21 was obtained using spherical glass particles so that it is advisable to measure H_M in a specific application. In addition although equation 20 can be used to predict V_{mSF}, better results are obtained using equation 1 and an experimental value for U_{mS}. To show this the experimental values for U_{mS}/U_{mF} from Table 1 are compared with those from equation 19 in Figure 3. Although the prediction of U_{mS} is satisfactory, there is a particle size effect unaccounted for.

CONCLUSIONS

1. Equations have been developed to predict the minimum spout-fluid flowrate, the annular velocity and pressure drop in the annulus in a liquid phase spout-fluid bed at V_{mSF}. The pressure drop data fit the equations quite well.

2. The equations reduce to those for a spouted bed when there is no annular flow and those equations have been experimentally verified [5].

REFERENCES

1. Littman, H., Vuković, D.V., Zdanski, F.K., Grbavčić, Ž.B., Can. J. Chem. Eng., 52, 174 (1974).

2. Ray, T.B., Sarkar, S., Indian J. Tech., 11, 1 (1973).

3. Nagarkatti, A., Chatterjee, A., Can. J. Chem. Eng., 52, 185 (1974).

4. Mamuro, T., Hattori, H.J., Chem. Eng. Japan, 1, 1 (1968).

5. Grbavčić, Ž.B., Vuković, D.V., Zdanski, F.K., and Littman, H., paper accepted for publication in Can. J. Chem. Eng.

6. Lefroy, G.A., Davidson, J.F., Trans. Instn. Chem. Engrs., 47, T120 (1969).

7. Wen, C.Y., Yu, Y.H., Chem. Eng. Progr. Symp. Ser. No. 62, 62, 100 (1966).

8. Richardson, J.F., in "Fluidization", Davidson, J.F., Harrison, D., Ed., Academic Press (1971), Chapter 2.

PART IV

Three Phase Fluidization

CONTRACTION OR EXPANSION OF THREE-PHASE FLUIDIZED BEDS

NORMAN EPSTEIN AND DOUGLAS NICKS

NOMENCLATURE

a	wall effect factor in equation (7) = $a(d_p/D, Re_o)$ [3,5]
D	column diameter
d_p	particle diameter
f	function in equation (11) equal to j_2
g	acceleration of gravity
h	function in equation (12) equal to ε
j_i	average superficial velocity of phase i
k	ratio of wake volume to bubble volume
n	exponent in equation (7) = $n(Re_o, d_p/D)$ [3]
Re_o	free settling particle Reynolds number = $d_p V_o/\nu_1$
V_b	terminal free rise velocity of gas bubbles in liquid medium
V_o	terminal free settling velocity of solid particles in liquid medium
v_i	average linear velocity of phase i = j_i/ε_i
v_{1f}	average linear velocity of non-wake liquid
v_{21}	relative velocity between gas and non-wake liquid
ε	bed voidage = $1 - \varepsilon_3$
ε_i	average volume fraction (i.e. holdup) of i
ε_{1f}	average liquid holdup in particulate region
ν_1	kinematic viscosity of liquid
ψ	numerator of equation (14)

Subscripts

f   liquid-fluidized, or particulate, region

k   wake region

i   phase or region

1   liquid phase

2   gas phase

3   solid phase

Superscripts

denotes differentiation with respect to v_{1f}

ABSTRACT

Some beds contract and others expand when gas bubbles are introduced into the bottom of a liquid-fluidized bed. A quantitative criterion for predicting which of these two events will occur in any particular case is derived and tested against experimental data.

THEORY

A recent paper [1] has shown that the holdup characteristics of a three-phase (gas-liquid-solid) fluidized bed in which the solids are wetted and supported by the liquid is best represented by a wake model in which the solids content of the liquid wakes behind the gas bubbles is assumed negligible compared to the solids content of the remaining liquid (within the liquid-fluidized or "particulate" region). More direct experimental evidence for wakes which are nearly solids-free was cited in the published discussion [2] of the above paper.

The governing equations for this simplified solids-free wake model, assuming steady state, are as follows. By definition,

$$\varepsilon_1 + \varepsilon_2 + \varepsilon_3 = 1 \qquad\qquad (1)$$

Since the liquid divides itself between the solids-free wakes and the particulate region of liquid-fluidized solids,

$$\varepsilon_1 = \varepsilon_k + (1 - \varepsilon_2 - \varepsilon_k)\varepsilon_{1f} \qquad (2)$$

The ratio of wake volume to gas bubble volume is given by

$$\varepsilon_k/\varepsilon_2 = k(\varepsilon_2, \varepsilon_3) \qquad (3)$$

A gas balance yields

$$j_2 = v_2\varepsilon_2 \qquad (4)$$

while a liquid balance gives

$$j_1 = \varepsilon_k v_2 + (1 - \varepsilon_2 - \varepsilon_k) \varepsilon_{1f} v_{1f} \qquad (5)$$

The velocity of the gas bubbles relative to that of the liquid in the particulate region is given by

$$v_{21} = v_2 - v_{1f} \qquad (6)$$

while the interstitial velocity of the liquid in the particulate region, assuming uniform-sized solids, can in general be represented by an equation of the Richardson-Zaki [3] type:

$$v_{1f}\varepsilon_{1f} = aV_o\varepsilon_{1f}^n \qquad (7)$$

Algebraic combination of equations (1)-(6) leads to

$$\varepsilon = 1 - \varepsilon_3 = \frac{j_1 - j_2 k}{v_{1f}} + \frac{j_2 (1+k)}{v_{1f}+v_{21}} \qquad (8)$$

while combination of equations (3)-(7) gives

$$(j_1 - j_2 k)\left(\frac{aV_o}{v_{1f}}\right)^{\frac{1}{n-1}} = v_{1f} - \frac{j_2 v_{1f} (1+k)}{v_{1f}+v_{21}} \qquad (9)$$

Equation (9) can be re-arranged to give j_2 explicitly:

$$j_2 = j_2 (v_{1f}, v_{21}, k, aV_o, n, j_1) \qquad (9a)$$

Substitution of equation (9a) into equation (8) yields

$$\varepsilon = \varepsilon (v_{1f}, v_{21}, k, aV_o, n, j_1) \qquad (10)$$

The objective of the present study is to predict whether bed expansion or bed contraction occurs on first introducing gas into a liquid-fluidized bed of solids. It is thus desired to evaluate $d\epsilon/dj_2$ at $j_2 = 0$. For present purposes it is permissible to assume that the wake-to-bubble volume ratio, k, is a known constant and that v_{21} may be determined from the appropriate two-phase gas-liquid equation if the bubble size is specified. Assume also that j_1, a, n and V_o are fixed, which is equivalent to saying that we start with a liquid-fluidized bed of specified properties and degree of expansion. By incorporating the constant parameters into the functionalities of equations (9a) and (10), these become respectively

$$j_2 = f(v_{1f}) \tag{9b}$$

and

$$\epsilon = h(v_{1f}) \tag{10a}$$

Differentiation of equation (9b) with respect to v_{1f} then gives

$$\frac{dj_2}{dv_{1f}} = f'(v_{1f}) \tag{11}$$

while differentiation of equation (10a) with respect to v_{1f} yields

$$\frac{d\epsilon}{dv_{1f}} = h'(v_{1f}) \tag{12}$$

To evaluate $d\epsilon/dj_2$, apply the chain rule

$$\frac{d\epsilon}{dj_2} = \frac{d\epsilon}{dv_{1f}} \cdot \frac{dv_{1f}}{dj_2} = \frac{h'}{f'} \tag{13}$$

by dividing equation (12) by equation (11). The result, evaluated at $j_2 = 0$, is

$$\left.\frac{d\epsilon}{dj_2}\right|_{j_2=0} = \frac{(\frac{n}{n-1} + k) v_1 - [(1+k)j_1 + \frac{kv_{21}}{n-1}]}{(\frac{n}{n-1}) v_1 (v_1 + v_{21})} \tag{14}$$

where $v_1 = j_1/\epsilon_1 = (aV_o)^{\frac{1}{n}} j_1^{\frac{n-1}{n}}$ via equation (7) in the case of zero gas flow (for which $\epsilon_{1f} = \epsilon_1$ and $v_{1f} = v_1$).

The magnitude and sign of the bed expansion on first introducing gas can be evaluated from equation (14), a negative sign denoting a contraction. If it is desired only to predict qualitatively whether an expansion or contraction will occur initially, it is necessary to consider only the numerator of equation (14). Since n always exceeds unity [3], the denominator is always positive; hence a positive numerator signifies a bed expansion and a negative numerator a bed contraction.

RESULTS

Equation (14) has been tested for all the pertinent experimental data of reference [4], some of which have been reported in reference [1]. The term v_1 in equation (14) was evaluated for each run as j_1/ϵ_1 from the experimentally measured value of ϵ_1 at zero gas flow (for which $\epsilon_1 = 1 - \epsilon_3$), rather than as $(aV_o)^{\frac{1}{n}} j_1^{\frac{n-1}{n}}$, which would be used in the absence of experimental data on ϵ_1 at $j_2 = 0$. The systems investigated are summarized in Table 1 and the results for each run in Table 2. The numerator of equation (14) is denoted by ψ. System 10, which involved no bubble wakes, will be excluded from this discussion till subsequently. Systems 1 - 9 include a total of 31 runs.

As a first approximation, a value of $k = 1$ was assumed. It is seen in Table 2 that the correct predictions are obtained on this basis in all 31 runs with the exception of the first run for System 8.

A somewhat better estimate of k at negligible gas holdup is suggested by reference [1]:

$$k = 3.5 (1 - \epsilon_3)^3 \qquad (15)$$

Although the previous derivation has assumed that k is independent of phase holdups, this assumption will be relaxed only for the purpose of calculating a value of k for each run by equation (15). It is seen in Table 2 that, on this basis, the correct predictions are now obtained for all 31 runs of Systems

Table 1 Properties of systems investigated [4]

Gas = air

D mm	System No.	Liquid	Solids	Solids spec.grav.	d_p mm	Re_o	n	V_o cm/sec	α
20.0	1	water	glass beads	2.824	1.08	191.4	2.70	17.7	0.749
	2	aq. glycerol #	glass beads	2.824	1.08	74.6	2.70	13.6	0.729
	3	water	sieved sand	2.578	0.458	34.0	3.95	7.06	0.733
50.8	4	water	glass beads	2.938	0.273	10.2	3.60	4.07	1.0
(2 inches)	5	water	glass beads	2.935	0.456	30.4	3.27	7.55	1.0
	6	water	glass beads	2.824	1.08	191.4	2.86	17.7	1.0
	7	water	glass beads	2.949	1.08	202.2	2.84	18.7	1.0
	8	water	lead shot	11.03	2.18	1762.5	2.28	80.8	0.794
	9	aq. PEG*	steel balls	7.756	3.18	15.5	4.24	29.4	1.0
	10	aq. PEG*	glass beads	2.824	1.08	0.23	4.93	1.47	1.0

Aqueous solution of glycerol

* Aqueous solution of polyethylene glycol

Table 2 Test of expansion-contraction predictions

System No.	j_1 cm/sec	ε_3 ($=1-\varepsilon_1$)	v_{21} cm/sec	$k = 1.0$		$k = 3.5\,(1-\varepsilon_3)^3$			Experimental observation
				ψ cm/sec	$d\varepsilon/dj_2\big|_{j_2=0}$ sec/cm	k	ψ cm/sec	$d\varepsilon/dj_2\big|_{j_2=0}$ sec/cm	
1	4.01	0.353	25.20	-6.80	-0.0220	0.948	-6.15	-0.0199	contraction
	4.81	0.300	23.52	-5.67	-0.0171	1.20	-8.03	-0.0242	contraction
	5.22	0.293	23.40	-5.09	-0.0141	1.24	-7.88	-0.0218	contraction
	6.02	0.247	22.18	-4.39	-0.0115	1.49	-9.82	-0.0256	contraction
	6.42	0.231	21.83	-4.07	-0.0102	1.59	-10.51	-0.0263	contraction
	6.82	0.223	21.70	-3.69	-0.0087	1.64	-10.60	-0.0250	contraction
2	1.67	0.249	21.08	-9.98	-0.1213	1.48	-15.67	-0.1904	contraction
	2.03	0.213	20.21	-9.27	-0.0993	1.70	-17.21	-0.1843	contraction
	2.38	0.178	19.43	-8.70	-0.0847	1.94	-18.95	-0.1846	contraction
	2.71	0.156	19.01	-8.29	-0.0732	2.10	-20.04	-0.1769	contraction
	3.04	0.127	18.45	-7.92	-0.0653	2.32	-21.66	-0.1786	contraction
3	4.06	0.276	22.53	-2.64	-0.0125	1.33	-4.65	-0.0220	contraction
	4.86	0.223	21.20	-2.28	-0.0099	1.64	-5.98	-0.0260	contraction
	6.03	0.175	20.25	-1.83	-0.0068	1.96	-7.19	-0.0267	contraction
4	1.25	0.237	32.70	-11.17	-0.1434	1.55	-17.87	-0.229	contraction
	1.87	0.163	29.51*	-9.76	-0.0994	2.05	-21.30	-0.217	contraction

*Bubble flow : $v_{21} = V_b/\varepsilon_1$; other runs in slug flow : $v_{21} = (0.2j_1 + 0.35\,\sqrt{gD})/\varepsilon_1$

Table 2 (continued) Test of expansion-contraction predictions

System No.	j_1 cm/sec	ε_3 ($\equiv 1-\varepsilon_1$)	v_{21} cm/sec	k = 1.0		k = 3.5 $(1-\varepsilon_3)^3$			Experimental observation
				ψ cm/sec	$d\varepsilon/dj_2\|_{j_2=0}$ sec/cm	k	ψ cm/sec	$d\varepsilon/dj_2\|_{j_2=0}$ sec/cm	
5	1.59	0.365	38.90*	-14.21	-0.0951	0.89	-12.42	-0.0832	contraction
	3.52	0.195	30.68*	- 9.88	-0.0448	1.82	-20.27	-0.0918	contraction
	4.55	0.148	28.99*	- 8.84	-0.0335	2.16	-22.74	-0.0861	contraction
	4.55	0.134	28.55*	- 8.85	-0.0346	2.27	-23.93	-0.0935	contraction
	5.71	0.078	26.79*	- 8.11	-0.0276	2.74	-27.80	-0.0945	contraction
6	7.02	0.262	33.47*	- 7.90	-0.0126	1.41	-14.25	-0.0227	contraction
7	7.65	0.240	32.50*	- 7.36	-0.0111	1.54	-15.59	-0.0236	contraction
	12.80	0.121	28.10*	- 3.83	-0.0040	2.38	-22.48	-0.0234	contraction
	12.80	0.116	27.94*	- 3.96	-0.0042	2.42	-23.13	-0.0244	contraction
8	8.26	0.583	59.23*	- 7.70	-0.0028	0.254	+18.20	+0.0065	expansion
	17.80	0.414	42.15*	+15.95	+0.0041	0.704	+21.98	+0.0056	expansion
	26.40	0.322	36.43*	+27.04	+0.0052	1.09	+25.60	+0.0049	expansion
	38.77	0.191	30.53*	+31.90	+0.0048	1.85	+19.40	+0.0029	expansion
9	13.82	0.135	31.78	- 0.563	-0.00056	2.26	-10.21	-0.0102	contraction
	18.84	0.062	30.40	- 0.693	-0.00052	2.89	-16.07	-0.0121	contraction
10	0.40	0.268	33.85	wakeless bubbles		0.00	+ 0.285	+0.0121	expansion
	0.86	0.154	29.40			0.00	+ 0.415	+0.0107	expansion

*Bubble flow : $v_{21} = V_b/\varepsilon_1$; other runs in slug flow : $v_{21} = (0.2j_1 + 0.35 \sqrt{gD})/\varepsilon_1$

1 - 9, with no exceptions. It should be noted that the value of k for the first run of System 8 (0.254) is by far the lowest of all the 31 values computed by equation (15), which presumably accounts for the inadequacy of the k = 1 assumption in this run.

It is of interest to note that for k = 0, as in low Reynolds number flow around the gas bubbles, the numerator of equation (14) reduces to

$$\psi = (\frac{n}{n-1} \cdot \frac{1}{\epsilon_1} - 1)j_1 \qquad (16)$$

which is always positive, since n always exceeds unity [3]. In other words, equation (14) correctly predicts that, in the absence of bubble wakes, bed expansion will always occur when gas is introduced to a liquid-fluidized bed. This is illustrated in Table 2 by System 10, for which the gas bubble Reynolds numbers were below that at which any appreciable wake forms [4].

Conclusive confirmation of equation (14) will require testing of data from other reported studies.

ACKNOWLEDGEMENTS

Continuing financial support from the National Research Council of Canada is gratefully acknowledged. Thanks for help in arriving at equations (8) and (9) are also owed to Professor T.J. Fitzgerald of Oregon State University.

REFERENCES

[1] Bhatia, V.K. and Epstein, N. "Three-Phase Fluidization : A Generalized Wake Model", Proceedings of the International Symposium on Fluidization and Its Applications, pages 380-392, Cepadues-Editions, Toulouse (1974).

[2] Ibid, pages 703-705.

[3] Richardson, J.F. and Zaki, W.N., Trans. Instn. Chem. Engrs. 32, 35 (1954).

[4] Bhatia, V.K., Ph.D. Thesis, University of British Columbia, Vancouver, Canada, 1972.

[5] Neuzil, L. and Hrdina, M., Coll. Czechoslov. Chem. Commun. 30, 752 (1965).

BUBBLE WAKE STRUCTURE IN
THREE-PHASE FLUIDIZATION

R. C. DARTON AND D. HARRISON

INTRODUCTION

In a liquid-fluidised bed the liquid hold-up, ε_ℓ, may be estimated from the properties of the materials and the superficial liquid
velocity, U_ℓ:

$$\varepsilon_\ell^n = U_\ell/u_t \qquad (1)$$

where u_t is the terminal velocity of the particles and n is a positive
constant (1). In a three-phase fluidised bed the liquid hold-up varies
with the gas hold-up, an interdependence quantified by Stewart (2)
and Efremov and Vakhrushev (3) in the following way. Let the flux of
liquid in the wakes of the bubbles be $\overline{k}U_g$, where $\overline{k}$ is the mean value
of (liquid-wake volume/bubble volume) and U_g is the superficial gas
velocity. Then in the particulate phase (that part of the bed which
is neither bubbles nor wakes) the superficial liquid velocity is
$(U_\ell - \overline{k}U_g)/(1 - \varepsilon_g - \overline{k}\varepsilon_g)$, and the liquid hold-up is $(\varepsilon_\ell - \overline{k}\varepsilon_g)/$
$(1 - \varepsilon_g - \overline{k}\varepsilon_g)$. Applying Eq.(1) to the particulate phase gives

$$\varepsilon_\ell = (U_\ell/u_t - \overline{k}U_g/u_t)^{1/n}(1 - \varepsilon_g - \overline{k}\varepsilon_g)^{1-1/n} + \overline{k}\varepsilon_g \qquad (2)$$

where ε_g and ε_ℓ are the volume fractions of the whole bed occupied by
gas and liquid respectively. Prediction of both ε_g and ε_ℓ requires a
second relation between ε_g and ε_ℓ, for example from the dependence of
gas slip velocity on gas hold-up.

We have previously suggested (4) an empirical expression for $\overline{k}$
(i.e. $1 + \overline{k} = 1.4(U_\ell/U_g)^{0.33}$) obtained by fitting data to Eq.(2); here
we attempt to predict the size of the liquid-wake by modelling the
fluid and particle flows.

The Bubble Wake

The wake of a bubble in a two-dimensional bed is observed to comprise: (a) particle-free liquid immediately below the bubble which is the "liquid-wake" of Stewart's model; and (b) a lower region of particles and liquid also apparently moving with the bubble.

(i)   The vortical wake. One possible description of the motion in a closed wake is Hill's spherical vortex, defined by the stream function $\psi = -\frac{3}{4}u_b x^2(1-r^2/a^2)$, where x is distance from the axis of symmetry and r distance from the vortex centre. The matching free stream velocity is $-u_b$ and the vortex radius a. Suppose the vortex is set up in the liquid phase and the particles remain fluidised, falling relative to the liquid with a velocity v_p. Then the stream function for the particles within the liquid vortex is $\psi_p = -\frac{3}{4}u'x^2(1-r^2/a'^2)$, where $u' = u_b(1-2v_p/3u_b)$, and the radius of the particle vortex is $a' = a(1-2v_p/3u_b)^{\frac{1}{2}}$. This is shown in Fig.1. In fact the vortex is induced by the rising bubble and the flows in Fig.1 must be distorted to allow for its presence as in Fig.2, which makes the geometry of Fig.2 uncertain. However the distance (a-a') should give an estimate of the height h of the liquid-wake, that is

$$h = a - a(1 - 2U_\ell/3\epsilon_\ell u_b)^{\frac{1}{2}} \qquad\qquad (3)$$

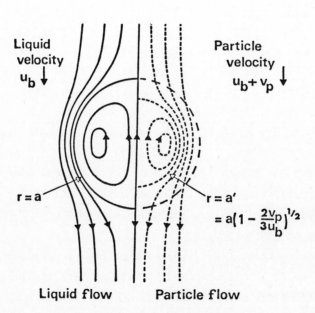

Figure 1. Particles falling through a Hill's vortex.

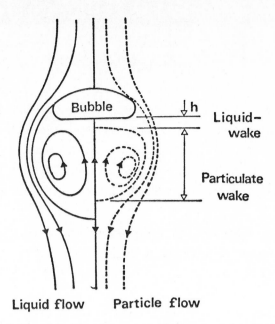

Figure 2. Solids and liquid motion near a bubble in a three-phase fluidized bed.

since v_p is the interstitial liquid velocity U_ℓ/ε_ℓ. When (a-a') is small $h/a \simeq U_\ell/3\varepsilon_\ell u_b$. The ratio h/a is k multiplied by a function of the bubble geometry.

(ii) The flat-topped wake. Alternatively we may suppose the flow towards the (flat) base of the bubble resembles axi-symmetric flow towards a stagnation point, described by $\psi = bx\sigma^2$ and $\psi_p = b(x + v_p/2b)\sigma^2$. Distance to the base of the bubble is x, and from the axis of symmetry σ. In this case the particulate wake has a flat top - Fig.3 - in agreement with observation, and $h = v_p/2b$. The strength of flow in the wake, b, is found by matching velocities at the rim of the bubble. Assuming potential flow around a sphere of radius a gives $b = 3u_b/2a$ and $h/a = U_\ell/3\varepsilon_\ell u_b$, as for the vortical wake when (a-a') is small. The same theory gives the result $h/a = U_\ell/2\varepsilon_\ell u_b$ for a flat-topped wake in a two-dimensional bed.

Comparison of Theory with Experiment

The following table gives approximate measurements from photographs of bubbles and wakes in two-dimensional beds.

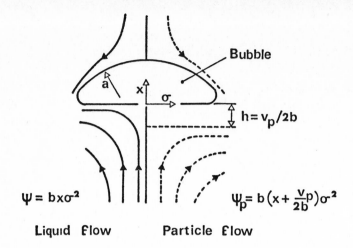

Figure 3. The flat-topped wake in 3-dimensions.

Fluidised bed	Bubble size cm³	h/a, measured	$h/a = U_\ell/2\varepsilon_\ell u_b$	Source of data
460µm ballotini	0.2	0.4	.17	Ref.5
U_ℓ = 2.5 cm/s	1.6	0.1	.09	
775µm sand	0.2	0.8	.18	
U_ℓ = 2.8 cm/s	0.5	0.2	.17	Ref.6
	1.0	0.15	.15	

Conclusions

The vortical wake model shows how a circulating flow of particles is maintained in the bubble wake. These particles are a primary cause of elutriation from the bed, which the model suggests will be small if the bubble size and velocity are small. The flat-topped wake model predicts the horizontal voidage discontinuity observed in practice, and its position fairly closely for large bubbles. Small bubbles (those not cap-shaped) are not well described by the theory: the strength of circulation in the wake is over-estimated and the size of liquid-wake underestimated. For cap bubbles of a given geometry, the liquid-wake volume/bubble volume ratio required by Stewart's model can be calculated.

REFERENCES

(1) Richardson, J.F. and Zaki, W.N. Trans.Instn Chem.Engrs 1954,
     32, 35

(2) Stewart, P.S.B. Ph.D. dissertation, University of Cambridge 1965

(3) Efremov, G.I. and Vakhrushev, I.A. Int.Chem.Engng 1970, 10, 37

(4) Darton, R.C. and Harrison, D. Chem.Engng Sci. In press

(5) Stewart, P.S.B. and Davidson, J.F. Chem.Engng Sci. 1964, 19, 319

(6) Rigby, G.R. and Capes, C.E. Can.J.Chem.Engng 1970, 48, 343.

THE EFFECTS OF SOLIDS WETTABILITY ON THE CHARACTERISTICS OF THREE PHASE FLUIDIZATION

E. R. ARMSTRONG, C. G. J. BAKER AND M. A. BERGOUGNOU

Three-phase (liquid-gas-solid) fluidization describes the process in which solid particles are simultaneously fluidized by a liquid and a gas. The former constitutes the continuous phase and the latter a discontinuous bubbling phase. Whereas liquid fluidized beds of small particles have been found to contract upon the introduction of gas; the reverse trend occurs in beds of large particles. For solids having properties similar to glass, the transition occurs at about 2.5 mm. The contraction has been attributed to the formation of wakes behind the fast moving gas bubbles which results in a quantity of liquid being carried through the bed at a much faster rate than that of the continuous phase. Consequently the average velocity of the bulk liquid phase is reduced and the bed contracts. In beds of large particles, the formation of wakes is presumably not as marked and thus expansion is observed at all times.

The properties of both the wakes and bubbles have been used by many authors to explain the complex behaviour of three-phase fluidized beds. It is the purpose of this study to determine the effect of solids wettability on the behaviour of three-phase fluidized beds of large particles and also to qualitatively determine the role of wakes and bubbles in such beds. For this purpose, 6 mm glass beads and teflon coated glass beads were employed as the wettable and non-wettable solids respectively. The liquid phase was water, the velocity of which was maintained at 0.275 ft/sec. The superficial velocity of the gas phase, air, was varied from 0.0 to

0.35 ft/sec. The experiments were carried out in a large two-dimensional bed of 26 in. x 1 in. cross-section and 84 in. high. The equipment has been described in detail elsewhere [1].

The effect of gas velocity, V_g on the individual gas, liquid and solids hold-ups, ε_g, ε_ℓ and ε_s, is shown in figure 1. As may be seen ε_s decreased with increasing gas velocity indicating the expected bed expansion. The teflon-coated beads exhibited a slightly smaller solids hold-up and hence a greater expansion. The most remarkable trend was observed in the gas hold-up which, as expected, increased with increasing gas rate. However, for the teflon-coated beads, it was much lower than that for the uncoated beads. This effect is shown more clearly in figure 2, in which the liquid and gas hold-ups are plotted against the bed height H. For a

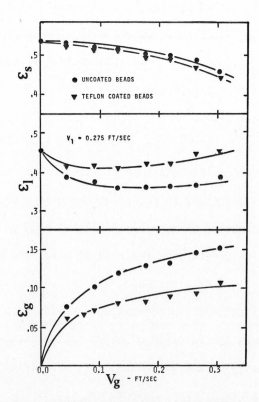

Figure 1. The effect of gas velocity on the individual phase hold-ups in beds of 6mm wettable and non-wettable solids.

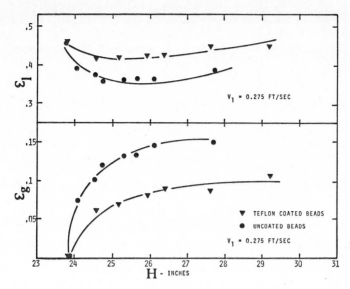

Figure 2. Liquid and gas hold-ups versus bed height.

given height, the teflon-coated beads exhibited a gas hold-up as much as 40% lower than that of the glass beads. Naturally, the reverse trend was observed with the liquid hold-up. Thus, the beds of teflon-coated beads exhibited increased expansion and, at the same time, a decreased gas hold-up.

Solids wettability, as the name implies, characterizes the ability of a liquid to wet a solid surface. The work of adhesion W, expressed by:

$$W = \sigma_{LV}\ (1 + \cos\theta) \qquad \text{---------------------} \qquad (1)$$

is the amount of energy required to separate a liquid from a solid. In this equation σ_{LV} is the liquid-gas surface tension and θ is the contact angle. As θ is increased, W decreases and the solid is said to become non-wettable. Consequently, the ability of a gas to contact the solid is increased. In three-phase fluidized beds of wettable solids, gas-solid contact is unlikely. However, in beds of non-wettable solids the reverse is true. Thus, by rendering the solids non-wettable, the formation of liquid wakes is probably inhibited and beds of small particles expand upon the

introduction of gas as observed by Bhatia [2]. The fact that increased
bed expansion was observed in the present study for non-wettable solids
would indicate that liquid wakes are also present in beds of large wettable
particles. These wakes are obviously less dominant than in small particle
beds since expansion is normally observed anyway.

The presence of wakes in beds of large wettable particles would explain why they exhibited a smaller expansion than those of non-wettable
solids. However, this does not adequately account for the decreased gas
hold-up. Beds of large particles have been termed bubble disintegrating
since the bubbles tend to break- up, or at least undergo reduced coalescence,
and this results in a large gas hold-up [1]. By rendering the solids non-
wettable, the increased gas- solid contact results in adherance of the bubbles to the solids and thus break-up is probably reduced. The result is
larger bubbles and hence a reduced gas hold-up. This agrees with visual
observation as shown in Plate 1.

Since solids wettability plays a key role in flotation processes, one
might assume that its effect in these studies would account for the increased
expansion. As shown in figure 3 a model based on this assumption showed
this to be erroneous. However, this effect would likely be of greater
importance in beds of small non-wettable solids.

Plate 1. Photograph of air bubbles adhering to teflon coated glass beads in a water atmosphere.

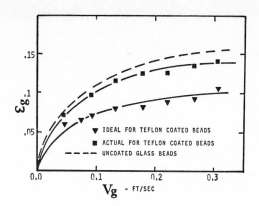

Figure 3. Ideal gas hold-up in bed of teflon coated glas beads.

This preliminary study has further illustrated the complex behaviour of three-phase fluidized beds. More detailed research will be necessary to gain a better understanding of the mechanisms involved.

REFERENCES

1. Kim S.D., Ph.D. Thesis, Univ. of Western Ontario, Canada (1974)

2. Bhatia, V.K. *et al.*, Ind. Eng. Chem. Process Des. Develop., Vol. 11, No. 1, (1972).

PERFORMANCE OF TURBULENT BED CONTACTOR: GAS HOLDUP AND INTERFACIAL AREA UNDER LIQUID STAGNANT FLOW

M. KITO, M. SHIMADA, T. SAKAI, S. SUGIYAMA, AND C. Y. WEN

NOMENCLATURE

a	:	gas-liquid interfacial area per unit bed volume	$[cm^2/cm^3]$
a^*	:	gas-liquid interfacial area per unit liquid volume	$[cm^2/cm^3]$
C_{Ai}, C_{AL}	:	concentration of dissolved gas at interface and in bulk of liquid, respectively	$[mole/cm^3]$
C_B	:	concentration of the reactant in bulk of liquid	$[mole/cm^3]$
$\overline{C}_B$	:	average concentration of the reactant	$[mole/cm^3]$
$\Delta C_B/\Delta t$	:	change in concentration of the reactant per unit time	$[mole/cm^3\ sec]$
D_A	:	diffusivity of dissolved gas in liquid	$[cm^2/sec]$
D_C	:	diameter of column	$[cm]$
D_P	:	diameter of spherical packing	$[cm]$
Fr	:	Froude number ($= U_G / \sqrt{g\ D_C}$)	$[-]$
g	:	acceleration of gravity	$[cm/sec^2]$
H	:	expanded bed height	$[cm]$
He	:	Henry's law constant	$[atm\ cm^3/mole]$
H_L	:	height of clear liquid	$[cm]$
H_S	:	static packing height	$[cm]$
K_G	:	over-all mass transfer coefficient based on gas component	$[mole/cm^2\ atm\ sec]$
k_G	:	gas-side mass transfer coefficient	$[mole/cm^2\ atm\ sec]$

k_L : liquid-side mass transfer coefficient with reaction [cm/sec]

k_L^o : liquid-side mass transfer coefficient without reaction [cm/sec]

k_r : reaction rate constant for second-order reaction [cm^3/mole sec]

$\overline{P}_A$: mean partial pressure of carbon dioxide [atm]

R_A : rate of absorption per unit area of interface [mole/cm^2 sec]

U_G : superficial gas velocity [cm/sec]

V_G, V_L, V_S : volume of gas, liquid and packings, respectively [cm^3]

We : Weber number ($= D_C U_G^2 \rho_L / \sigma$) [-]

z : number of moles of reactant reacting with each mole of A [-]

β : reaction factor difined by k_L / k_L^o [-]

ε_G : gas-holdup ($= V_G / (V_G + V_L + V_S) $) [-]

ε_L : liquid-holdup ($= V_L / (V_G + V_L + V_S) $) [-]

μ_L : viscosity of liquid [g/cm sec]

ρ_L : density of liquid [g/cm^3]

ρ_P : bulk density of spherical packing [g/cm^3]

γ : ratio difined by $\sqrt{D_A k_r C_B} / k_L^o$ [-]

σ : surface tension [dyne/cm]

Subscripts

A : substance transfered from A (gas:CO_2) - phase to B (liquid:NaOH)-phase

B : reactant in B - phase

        The turbulent bed contactor is a three phase fluidized
bed characterized by the use of an essentially nonflooding
packing consisting of spheres of low density and large diameter placed between retaining grids sufficiently far apart

to permit turbulent and random motion of packing. This
contacting device has many advantages over conventional
absorber, namely high capacity, high efficiency and non-
clogging[3,4]. There are, however, a number of problems
concerning the performance of the turbulent bed contactor
that must be solved before a proper design can be made.
Although mass transfer rates, effective interfacial areas
and liquid holdups in the turbulent bed contactor were investigated by Kossev et al[5], Strumillo et al[9] Wozniak et al[13]
and Chen et al[1], most of the results were not clarified
effects of process variables on the characteristic. The
purpose of this investigation is to determine the gas holdup,
the mass transfer coefficient and the interfacial area, and
their dependence on process variables such as gas flow rates,
the free opening of the supporting grid, the diameter and
density of packing and static packing heights.

EXPERIMENTAL

Equipment and method.   A flow diagram is shown in
Fig. 1. The test section of the equipment is 5 cm and 10 cm
in diameter and 150 cm high. A sieve plate and perforated
plate were used as the supporting grid. Their free opening
areas were 31.5 % and 1.27 %. Spheres used had effective
densities of 1 g/cm^3 and 0.59 g/cm^3 and had diameters of
1.1 cm, 2.65 cm and 2.87 cm. The static packing height was
varied from 5 cm to 20 cm. All experimental runs were
operated under the liquid stagnant flow. The superficial
gas velocity was varied from 5 cm/sec to 400 cm/sec. The

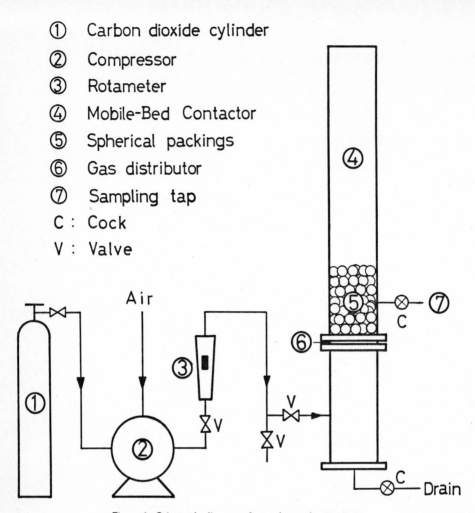

①   Carbon dioxide cylinder
②   Compressor
③   Rotameter
④   Mobile-Bed Contactor
⑤   Spherical packings
⑥   Gas distributor
⑦   Sampling tap
C :   Cock
V :   Valve

Figure 1. Schematic diagram of experimental apparatus.

bed was fluidized by the introduction of a gas flowing upwards. Table 1 lists the properties of liquid used in the
experiments on the gas holdup. Values of the volumetric
mass transfer coefficient and the interfacial area with
respect to the unit volume of liquid were obtained from experiments of absorption of CO_2 into the aqueous sodium
hydroxide solution. The gas was a mixture of 5 % CO_2 and
95 % atmospheric air. The liquid was an aqueous sodium

hydroxide solution of molality 0.0005 g-mol/cm^3.   All ex-

perimental runs were carried out near isothermal at the room

temperature.   The rate of absorption was calculated on the

basis of chemical analyses of liquid and gas phases.   The

conversion of sodium hydroxide was about 20 per cent.   The

holdups of gas, liquid and packing were determined by directly

measuring the height of aerated bed and that of bed without

aeration.   The average fractional gas holdup can be given as

$$\varepsilon_G = \frac{H_G}{H_G + H_L + H_S}$$

The overall coefficient can be calculated by

$$K_G a^* = (\Delta C_B / \Delta t) / 2\overline{p}_A \tag{1}$$

Procedure   The rate of absorption of CO_2 accompanied by

chemical reactions is given by Eq. (2).

$$R_A = k_L(C_{Ai} - C_{AL}) = \beta k_L^{\circ}(C_{Ai} - C_{AL}) \tag{2}$$

Table 1   Properties of liquids [at 20°C]

Material	Density [g/cm^3]	Surface tension [dyne/cm]	Viscosity [cp]
Ethanol	0.797	22.5	1.38
25wt%.Glycerol soln.	1.068	70.8	1.33
45wt%.Glycerol soln.	1.118	69.1	5.21
65wt%.Glycerol soln.	1.165	67.5	14.45
80wt%.Glycerol soln.	1.211	65	62
Methanol	0.790	22.3	0.58
Water	0.998	72.8	1.01

C_{AL} can be assumed zero for an irreversible reaction if
enough quantity of the reactive liquid component exists.
Thus Eq. (2) may be reduced to Eq. (3).

$$R_A = \beta k_L C_{Ai} = k_L{}^{\circ}C_{Ai} \tag{3}$$

Although the reaction between CO_2 and NaOH is essentially a
second-order reaction, this reaction may be treated as a
pseudo-first order reaction provided that[2]

$$5 < \gamma < 0.2 \quad \frac{C_B}{z\,C_{Ai}} \tag{4}$$

The coefficient for absorption accompanied by a pseudo-first
order reaction, k_L, is given as

$$k_L = (k_r C_B D_A)^{1/2} \tag{5}$$

and β in Eq. (2) may be approximated by γ. Since k_L is
independent of hydrodynamic conditions, the interfacial
area with respect to the unit volume of liquid, a*, and
with respect to the unit volume of tower, a, can be calculated
respectively, by

$$a^* = \left(\frac{1}{k_G} + \frac{He}{(k_r C_B D_A)^{1/2}} \right) K_G a^* \tag{6}$$

$$a = a^* \varepsilon_L \tag{7}$$

Thus, knowledge of k_r, He and D_A for the CO_2-NaOH system is required to calculate the interfacial area. In this study, the rate constant, k_r, was taken from the works of Nijsing et al[7] and Pinsent et al[8], while the Henry's law constant, He, and the diffusivity D_A, respectively, were obtained from equations suggested by van Krevelen et al[10] and Wales[12].

RESULTS

Gas Holdup

Fig. 2 shows the effect of the superficial gas velocity on the gas holdup as a function of physical properties of

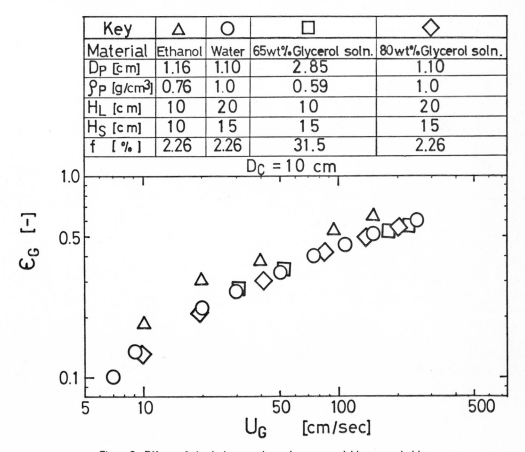

Key	△	○	□	◇
Material	Ethanol	Water	65wt% Glycerol soln.	80wt% Glycerol soln.
D_P [cm]	1.16	1.10	2.85	1.10
ρ_P [g/cm³]	0.76	1.0	0.59	1.0
H_L [cm]	10	20	10	20
H_S [cm]	10	15	15	15
f [%]	2.26	2.26	31.5	2.26

Figure 2. Effects of physical properties and process variables on gas holdup.

liquid, the packing density and the free opening of the supporting grid. As can be seen from Fig. 2, ε_G is nearly
independent of the packing density, the free opening of the
supporting grid and liquid viscosity, and dependent on surface
tension and the superficial gas velocity in the fully fluidized mobile bed. As can be seen from Fig. 3, ε_G is also
independent of the ratio of the diameters of tower and packing. The effect of the number of packing held in the bed

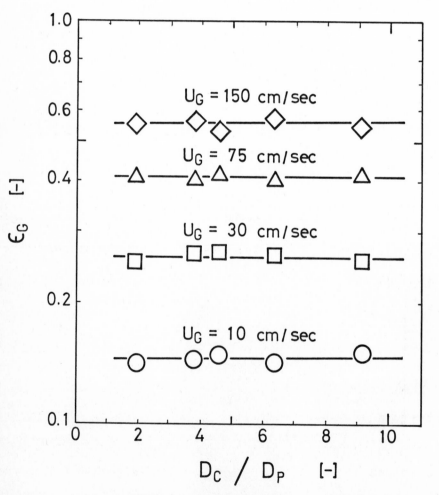

Figure 3. Effect of D_C/D_p on ε_G.

Key	△	○	□	◇
Material	Ethanol	Water	25wt% Glycerol soln.	65wt% Glycerol soln.
D_P [cm]	1.16	1.10	1.10	1.10
ρ_P [g/cm³]	0.76	1.0	1.0	1.0
H_L [cm]	10	10	10	10
H_S [cm]	10	15	15	15
f [%]	2.26	31.5	31.5	31.5

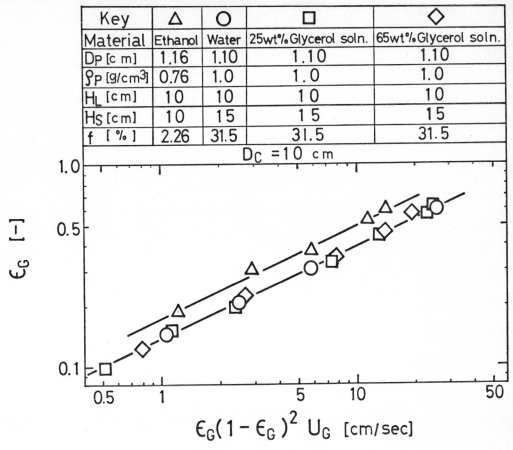

Figure 4. ϵ_G vs $\epsilon_G (1 - \epsilon_G)^2 U_G$.

on ϵ_G is negligible. From the results obtained, conceivable factors affecting the gas holdup could be U_G, σ, g and ρ_L. By dimensional analysis, Eq. (8) can be obtained.

$$\epsilon_G = C \, (We)^a (Fr)^b \qquad (8)$$

Although the relation between ϵ_G and U_G shown in Fig. 2 has not given a linear relationship, an empirical plot of the values of ϵ_G against $\epsilon_G (1-\epsilon_G)^2 U_G$ on a log-log paper gives straight lines with a slope of 0.44 as shown in Fig. 4. Therefore, Eq. (8) could be represented as follows:

$$\frac{\varepsilon_G}{\{\varepsilon_G(1-\varepsilon_G)^2\}^{0.44}} \;=\; C\,(We)^a(Fr)^b \qquad (9)$$

The values of a and b in Eq. (9) can be determined from an empirical plot of the values of $\varepsilon_G/\{\varepsilon_G(1-\varepsilon_G)^2\}^{0.44}$ against

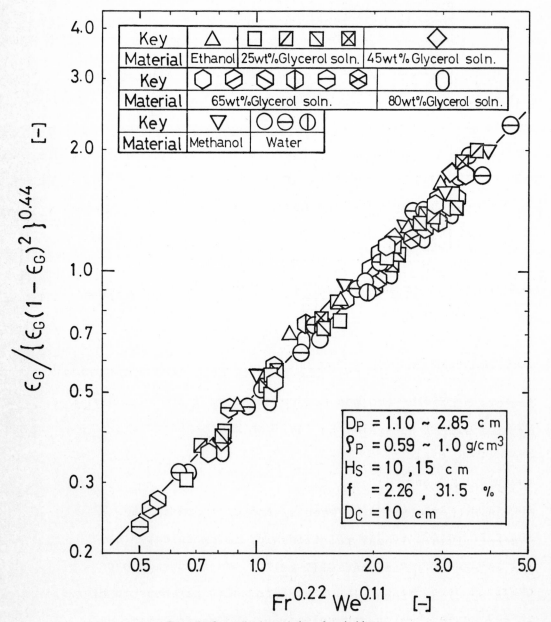

Figure 5. Generalized correlation of gas holdup.

the Weber number or Froude number on a log-log paper.

$\varepsilon_G/\{\varepsilon_G(1-\varepsilon_G)^2\}^{0.44}$ is proportional to the Weber number to the

0.11 power and to the Froude number to the 0.22 power. The

value of C in Eq. (9) determined from the log-log plot of

$\varepsilon_G/\{\varepsilon_G(1-\varepsilon_G)^2\}^{0.44}$ against the product of $(We)^{0.11}$ and $(Fr)^{0.22}$

is 0.5. Thus, Eq. (9) becomes

$$\frac{\varepsilon_G}{\{\varepsilon_G(1-\varepsilon_G)^2\}^{0.44}} = 0.5 \ (We)^{0.11}(Fr)^{0.22} \quad (10)$$

This correlation is valid for $2.0 < We \ Fr < 4.8 \times 10^4$.

In Fig. 5, all the data are plotted in accordance with Eq.

(10). Eq. (10) may be useful for the prediction of gas hold-

up in fully fluidzed mobile beds. The gas holdup in aqueous

solutions of electrolytes such as sodium hydroxide is larger

than that in nonelectrolyte solutions. For electrolyte

solutions, it is suggested that 0.6 is taken as the value

of coefficient in Eq. (10).

MASS TRANSFER AND INTERFACIAL AREA

Figs. 6-12 show data obtained in the absorption runs. Fig. 6

shows the plot of $1/K_G a^*$ vs $He/(D_{CO_2} \ C_{NaOH} \ k_r)^{1/2}$ which can

be used to evaluate a^* and K_G. The linear relationship be-

tween $1/K_G a^*$ and $He/(D_{CO_2} \ C_{NaOH} \ k_r)^{1/2}$ shown in the figure

indicates that Eq. (5) holds for these runs and the inter-

facial area of gas-liquid contact in the mobile bed can be

calculated by the aforementioned method. The values of $K_G a$

obtained have the same magnitude as that obtained by Kossev

et al[5] under higher gas velocities. However, they are

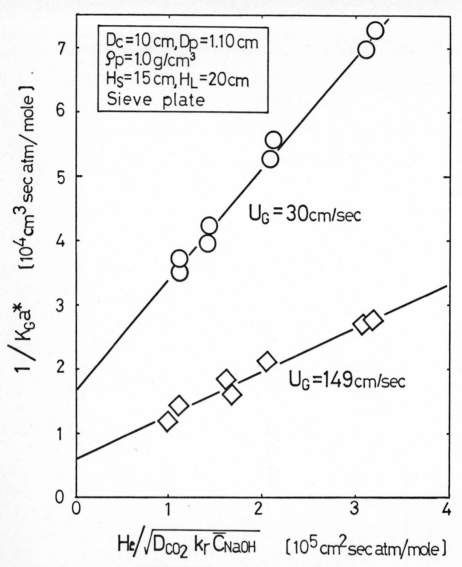

Figure 6. Plot of $1/K_G a^*$ against $He/\sqrt{D_{CO_2}\, k_r\, \bar{C}_{NaOH}}$ for the absorption of carbon dioxide in aqueous NaOH solution.

smaller than that obtained by Wozniak et al[13]. $K_G a$ is seen
to decrease as the gas velocity is increased ($U_G > 200$ cm/sec).
The decneasing trend of $K_G a$ is agreement with that reported
by Levesh et al[6]. As the results of the absorption of CO_2
with the aqueous sodium hydroxide solution under a low CO_2
pressure obtained by Vidwan's et al[11] shows that the gas-

film resistance compared with that of liquid phase cannot
be neglected, the calculation of a* based on Eq. (6) should
be taken account of k_G. In this work, k_G was calculated to
be 1.04 x 10^{-5} [g-mol/cm^2·atm·sec]. This value is roughly of
the same magnitude as that obtained by Wozniak et al[13] in
a fully fluidized mobile bed. Fig. 7 plots the interfacial
areas calculated based on Eqs. (6) and (7) as a function of
the superficial gas velocity indicating the effect of the
free opening of the supporting grid on the interfacial areas.
a* seems to increase with increasing U_G and approches a
certain value in the range of high gas velocity. On the

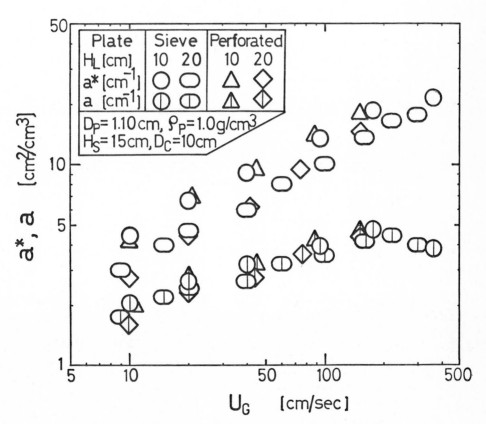

Figure 7. Effect of free opening of supporting grid on interfacial area.

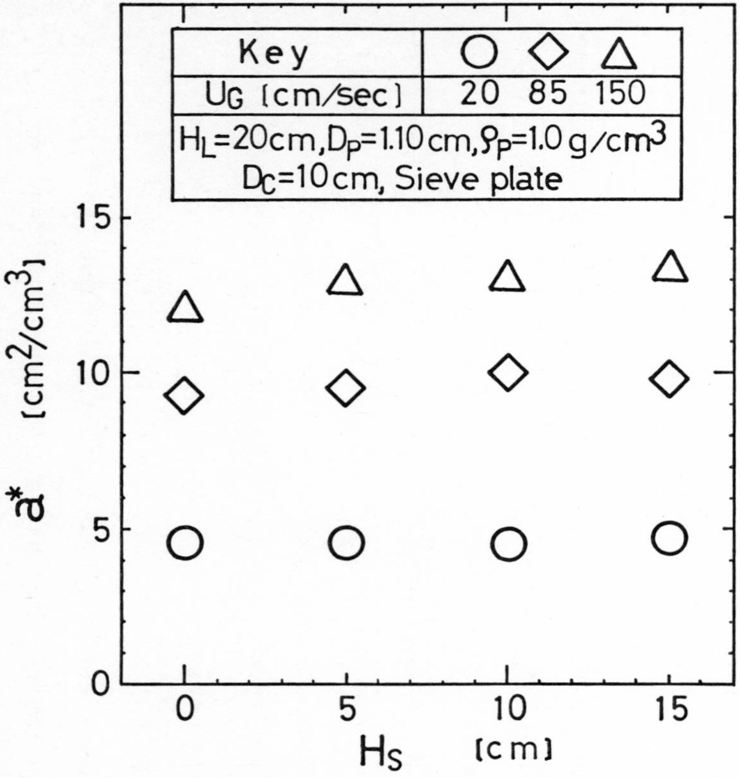

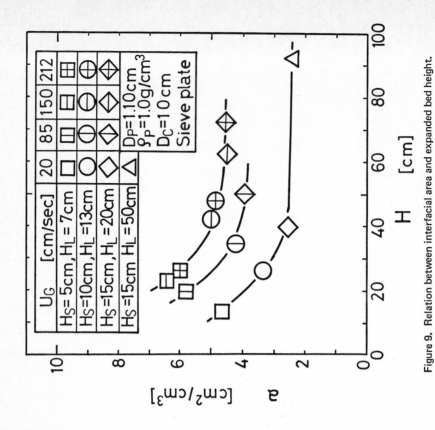

Figure 8. Effect of static packing height on interfacial area.

Figure 9. Relation between interfacial area and expanded bed height.

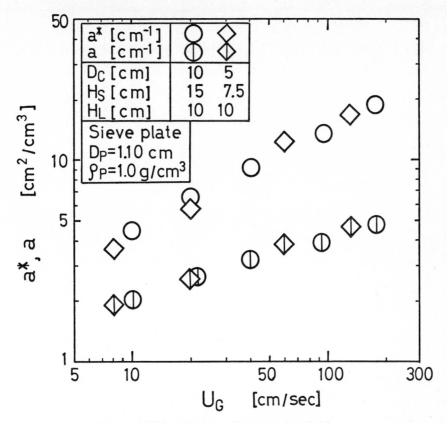

Figure 10. Effect of column diameter on interfacial area.

other hand the value of a seems to increase up to 200 cm/sec
and decreases in higher gas velocity than 200 cm/sec.   The
interfacial area seems to be nearly independent of the free
opening of the supporting grid.   The value of a is   estimated
to be roughly proportional to the 0.4 power of U_G.   Fig. 8
shows the effect of the static packing height on the interfacial area.   This figure indicates that in the range of the
static packing height Hs < 5 cm (V_S/V_L < 0.125), a* is
slightly dependent on Hs, but at Hs = 10 ∿ 15 cm (0.27 < V_S/V_L
< 0.425) a* is practically independent of Hs.   In the range
of above U_G = 150 cm/sec, a* is slightly dependent on the

ratio (V_S/V_L), while a decreases when the ratio $(V_S/V_L) > 1$.
Fig. 9 shows the effect of the expanded bed height, H, on the
interfacial area. The ratio (V_S/V_L) were 0.425 and 0.224.
As can be seen from Fig. 9, a decreases with the increasing
of H, but in H > 40 cm, a approches the constant value at
the superficial gas velocity. When the static liquid height,
H_L, is above 20 cm, a is almost a constant value. In these
cases, the expanded bed height could be as high as 20 to
30 cm, while the static bed height is roughly half of H.
Fig. 10 shows the effect of Dc on the interfacial area. The
values of a along the radial direction may be assumed uniform.

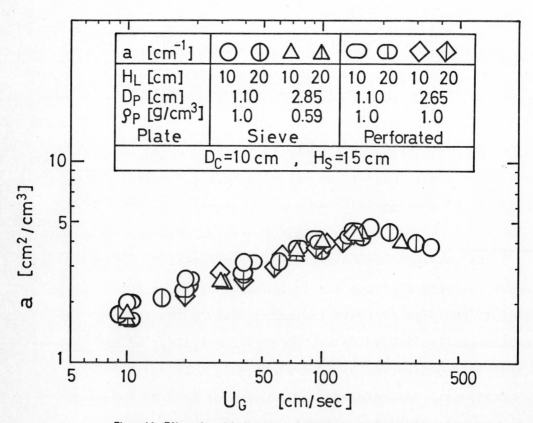

Figure 11. Effect of particle-diameter and -density on interfacial area.

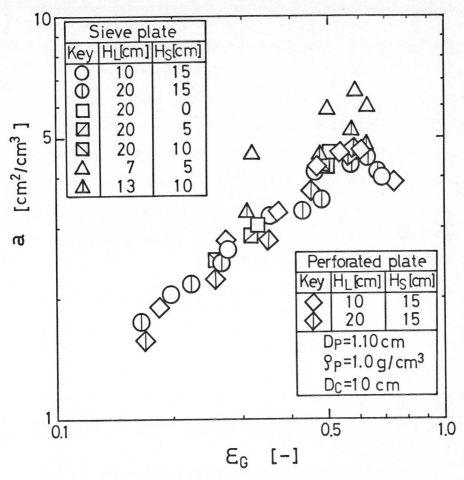

Figure 12. Relation between interfacial area and gas holdup.

Effects of the diameter and density of packings are shown in
Fig. 11. a seems to be independent of the diameter and density of packings under the present operating conditions.
Fig. 12 shows an empirical plot of the value of a against ε_G.
The value of a in the static bed height roughly 10 cm
shown in the figure is the different value which is larger
than that in the static bed height above roughly 20 cm with
the same ε_G. The values of a in present operating conditions
are seen to be proportional to ε_G up to ε_G = 0.6.

CONCLUSION

From the results obtained from the investigation of Turbulent
Bed Contactor under the condition of liquid stagnant flow,
the following conclusion can be drawn.

1. In the fully fluidized mobile bed, the free opening of
    the supporting grid, the density and diameter of packings
and the diameter of tower affect very little if any on the
gas holdup. The gas holdup in aqueous solutions of electrolytes is slightly larger than that in nonelectrolyte solutions.
ε_G can be represented by Eq. (10).

2. The effective interfacial area per unit bed volume in-
    creases with the increasing gas velocity up to 200 cm/sec,
but decreases at a higher gas velocity. The bed is operated
best under the conditions such that the expanded bed height
is 20 to 30 cm and the static bed height is half of the
expanded bed height, provided that ε_G is nearly 0.6.

LITERATURE CITED

1)   Chen, B.H. and W.J.M.Douglas ; *Can.J.Chem.Eng.*, **46**, 245 (1968)

2)   Danckwerts, P.V. ; " *Gas - Liquid Reactions* ", McGraw-Hill, New York
      (1970)

3)   Douglas, H.R., I.W.A.Snider and G.H.Tomlinson ;*Chem.Eng.Progr.*, **59**,
      85 (1963)

4)   Douglas, W.J.M. ; *Chem.Eng.Progr.*, **60**, 66 (1964)

5)   Kossev, A., G.Peev and D.Elenkov ; *Verfahrenstechnik*, **5**, 340 (1971)

6)   Levsh, I.P., M.I.Niyasov, N.I.Krainev and F.F.Ganikhanova ; *Int.Chem.
      Eng.*, **8**, 379 (1968)

7)   Nijsing, R.A.T.O., R.H.Hendriksz and H.Kramers ; *Chem.Eng.Sci.*, **10**,
      88 (1959)

8)   Pinset, B.R.W., L.Pearson and F.J.W.Roughton ; *Trans. Faraday Soc.,*
     52, 1512 (1956)

9)   Strumillo, C., J.Adamiec, T.Kudra ; *Int.Chem.Eng.,* 14, 652 (1974)

10) van Krevelen, D.W. and P.J.Hoftijzer ; *Chim.Ind. XXIeme Congr.Int.*
     Chim.Ind., 168 (1948)

11) Vidwans, A.D. and M.M.Sharma ; *Chem.Eng.Sci.,* 22, 673 (1967)

12) Wales, C.E. ; *A.I.Ch.E.J.,* 6, 1166 (1966)

13) Wòzniak, M. and K.Østergaard ; *Chem.Eng.Sci.,* 28, 167 (1973)

DISPERSION OF SOLID PARTICLES ON A MULTISTAGE THREE-PHASE FLUIDIZED BED

HIROSHI KUBOTA AND TSUNEO SEKIZAWA

NOMENCLATURE

Ar	:	fractional open area of partition plate	[-]
C	:	concentration of solid particles	$[g/cm^3]$
D_T	:	column diameter	[cm]
d_0	:	hole diameter of partition plate	[cm]
$\bar{d}_p$	:	average diameter of suspended solid parti- cles	[cm]
E_L	:	longitudinal dispersion coefficient of liquid in stage	$[cm^2/sec]$
E_{L0}	:	E_L in the column without partition plate	$[cm^2/sec]$
K	:	parameter defined by Eq. (13)	[-]
K_1, K_2	:	parameters defined by Eq. (6)	[-]
ℓ	:	longitudinal length along the column	[cm]
$\Delta \ell$	:	distance between partition plates	[cm]
u_G	:	superficial gas velocity	[cm/sec]
u_L	:	superficial liquid velocity	[cm/sec]
u_L'	:	superficial velocity of liquid backflowed through partition plate	[cm/sec]
β	:	fractional area of liquid ascending part in plate	[-]
δ	:	plate thickness	[cm]

κ :   back flow ratio defined by u'_L/u_L [-]

φ :   gas holdup in stage   [-]

φ :   gas holdup in hole of partition plate   [-]

μ :   viscosity of liquid   [g.cm/sec]

The longitudinal concentration distribution of
suspended solid particles on a three-phase fluidized bed
is significantly controlled by presence of partition
plates.   A model, which describes the exchange of solid
particles through the perforated partition plate, was
presented.   The model interpretes well the characteristic
concentration difference of suspended solid particles
through partition plate experimentally observed: the
stepwise profile of the solid concentration at low liquid
flow rate and the saw-toothed profile at high liquid flow
rate.   A model parameter was determined under varied
experimental conditions.

1. PREFACE

The three-phase fluidized beds are often used as
chemical reaction systems.   The solid particles suspended
may be reactants, products or catalysts for reactions.
For example, they are catalysts in hydrogenation of
hydrocarbons, and products in biochemical cultivation of
microorganisms.   In these reaction systems, to approach
the preferable plug flow pattern of liquid, perforated
plates are often placed in longitudinal direction to partition the bed space.   Lack of information on the behavior of

suspended solid particles, however, makes unable us to predict the concentration profile of solid particles along the longitudinal direction of the bed. These informations will be essential for the rational design of the chemical reactors. Suganuma and Yamanishi's report[4] may be only one quantitative study in this field so far presented, but the results was not analyzed to be used for design purposes.

The present authors[2][5] studied on the behavior of suspended solid particles in a single stage three-phase fluidized bed. Recently they[3] reported also the informations of mixing characteristics of liquid in the multi-stage bubble column with perforated partition plates, which is gas-liquid two phase system. Here the authors will extend these works to have useful informations predicting the concentration profile of suspended solid particles in the multistage three-phase fluidized bed.

2. MODELING OF BEHAVIOR OF SUSPENDED SOLID PARTICLES

Consider a three-phase fluidized system in which gas and liquid with suspended solid particles flow upward concurrently.

2.1 Liquid Flow on Multistage Gas-liquid System

The authors' previous study[3] shows that the longitudinal liquid mixing characteristics in a multistage bubble column is well expressed by combining the backflow through perforated partition plate and the longitudinal dispersion in each stage between partition plates. The value of the backflow ratio of liquid through the partition

plate $\mathcal{K}$, which was defined as the ratio of the backflow
rate u_L' to the net flow rate of liquid u_L , was empirically
correlated with column configurations and flow conditions
in the following equation.

$$K = \frac{u_L'}{u_L} = \frac{7.15 \; u_G^{0.2} Ar}{1 + 0.01 \; u_G/Ar} \left(\frac{1}{u_L} - \frac{0.552}{u_L^{0.85} Ar^{0.2}} \right)$$

$$X \; \frac{1.48}{\exp \left\{ \frac{1.5}{d_0} \left(\mu^{0.5} + \frac{0.01 \, \delta^{0.2}}{d_0^2} \right) \right\}} \tag{1}$$

where d_0 and δ should be mentioned in cm, u_G and u_L in
cm/sec, and μ in g/cm·sec.

Eq.(1) is thus dimensionally nonhomogeneous, but
will be practically used, since the size of column and the
distance between partition plates are independent of $\mathcal{K}$.

2.2 Behavior of Suspended Solid Particles in Stage

According to the one-dimensional dispersion model,
mass balance of suspended solid particles under steady
state conditions of fluid flow gives the following
equation.[5]

$$-(1-\varphi) \; E_P \; \frac{dC}{d\ell} + u_L C - (1-\varphi) \; v_f C = u_L C_e \tag{2}$$

In the left side of this equation, the first term
is mass flux of solid dispersed with liquid mixing, the
second is that accompanied with liquid flow, and the third
is that settling against liquid flow. The right side term

is thus the net mass flux of solid across a section of the bed.

One of the present authors[2,5] has observed the value of longitudinal dispersion coefficient E_p and the settling velocity of suspended solid particles v_f in a three-phase system. It was concluded that E_p was the same as the liquid dispersion coefficient E_L. And the settling velocity v_f empirically determined[2] was expressed by

$$v_f = 1.45 \ v_{ft}^{0.65} \ / \ F(\varepsilon) \tag{3}$$

where $F(\varepsilon)$ is a voidage function which is taken as $F(\varepsilon)$ $= \varepsilon^{-4.65}$, ε is liquid volume fraction in slurry, and v_{ft} is the terminal velocity of a single particle with the averaged diameter.

These informations will be directly applied in each stage of a multistage system. In the latter case, the value of the dispersion coefficient E_L has been empirically correlated[3] with the dispersion coefficient obtained in the single-stage bed E_{LO} by

$$\frac{E_L}{E_{LO}} = \frac{u_G^{0.1}}{1 + 0.5 \ (D_T/\Delta\ell)^{0.2}} \tag{4}$$

where u_G is used in cm/sec. For the values of E_{LO} the previous experimental study[1] would be referred.

2.3 Exchange of Solid Particles through Partition Plate

From the investigation of liquid exchange through the perforated partition plate in a gas-liquid multistage

system, the authors[3] have concluded that backflow through
the plate is caused by the gas-lift action. Gas bubbles,
which ascend continuously through some parts of holes of
the plate, carry liquid to the upper stage from the lower.
The amount of liquid flowing up over the net flow, therefore,
must descend also continuously through the other part of
holes of the plate. Besides this liquid exchange it is now
assumed that the ascending and the descending flows of
suspended solid particles with liquid in holes of the plate
have no dispersion in direction of flow. This allows that
the ascending flow of slurry will have the same concentration of solid as that of slurry at just underneath the
plate c_{i-1}^0. For descending flow of slurry the same will be
assumed.

    Mass balance through a plane across the plate, there-
fore, derives the following equation (refer Fig. 1).

$$\left\{ u_L + u_L' - v_f Ar \quad (1- y') \right\} c_{i-1}^0$$
$$- \left\{ u_L' + v_f Ar(1-\beta) \right\} c_i = u_L c_e \qquad (5)$$

In this equation β is the fractional open area of
the plate effective for liquid ascending and y' is the gas
holdup in holes of the plate there. It is also assumed that
there is no gas bubble in holes used for liquid descending.
The first term of the left side of Eq.(5), therefore, is
the flow rate of suspended solid ascending through the plate
and the second term descending.

    Since the values of β and y' can not be evaluated,
the following new parameters K_1 and K_2 are introduced.

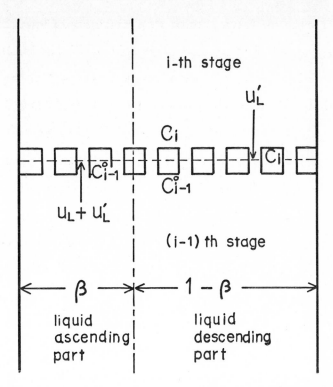

Figure 1. Schematic model of back flow of liquid with suspended solid particles through partition plate.

$$K_1 = (1 - \mathcal{G}'), \qquad K_2 = 1 - \beta \qquad\qquad (6)$$

Then rearranging Eq. (5) leads Eq. (7)

$$c^0_{i-1} = \frac{(K + v_f \, Ar \, K_2 / u_L) \, c_i + c_e}{1 + K - v_f \, Ar \, K_1 / u_L} \qquad\qquad (7)$$

where $K = u'_L / u_L$ is the back flow ratio of liquid through the plate.

When liquid flow through the column is absent or so small that one can approximate $u_L = 0$, Eq. (7) is reduced to

$$c^0_{i-1} = \frac{u'_L + v_f \, Ar \, K_2}{u'_L - v_f \, Ar \, K_1} \, c_i \qquad\qquad (8)$$

Since c^0_{i-1} is always larger than C_i, the experimentally observed stepwise concentration profile of soild particles
is explained.

If liquid flow rate u_L is large enough, Eq.(7) is

$$c^0_{i-1} = \frac{K}{1 + K} C_i + \frac{1}{1 + K} C_e \qquad (9)$$

c^0_{i-1} is always smaller than C_i , and therefore the observed
saw-toothed solid concentration profile is interpreted.

From Eq.(7) it is also derived the following.   When
the backflow ratio $K \ll 1$,

$$c^0_{i-1} \simeq C_e \qquad (10)$$

The solid concentration at the outlet of each stage will
almost equal that at the column end C_e , which is the same
as the feed concentration under steady state flow conditions.
On the other hand, if $K \gg 1$,

$$c^0_{i-1} \quad C_i \qquad (11)$$

This means that the exchange of liquid through the plate is
so large that the solid concentration difference will be
negligible.

3. EXPERIMENTAL

The schematic diagram of the experimental apparatus
used is shown in Fig. 2.   A vertical multistage column, made
of prexiglass, 10 cm in dia., was composed of several sections
coupled with flanges.   The perforated plates were mounted

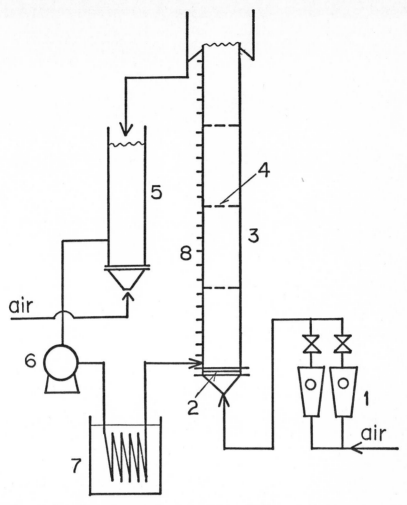

Figure 2.   Flow diagram of experimental system: 1. rotameter; 2. gas distributor; 3. plexiglass column; 4. partition plate; 5. mixing tank; 6. pump, 7. thermostat; sampling tap.

between flanges, and thus the column was partitioned.   The
dimensions of the plates, suspended solid particles used and
the experimental conditions are summarized in Table 1.
Air and tap water with suspended solid particles were introduced concurrently from the bottom of the column.   The slurry
flowed out of the column was well mixed in a tank and then
recycled into the inlet of the column.   The temperature of
the slurry was maintained constant 20°C.

Table 1 Dimensions of experimental apparatus used and experimental conditions

Column :

    Column diameter D_T [cm]   10

    Plate spacing   [cm]   50

Partitioned plate :

    Hole diameter, d_0 [cm]   0.2   0.5   1.0   1.5   2.0

    Thickness,   [cm]   0.5   1.0

    Free area fraction, Ar   [-]   0.0700   0.202

Suspended Solid particles :

    Material   glass beads

    Mean diameter, $\bar{d}_p$ [cm]   0.089, 0.106, 0.128, 0.153

    Concentration, C [g/cm^3] 0.004   0.4

Flow conditions :

    Gas ; Air

    Liquid ; Tap water   20°C

    Superficial gas velocity, u_G [cm/sec] 1.79 10.8

    Superficial liquid velocity, u_L [cm/sec] 0 1.8

In order to determine the concentration of solid particles in slurry, the samples were withdrawn at different longitudinal positions along the column. After weighing the sample, solid particles were separated from liquid on a glass filter, dried and weighed to obtain the solid concentration.

It wsa assertained that the solid concentration of the slurry sampled across the radial position in the column was sufficiently uniform.

4. EXPERIMENTAL RESULTS AND DISCUSSIONS

4.1 Experimental Data and Their Analysis

Typical examples of experimentally obtained concentration profiles of suspended solid particles are illustrated in Fig. 3. Fig. 3(a) corresponds to the case of $u_L = 0$. As predicted from Eq. (8), the stepwise profile is obtained. Increase of u_L decreases the solid concentration difference through the plate and finally the concentration at the upper surface of the plate becomes smaller than that at the lower surface, as shown in Fig. 3(b). The further increase of u_L leads the typical saw-toothed profile shown in Fig. 3(c). This is also interpreted by Eq. (9).

From the experimental data such shown in Fig. 3, the parameters K_1 and K_2 which are included in Eq. (7) might be obtained. Since it is difficult, however, to determine them separately, a sole parameter K defined below is introduced.

$$K \, \bar{C}_i = K_2 C_i + K_1 C_{i-1}^0 \tag{12}$$

As the value of $\bar{C}_i$ the arithmetic mean between C_i and C_{i-1}^0 may be approximately used. And thus

$$K \simeq 1 - \beta \mathcal{Y}' \tag{13}$$

Eq. (7) is now reduced to

$$C_{i-1}^0 = \frac{\mathcal{K} C_i + v_f A r K \bar{C}_i \, / \, u_L + C_e}{1 + \mathcal{K}} \tag{14}$$

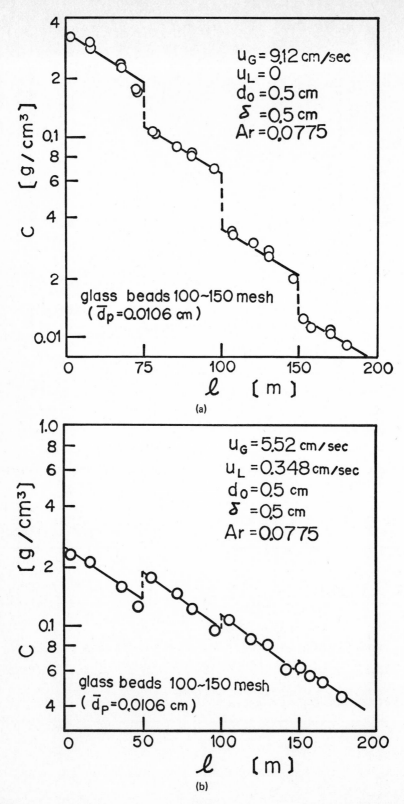

Figure 3. Typical concentration profiles of suspended solid particles in a multistage bubble column; (a) at $u_L = 0$; (b) at $u_L = 0.348$ cm/sec.

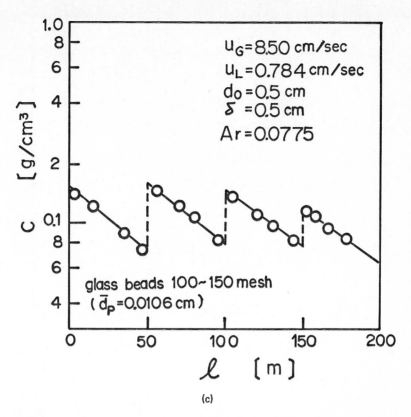

(c)

Figure 3. (continued) Typical concentration profiles of suspended solid particles in a multistage bubble column; (c)
at $u_L = 0.784$ cm/sec.

Rearranging Eq.(14) allows the evaluation of values of K
from the experimental data by

$$K = \frac{(c^0_{i-1} - c_i) + c^0_{i-1} - c_e}{v_f Ar \bar{c}_i / u_L}$$ (15)

where c^0_{i-1} and c_i are determind by extraporations of the
observed concentration distributions to the upper and to the
lower surfaces, respectively, of the partition plate.

When $u_L = 0$, by modifying Eq.(15), the value of K
is determined by

$$K = \frac{u'_L (c^0_{i-1} - c_i)}{v_f Ar \bar{c}_i}$$ (16)

The value of K evaluated in the manner mentioned above will be affected by gas velocity u_G , but may be independent of liquid velocity u_L , hole diameter of plate d_0 , plate thickness δ , average diameter of suspended solid particles $\overline{d}_p$ and fractional open area Ar.

The values of K determined from the data obtained under varied experimental conditions are shown in Fig. 4 to 8. As seen in these results the value of K obtained in most of experimental conditions are rather larger than unity.

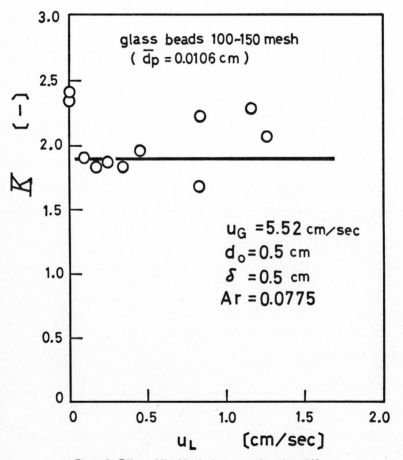

Figure 4. Effect of liquid velocity u_L on the value of K.

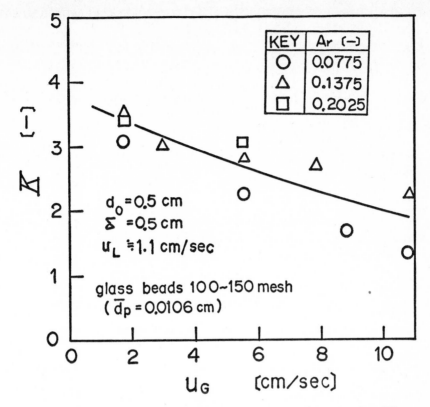

Figure 5. Effect of gas velocity u_G and fractional open area Ar on the value of K.

However, Eq. (13) suggests that K should be less than unity. The main reason of this disagreement might be that the actual settling velocity v_f at the gas ascending part is much higher than that predicted from Eq. (3), where the segregation of solid particles from liquid phase may cause owing to the high volume fraction of gas phase. One should consider thus that K is the empirical parameter which includes such unknown factors.

Fig. 4 shows that the values of K are almost independent of u_L as expected, but rather large at $u_L = 0$. The transition of the profile from the stepwise to the sawtoothed appears at $u_L \simeq 0.18$cm/sec. It is not clear why

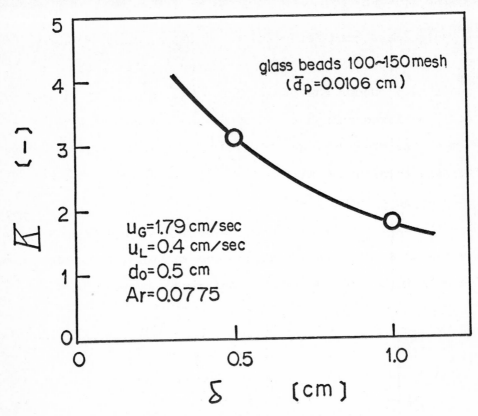

Figure 6. Effect of plate thickness δ on the value of K.

the value of K at $u_L = 0$ is large. There may be some difference in the end effect between the case of giving the typical
stepwise profile and that of giving saw-toothed one. Eq. (13)
predicts that the increase of u_G will decrease K, since β
and $\mathcal{Y}$ decrease with increase of u_G. This is illustrated
in Fig.5. The values of K are scattered but it is seen that
the effect of the fractional open area of plate Ar on K is
minor as expected. The effect of plate thickness δ on K
is seen in Fig.6. When the fundamental equation (5) is
derived, no longitudinal dispersion of solid particles in
holes of plate is assumed. This will not be essentially

correct so that deviation from this simple assumption seems to appear. Mass transfer owing to concentration differnce between the upper and the lower surfaces of plates may not be neglected. Under the conditions corresponding to Fig.6, where u_L is large enough to have the saw-toothed profiles, such mass transfer will cause the decrease of concentration difference through the plate, and thus higher values of K are given for thinner plates.

The effect of size of suspended solid particles on the value of K does not appear as seen in Fig.7, but the difference in hole diameter of plate gives appreciable

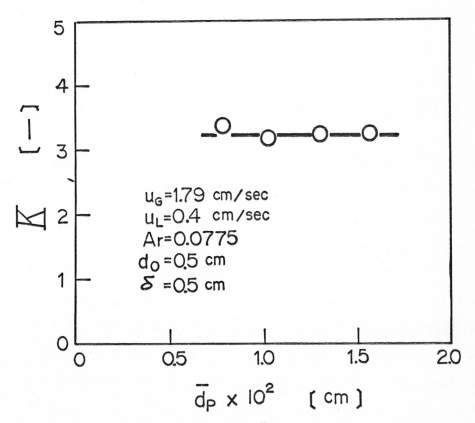

Figure 7. Effect of particle size $\bar{d}_p$ on the value of K.

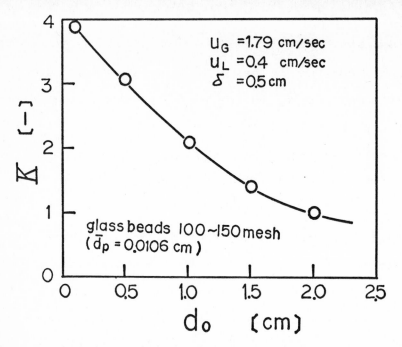

Figure 8. Effect of hole diameter d_0 on the value of K.

effect on the value of K, as seen in Fig.8. This latter
effect can be explained by high gas holdup underneath the
plate. In order to have detailed informations, further
experimental studies will be needed.

4.2 Determination of Whole Profiles of Solid Concentration

The concentration profile of suspended solid particles, under steady state of concurrent flow of gas and
liquid, along the whole length of the multistage bed, in
which the number of stages are n, will be determined in the
following manner.

1) First, the concentration at the top of the column
c_n^0 is taken as that of the feed slurry c_e.

2) The concentration profile in the top n-th stage
is determined by integration of Eq. (2) with the boundary

condition of $C = C_n^0$ at the outlet. The difference in concentration in a stage is not so large that the value of v_f will
be taken as a constant. There, Eq.(2) is easily integrated. The value of E_p in stage will be the same as that
of E_L given by Eq.(4). The value of C_n at the inlet of
this n-th stage is thus determined.

3) The concentration at the outlet of the (n-1)-st
stage C_{n-1}^0 is determined from Eq.(14). To obtain the value
of K in Eq.(14), which depends on column configurations and
flow conditions, the informations obtained here will usefully
be referred.

4) The same procedures as those of 2) and 3) are
iterated to the inlet of the first stage. Once the concentration profile along whole length of the column is given,
the total holdup of solid will be easily calculated.

4.3 Effect of Partition Plates of Solid Suspension in Column

Now summarize how the existance of the partition
plate affects the concentration profile of suspended solid
particles in the column.

When the superficial liquid velocity u_L is zero or
very small, as seen above the presence of partition plates
gives the stepwise profile of solid particles. This avoids
uniform distribution and gives large solid holdup. In the
continuous operation in this range, therefore, higher mean
residence time of solid particles than that in the column
without partition plate will result.

When u_L is large, the partition plates contribute to the uniform distribution and give small solid holdup. When the back flow ratio of liquid K is small, the effect of the plates is evident, but as increase in K the effect becomes minor.

Since the concentration profile changes widely from the stepwise to the saw-toothed as increase of u_L, there will be a critical value of u_L, in which concentrations at the upper and the lower surfaces of the plate C_i and C_{i-1}^0 approach to the same value. This value of u_L is easily obtained from Eq(14) as

$$u_L = v_f ArK / (1 - C_e / C_i) \qquad (17)$$

It will be noted that this critical value of u_L is independent of the back flow ratio K .

It should be added to note that the fractional open area of plate Ar affects the solid concentration profile with not only its contribution to the back flow ratio K but also its appreciable contribution to the settling of solid particles.

5. CONCLUSIONS

(1) A model, which describes the exchange of solid particles through the perforated partition plate in the multistage three-phase fluidized bed, was presented.

(2) The model interpretes well the transition from the stepwise type to the saw-toothed type profile of solid concentration experimentally obtained.

(3)   A parameter of the model was determined under varied experimental conditions.

LITERATURE CITED

1)   Aoyama, Y., Ogushi, K., Koide, K., Kubota, H.; J. Chem. Eng. Japan, $\underline{1}$, 158 (1968)

2)   Imafuku, K., Wang, T. Y., Koide, K., Kubota, H.; ibid., $\underline{1}$, 153 (1968)

3)   Sekizawa, T., Kubota, H.; ibid., $\underline{7}$, 441 (1974)

4)   Suganuma, T., Yamanishi, T.; Kagaku Kōgaku (Chem. Eng., Japan), $\underline{31}$, 1006 (1967)

5)   Yamanaka, Y., Sekizawa, T., Kubota, H.; J. Chem. Eng. Japan, $\underline{3}$, 264 (1970)

HEAT TRANSFER AND HYDRODYNAMIC STUDIES ON THREE PHASE FLUIDIZED BEDS

E. R. ARMSTRONG, C. G. J. BAKER AND M. A. BERGOUGNOU

The hydrodynamics of three-phase fluidization are relatively well understood and numerous publications are available on this topic. However, heat transfer studies in such beds are essentially non-existent and the ones that are available cover such a limited range of variables that mean-ingful conclusions cannot be made. The present study was therefore under-taken to expand knowledge in this area. Measurements of the individual phase hold-ups were also performed. These were in general agreement with previously published data. To compliment the work carried out in three-phase fluidized beds, studies on liquid- gas and liquid-solid beds were also performed.

The experiments were conducted in a large three-dimensional column 9.5 in. in diameter and 108 in. high. The heat transfer surface was con-centrically located within the column and consisted of a brass sheath 2.5 in. in diameter by 10 in. long and housed four 1500 watt electric heaters. The heat transfer coefficient was measured for four particle sizes, namely, 1/2, 1, 3 and 5 mm. Water and air formed the liquid and gas phases. The superficial velocity of the former was varied from 0.027 to 0.414 ft/sec and that of the latter from 0.0 to 0.78 ft/sec.

Plots of the heat transfer coefficient h against gas velocity V_g are shown in figure 1 for three phase fluidized beds of 3 and 5 mm glass beads. Similar plots for the 1/2 and 1 mm glass beads are shown in figure 2. As

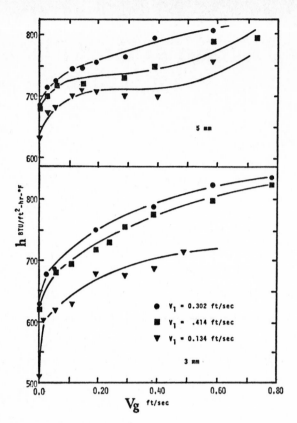

Figure 1. The effect of gas velocity on the heat transfer coefficient in three-phase fluidized beds of 3 and 5mm glass beads.

may be seen introducing gas into a liquid-solid bed initially resulted in an increase in the heat transfer coefficient. At higher gas rates this increase became less pronounced and h appeared to approach a maximum. Similar trends were observed for all particle sizes and liquid velocities studied. It should be noted that this trend is identical to that observed in liquid-gas beds.

As may be seen in figures 1 and 2, a maximum was observed in plots of heat transfer coefficient against liquid velocity. Since, in our opinion, this is not due to the change in liquid velocity but rather to the change in bed porosity, a plot of h against $(\varepsilon_{\ell} + \varepsilon_{g})$ is shown in figure 3. As may be seen, h increased with increasing porosity, went through

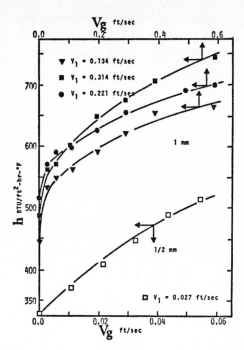

Figure 2. The effect of gas velocity on the heat transfer coefficient in three-phase fluidized beds of ½ and 1mm glass beads.

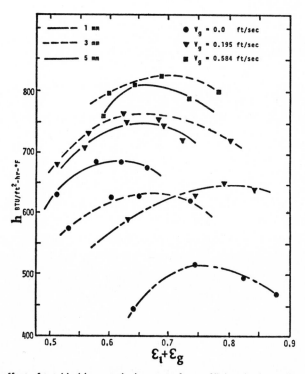

Figure 3. The effect of total hold-up on the heat transfer coefficient in three-phase fluidized beds.

a maximum, and then decreased. The same trend was observed for all particle sizes and gas rates employed. It is interesting to note the close
resemblence between the trends for the two and three-phase fluidized beds.
The former is illustrated in figure 3 for gas velocities equal to zero.
At a given gas rate in the three phase beds, the maxima in these curves
appeared to be shifted to lower porosities on increasing the particle size
from 1 to 3 mm. There was little shift on going from 3 to 5 mm. Unfortunately, due to equipment limitations, the effect of porosity on the
1/2 mm glass beads could not be studied.

It would therefore appear that particle size has a marked effect on
the heat transfer coefficient in three- phase beds. This is further
illustrated in figure 4 which shows plots of h against the particle size,
D_p. Introducing solids into a liquid-gas bed $(D_p = 0)$, in general, re-

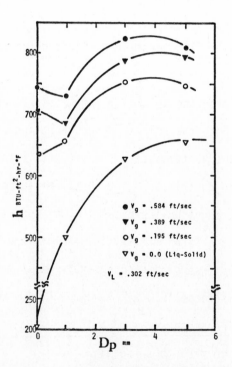

Figure 4. The effect of particle size on the heat transfer coefficient in three-phase fluidized beds.

sulted in an increase in h. A notable exception to this was observed at higher gas rates where the heat transfer coefficient went through a local minimum. Thus, at these high gas rates, h was smaller in three-phase beds than in the corresponding liquid-gas beds up to a particle size of about 1.5 mm. Thereafter h increased with D_p. However, for particle sizes larger than 3 mm, h became essentially independent of this parameter. Similar trends were observed at all liquid and gas velocities. However, at the lower liquid rates (not shown) the local minimum occurred at lower gas velocities. Also shown in figure 4 is the heat transfer coefficient for liquid-solid beds. As may be seen values of h for liquid-gas and three-phase beds were, in general, greater than those in liquid-solid beds. An exception to this was observed for liquid-gas beds at relatively low gas rates which exhibited values of h smaller than those in liquid-solid beds. It is interesting to note that for particle sizes greater than about 1.5 mm the trends shown in figure 4 are all essentially identical.

From the studies in liquid-gas and liquid-solid beds it was observed that the presence of either gas or solids significantly increased the heat transfer coefficient over the value for the liquid alone. In three-phase beds it appears that both the solids and gas contributed towards the increase in h. Thus, the heat transfer coefficient in three-phase beds may be thought of as being contributed to by both the gas and the particles. These contributions are not unlike those in the respective two-phase beds. That is, h experiences a large initial increase on the introduction of gas but becomes relatively insensitive to gas rate at high values of the latter. The solids cause an increase in h which is a function of both their size and concentration. The enhancement of h in three-phase beds is thus the result of a combination of both these effects.

1975 ENGINEERING FOUNDATION CONFERENCE

INTERNATIONAL CONFERENCE ON FLUIDIZATION

Asilomar Conference Grounds

Pacific Grove, California

June 15–20, 1975

S. B. ALPERT   Electric Power Research Institute, 3412 Hillview Avenue, Palo Alto, California

ROGER E. ANDERSON   Aerojet Liquid Rocket Company, P.O. Box 13222, Sacramento, California 95813

H. ANGELINO   Institute Genie Chimique, Chemin de La Loge, Toulouse 31400, France

DAVID H. ARCHER   Westinghouse Research Laboratories, Beulah Road, Pittsburgh, Pennsylvania 15235

M. M. AVEDESIAN   Noranda Research Centre, 240 Hymus Boulevard, Pointe Claire, Quebec, Canada

SURESH P. BABU   Institute of Gas Technology, 3424 South State Street, Chicago, Illinois 60559

C. G. J. BAKER   The University of Western Ontario, London, Ontario N6A 5B9, Canada

JOHN A. BAZAN   Foster Wheeler Energy Corporation, John Blizard Research Center, 12 Peach Tree Hill Road, Livingston, New Jersey 07039

S. R. BECK   Atlantic Richfield Company, P.O. Box 2819, Dallas, Texas 75221

JOHN M. BEGOVICH   Oak Ridge National Laboratory, P.O. Box X, Oak Ridge, Tennessee 37830

L. A. BEHIE   Domtar Ltd., Research Centre, Senneville, Quebec H9X 3L7, Canada

M. A. BERGOUGNOU   University of Western Ontario, London, Ontario N6A SB9, Canada

OSZKAR BORLAI   MTA Muszaki Kemiai Kutato, Intezet, Veszprem, Hungary, Torokvesz ut 12/a, Budapest 10022, Hungary

J. S. M. BOTTERHILL   University of Birmingham, Department of Chemical Engineering, Birmingham B15 2TT, England

COLIN I. BRADLEY   British Gas Corporation, Midlands Research Station, Wharf Lane, Solihull, Warwickshire, England

T. E. BROADHURST   Imperial Oil Enterprises Ltd., Research Department, Sarnia, Ontario, Canada

GARY L. BROWN   Union Carbide Corporation, P.O. Box 8361, South Charleston, West Virginia 25303

RICHARD W. BRYERS   Foster Wheeler Energy Corporation, John Blizard Research Center, 12 Peach Tree Hill Road, Livingston, New Jersey 07039

P. H. CALDERBANK   University of Edinburgh, Chemical Engineering Department, Edinburgh, Scotland

G. S. CANADA   General Electric Corporation R & D, P.O. Box 43, Schenectady, New York 12301

ERWIN L. CARLS   Argonne National Laboratory, 9700 South Cass Avenue, Argonne, Illinois 60439

K. E. CARMICHAEL   Union Carbide Corporation, P.O. Box 8361, Technical Center, South Charleston, West Virginia 25303

J. CARVALHO   University of Cambridge, Department of Chemical Engineering, Pembroke Street, Cambridge CB2 3RA, England

JOHN CHEN   Lehigh University, Mechanical Engineering, Building 19, Bethlehem, Pennsylvania 18015

HERMAN C. T. CHENG   DuPont Experimental Station, Wilmington, Delaware 19810

CHARLES K. CHOI   Garrett Research & Development, 1855 Carrion Road, La Verne, California 91750

DONALD K. CLARKE   Stearns-Roger Incorporated, P.O. Box 5888, Denver, Colorado 80217

SANDFORD S. COLE   Director, Engineering Foundation, 345 East 47th Street, New York, New York 10017

R. C. DARTON   ICI Research Fellow, University of Cambridge, Pembroke Street, Cambridge CB2 3RA, England

J. F. DAVIDSON   University of Cambridge, Pembroke Street, Cambridge CB2 3RA, England

LAWRENCE T. DENK   M. W. Kellogg Company, 1300 Three Greenway Plaza East, Houston, Texas 77046

SHELTON EHRLICH   Pope, Evans, and Robbins, 320 King Street, Suite 503, Alexandria, Virginia 22314

NORMAN EPSTEIN   Department of Chemical Engineering, University of British Columbia, Vancouver 8, British Columbia, Canada

ANDREW ERDMAN, JR.   Dorr-Oliver, 66 Jack London Square, Oakland, California 94607

L. T. FAN   Department of Chemical Engineering, Kansas State University, Manhattan, Kansas 66506

THOMAS FITZGERALD   Program Director, Chemical Processes, National Science Foundation, 1800 G Street, N.W., Washington, D.C. 20550

STEPHEN FREEDMAN   U.S. ERDA, 2100 M Street, N.W., Room 126, Washington, D.C. 20037
MASAHISA FUJIKAWA   Hokkaido University, Kita 13, Nishi 8, Sapporo, Japan
ROBERT J. GARTSIDE   Stone & Webster Engineering Corporation, P.O. Box 2325, Boston, Massachusetts 02107
DEREK GELDART   Bradford University, Great Horton Road, Bradford Yorks, BD7 1DP, England
M. J. GLUCKMAN   City University of New York, Steinman Hall, 140 Street & Convent Avenue, New York, New York 10031
NELLO DEL GOBBO   National Science Foundation, 3943 Wilcoxson Drive, Fairfax, Virginia 22030
W. R. A. GOOSSENS   S.C.K./C.E.N., Boeretang 200, MOL, Belgium 2400
JACK GORDON   MITRE Corporation, Westgate Research Park, McLean, Virginia 22101
JOHN R. GRACE   McGill University, P.O. Box 6070, Montreal H3C 3G1, Canada
ROBERT A. GRAFF   City College of New York, Department of Chemical Engineering, New York, New York 10031
EARL H. GRAY   Phillips Petroleum Company, Building 94G, Phillips Research Center, Bartlesville, Oklahoma 74004
JOHN L. GUILLORY   Envirotech Corporation, One Davis Drive, Belmont, California 94002
CHAIM GUTFINGER   Associate Professor of Mechanical Engineering, Technion, I.I.T., Haifa, Israel
JOSEPH GWOZDZ   Copeland Systems Incorporated, 2000 Spring Road, Suite 300, Oak Brook, Illinois 60521
JOHN S. HALOW   Exxon Research & Engineering Company, P.O. Box 101, Florham Park, New Jersey 07932
DAVID HARRISON   University of Cambridge, Department of Chemical Engineering, Pembroke Street, Cambridge CB2 3RA, England
JOHN HART   PPG Industries, P.O. Box 1000, Lake Charles, Louisiana 70601
T. D. HEATH   Dorr-Oliver, Incorporated, 77 Havermeyer Lane, Stamford, Connecticut 06904
CLAIR E. HILDEBRAND   Allied Chemical Company, P.O. Box 2105 R, Morristown, New Jersey 07960
W. C. A. HOLTKAMP   SASOL, P.O. Box 1, Sasolburg, South Africa
MASAYUKI HORIO   West Virginia University, Department of Chemical Engineering, Morgantown, West Virginia 26506
SVEND HOVMAND   NIRO Atomizer, 305 Gladsaxevej, Soborg 2860, Denmark
C. L. JOHNES   Esso Petroleum Company, Victoria Street, London, England
DAVID H. JONES   The Badger Company, Incorporated, One Broadway, Cambridge, Massachusetts
ALBERT A. JONKE   Argonne National Laboratory, 9700 South Cass Avenue, Argonne, Illinois 60439
M. R. JUDD   University of Natal, Department of Chemical Engineering, King George V Avenue, Durban 4001, South Africa
DAVID JUNGE   Research Associate, Department of Mechanical Engineering, Oregon State University, Corvallis, Oregon 97330
KUNIO KATO   Department of Chemical Engineering, Gunma University, Tenjin-cho, Kiryu-shi, Gunma 376, Japan
DALE L. KEAIRNS   Westinghouse Research Laboratory, Beulah Road, Pittsburgh, Pennsylvania 15235
MELISSA KEENBERG   Engineering Foundation, 345 East 47th Street, New York, New York 10017
PATRICK G. KELLEY   Tennessee Valley Authority, 503 Power Building, Chattanooga, Tennessee 37401
GEORGE H. KESLER   Engineering Consultant, 2758 South Olympia Circle, Evergreen, Colorado 80439
MASAO KITO   Department of Chemical Engineering, Gunma University, Tenjin-cho, Kiruyshi, Gunma 376, Japan
TED M. KNOWLTON   Institute of Gas Technology, 4201 West 36 Street, Chicago, Illinois
EIICHI KOJIMA   Department of Chemical Engineering, Kansas State University, Manhattan, Kansas 66506
C. R. KRISHNA   Brookhaven National Laboratory, Building 526, Upton, New York 11973
KIROSHI KUBOTA   Tokyo Institute of Technology, 1-26-3 Aobadai Midri-ku, Yokohama 227, Japan
DAIZO KUNII   University of Tokyo, 7-3-1. Hongo, Bunkyo-ku, Tokyo 113, Japan
CHARLES E. LAPPLE   Stanford Research Laboratory, 333 Ravenswood Avenue, Menlo Park, California 94025
BERNARD S. LEE   Institute of Gas Technology, 3424 South State Street, Chicago, Illinois 60616
DAVID LEE   PPG Industries, P.O. Box 1000, Lake Charles, Louisiana 70601
RICHARD H. C. LEE   The Aerospace Corporation, P.O. Box 92957, Los Angeles, California 90009
PIERRE LE GOFF   Centre de Cinetique du CNRS, Route de Vadoeuvre, Villers le Nancy 54600, France
L. S. LEUNG   Queensland University, Chemical Engineering Department, Brisbane 4067, Australia
OCTAVE LEVENSPIEL   Oregon State University, Chemical Engineering Department, Corvallis, Oregon 97331
EDWARD K. LEVY   Lehigh University, Packard Laboratory, Building No. 19, Bethlehem, Pennsylvania 18015
HOWARD LITTMAN   RPI, 123 Ricketts Building, Troy, New York 12181
DWIGHT N. LOCKWOOD   Graduate Student, Oregon State University, 2365 South West Pickford Street, Corvallis, Oregon 97330

MIKE MAAGHOUL Electric Power Research Institute, 3412 Hillview Avenue, Palo Alto, California 94304
YVES MARTINI Graduate Student, University of Western Ontario, London, Ontario N6A 5B9, Canada
LEOPOLDO MASSIMILLA University of Naples, Facolta Inqegneria, Piazzale Tecchio, Naples, Italy
KISHAN B. MATHUR University of British Columbia, Department of Chemical Engineering, Vancouver, B.C. V6T 1W5, Canada
JOHN M. MATSEN Exxon Research & Engineering Company, P.O. Box 101, Florham Park, New Jersey 07932
CORNELIUS L. McNALLY M. W. Kellogg Company (R & D), P.O. Box 79513, Houston, Texas 77079
OTTO MOLERUS University Erlangen-Nurnberg, 852 Erlangen, Martensstr. 9, West Germany
SHIGEKATSU MORI Nagoya Institute of Technology, Gokiso-cho, Showa-ku, Nagoya, Japan
SHIGEHARU MOROOKA University of Kentucky, 3300 Montavesta Road No. A51, Lexington, Kentucky 40502
G. MOSS Esso Petroleum Company, 78 Harpes Road, Oxford, England
HERMAN NACK Battelle-Columbus Laboratories, 505 King Avenue, Columbus, Ohio 43201
KOZO NAKAMURA Department of Agriculture, University of Tokyo, Yayoi 1-1-1-, Bunkyo-ku, Tokyo 113, Japan
ALLEN S. NEULS Allied Chemical Corporation, 550 2nd Street, Idaho Falls, Idaho 83401
RALPH H. NIELSEN Teledyne Wah Chang Albany, P.O. Box 460, Albany, Oregon 97321
RICHARD C. NORTON Stone & Webster Engineering Corporation, 225 Franklin Street, Boston, Massachusetts 02107
M. S. NUTKIS Exxon Research & Engineering Company, Box 8, Linden, New Jersey 07036
KATSUYA OHKI Tokyo Institute of Technology, 2-12-1 Ookayama, Meguro-ku, Tokyo 152, Japan
KNUD OSTERGAARD Technical University of Denmark, Department of Chemical Engineering, Building 229, DTH, 2800 Lyngby, Denmark
W. BRENT PALMER Allied Chemical Corporation, 550 2nd Street, Idaho Falls, Idaho 83401
ROBERT PFEFFER City College of New York, Department of Chemical Engineering, 140 Street & Convent Avenue, New York, New York 10031
BERT PHILLIPS NASA Lewis Research Center, 21000 Brookpark, Cleveland, Ohio 44135
OWEN POTTER Monash University, 835 Riversdale Road, Camberwell 3124, Victoria, Australia
B. B. PRUDEN Department of Chemical Engineering, University of Ottawa, Ontario K1N 6N5, Canada
DHARAM V. PUNWANI Institute of Gas Technology, 3424 South State Street, Chicago, Illinois 60616
D. L. PYLE Imperial College, Department of Chemical Engineering, Prince Consort Road, London SW7 2BY, England
WILLIAM E. REASER Assistant Director of Conferences, Engineering Foundation, 345 East 47 Street, New York, New York 10017
LOTHAR REH Lurgi Chemie and Huetten, Technik GMBH, Gervinusstr 17-19, 6 Frankfurt/Main, West Germany
LOUIS F. RICE C. F. Braun & Company, 1000 South Fremont Avenue, Alhambra, California 91801
K. RIETEMA Department of Physical Technology, Technical University, F.T.-hal, P.O. Box 513, Eindhoven, Netherlands
GILLES ROCHE Graduate Student, Ecole Polytechnique de Montreal, Park Avenue 6028 No. 11, Montreal (PQ), Canada
P. N. ROWE Ramsay Memorial Professor, University College, Torrington Place, London WC E 7JE, England
LAWRENCE A. RUTH Exxon Research & Engineering Company, P.O. Box 8, Linden, New Jersey 07036
S. C. SAXENA University of Illinois, Box 4348, Chicago, Illinois 60680
B. SCARLETT Loughborough University, Loughborough, Leicestershire, England
A. R. SCHAEDAL Field Services Engineer, Copeland Systems, Incorporated, 2000 Spring Road, Oakbrook, Illinois 60521
KLAUS W. SCHATZ Mobil Research & Development Corporation, Research Department, Paulsboro, New Jersey 08066
T. SHINGLES SASOL, P.O. Box 1, Sasolburg, South Africa
TAKASHI SHIRAI Tokyo Institute of Technology, O-okayama, Meguro-ku, Tokyo 152, Japan
A. E. SKRZEC Stauffer Chemical Company, East Research Center, Dobbs Ferry, New York 10522
LARRY M. SOUTHWICK C. F. Braun & Company, Alhambra, California 91802
ARTHUR M. SQUIRES City College of New York, 245 West 104 Street, New York, New York 10025
FRED W. STAUB General Electric Company, I River Road, Schenectady, New York 12345
CHARLES V. STERNLING Shell Development Company, 3737 Bellaire Boulevard, Houston, Texas 77001
J. D. A. STONES SASOL, 16 Goeie Hoop, Retief Street, Sasolburg, South Africa
KAZUO SUGIMOTO Kurita Water Industries, C. Ito & Company, Incorporated, 270 Park Avenue, New York, New York 10017

STEPHEN SZEPE   University of Illinois at Chicago Circle, Box 4348, Chicago, Illinois 60680

MICHEL TASSART   Graduate Student, Ecole Polytechnique de Montreal, Park Avenue 6028 No. 1, Montreal (P.Q.), Canada

A. I. THOMPSON   Imperial Chemical Industries, Mond Division, P.O. Box 14, The Heath, Runcorn, Cheshire WA6 91Z, England

HOSHENG TU   Standard Oil Company (Ohio), 4440 Warrensville Center Road, Cleveland, Ohio 44128

F. VERGNES   Centre de Cinetique du CNRS, Route de Vandoeuvre, Villers les Nancy, 54500, France

DIDIER VIEL   S.N.I. Aerospatiale, Route de Verneuil, Les Mureaux 78130, France

STERLING N. VINES   Pre-doctoral student, University of Virginia, 25 University Circle, Charlottesville, Virginia 22903

GORDON L. WADE   Combustion Power Company, 1346 Willow Road, Menlo Park, California 94025

PARVEZ H. WADIA   Union Carbide Corporation, P.O. Box 8361, Building 740-5320, South Charleston, West Virginia 25303

JACK S. WATSON   Oak Ridge National Laboratory, P.O. Box X, Oak Ridge, Tennessee 37830

DAVID F. WELLS   E.I. duPont de Nemours Incorporated, Engineering Department, Wilmington, Delaware 19898

C. WEN   West Virginia University, Evansdale Campus, Morgantown, West Virginia 26506

JOACHIM WERTHER   University of Erlangen, Martensstrasse 9, D 8520 Erlangen, West Germany

BASIL WHALLEY   Department of Energy Mines & Resources, 562 Booth Street, Ottawa, Ontario K1A OG1, Canada

MARVIN E. WHATLEY   Oak Ridge National Laboratory, P.O. Box Y, Oak Ridge, Tennessee 37830

WEN-CHING YANG   Westinghouse Electric Corporation, Research & Development Center, Pittsburgh, Pennsylvania 15235

JOSEPH YERUSHALMI   Chemical Engineering Department, City College of New York, 140 Street & Convent Avenue, New York, New York 10031

F. A. ZENZ   Particulate Solid Research, Incorporated, P.O. Box 205, Garrison, New York 10524

INDEX

Numbers in *italics* refer to pages in Volume I; numbers in **bold type** refer to Volume II.

113569